Advanced Principles of Environmental Soil Science

Advanced Principles of Environmental Soil Science

Edited by Henry Wang

SYRAWOOD
PUBLISHING HOUSE
New York

Published by Syrawood Publishing House,
750 Third Avenue, 9th Floor,
New York, NY 10017, USA
www.syrawoodpublishinghouse.com

Advanced Principles of Environmental Soil Science
Edited by Henry Wang

International Standard Book Number: 978-1-68286-752-5 (Hardback)

Cataloging-in-Publication Data

Advanced principles of environmental soil science / edited by Henry Wang.
    p. cm.
Includes bibliographical references and index.
ISBN 978-1-68286-752-5
1. Soils--Environmental aspects. 2. Soil science. 3. Soil ecology. I. Wang, Henry.
S596 .A38 2019
631.4--dc23

TABLE OF CONTENTS

PREFACE

In my initial years as a student, I used to run to the library at every possible instance to grab a book and learn something new. Books were my primary source of knowledge and I would not have come such a long way without all that I learnt from them. Thus, when I was approached to edit this book; I became understandably nostalgic. It was an absolute honor to be considered worthy of guiding the current generation as well as those to come. I put all my knowledge and hard work into making this book most beneficial for its readers.

The study of the interaction of humans with the pedosphere, hydrosphere, lithosphere, biosphere and the atmosphere falls under the domain of environmental soil science. The field also delves into diverse aspects of surface water quality, erosion control, septic drain field site assessment, vadose zone functions, land treatment of wastewater, storm water, etc. The restoration of wetlands, remediation of contaminated soils, soil degradation, bioremediation and nutrient management are also covered in this field. This book is a compilation of chapters that discuss the most vital concepts and emerging trends in the field of environmental soil science. Such selected concepts that redefine this field have been presented in this book. With state-of-the-art inputs by acclaimed experts of this field, this book targets students and professionals.

I wish to thank my publisher for supporting me at every step. I would also like to thank all the authors who have contributed their researches in this book. I hope this book will be a valuable contribution to the progress of the field.

Editor

Sustainable Management of Calcareous Saline-Sodic Soil in Arid Environments: The Leaching Process in the Jordan Valley

Mufeed Batarseh[1,2]

[1]*Chemistry Department, Mu'tah University, P.O. Box 7, Mu'tah 61710, Jordan*
[2]*Abu Dhabi Polytechnic, P.O. Box 111499, Abu Dhabi, UAE*

Correspondence should be addressed to Mufeed Batarseh; mufeed.batarseh@adpoly.ac.ae

Academic Editor: Rafael Clemente

A leaching experiment of calcareous saline-sodic soil was conducted in Jordan Valley and aimed to reduce the soil salinity $\leq$ 4.0 dS m^{-1}. The quantification of salt removal from the effective root zone was done using three treatment scenarios. Treatment A contained soil amended with gypsum leaching with fresh water (EC = 1.1 dS m^{-1}). Treatments B and C contained nonamended soil, but B was leached with fresh water only while treatment C's soil was washed with saline agricultural drainage water (EC = 8 dS m^{-1}) at the start of the experiment and continued with fresh water to reach the desired soil salinity. All treatments were able to reduce the soil salinity to the desired level at the end of the experiment; however, there were clear differences in the salt removal efficiencies among the treatments which were attributed to the presence of direct source of calcium ion. The soil amended with gypsum caused a substantial decline in soil salinity and drainage water's electrical conductivity and drained the water twice as fast as the nonamended soil. It was found that utilizing agricultural drainage water and gypsum as a soil amendment for calcareous saline-sodic soil reclamation can beneficially contribute to sustainable agricultural management in the Jordan Valley.

1. Introduction

Saline-sodic soil can significantly reduce the productivity of affected agricultural land. Saline soil has an electrical conductivity (EC) of the saturation soil extract of more than 4 dS m^{-1} at 25°C [1, 2]. Soil salinity is a major problem facing sustainable development in the agricultural sector in the Jordan Valley and adversely affects soil productivity. Many arid and semiarid regions in the world contain soil and water resources that have high salinity rates not suitable for most common economic crops [3]. Jordan, with an area of 89,556 km^2 of which more than 80% is desert, has more than 180,000 ha of arable land affected by soil salinity [4]. The main sources of soil salinity are either natural (due to parent soil materials, salt deposits, and climatic conditions) or caused by long-term use of fertilizers, low rainfall, and the use of high-salt irrigation water. In the Jordan Valley, these sources are predominantly found on a large portion of land surrounding the Dead Sea area and are naturally high in saline due to the declination of the Dead Sea's water level. In other areas in the Middle and Northern parts of the valley, soil salinity is due to upward movement irrigation practices in hot, dry climatic conditions where dissolved salts accumulate in the root zone [5, 6]. Removing salts from the root zone can be done by using various methods. Soil leaching is well known to be the most effective procedure, but this requires a large amount of water. Therefore, it is not recommended for a country, such as Jordan, with scarce water resources. Biological methods, such as salt tolerant plants (halophytes) and certain bacterial strains, were reported to be effective in removing salts from the root zone [2, 7]. Poor irrigation water quality along with saline-sodic soil has reduced plant yield and represents a serious threat to soil productivity. Nevertheless, information on the effects of these methods on soil reclamation is scarce in this region.

Reuse of saline water and amelioration of salt-affected soil is becoming an increasingly vital tool to improve crop production and has created extra farm land which has saved fresh water resources. Additionally, many areas of the world have soil with problems of high salinity and water logging. A large number of subsurface/surface drainage systems have been installed to lower the water table and leach salts from saline-sodic soil. Poor quality water can be used for the leaching of saline-sodic soil in areas where there are limited fresh water resources and salt-affected soils.

Sodic soil contains an excess of exchangeable Na^+ on soil colloids and has soluble carbonates which results in higher soil pH. The amendments to saline-sodic soil are materials that will furnish divalent cations such as Ca^{+2} to replace exchangeable Na^+. The exchange of Ca^{+2} with Na^+ in soil results in flocculation of soil particles, the restoration of porous structures, and the increase in soil permeability. The reclamation of sodic soil also requires water high in electrolytes to pass through the soil profile having leaving divalent ions, usually Ca^{+2}, in the soil and washing exchanged Na^+ ions away from the root zone (Hanay et al. 2004) [8].

Saline-sodic soil containing calcium carbonate ($CaCO_3$) is common in arid and semiarid regions (Oster et al. 1996). Since the $CaCO_3$ dissolution rate is too slow to supply enough Ca^{+2} for ion exchange, an acid or acid former, such as gypsum (Wong et al. 2009), sulfuric acid (Sadiq et al. 2007), or organic matter (Li and Keren 2009), can be applied as a soil amendment to enhance $CaCO_3$ dissolution. Gypsum is the most prevalent agricultural soil amendment used for reclamation of sodic soil. It decreases the damaging effects of high sodium content in irrigation water because of its high solubility, ease of use, and low cost [8]. Several studies have shown that a high application of gypsum to reclaim saline-sodic soils increases the removal of excess Na^+ from soil and causes a significant reduction in electrical conductivity and sodium adsorption ratio within the soil (Keren 1996; Sadiq et al. 2003; Hamza and Anderson et al. 2003; Hanay et al. 2004). Therefore, the objective of this study was to reclaim saline-sodic soils using gypsum-amended soil, fresh water, and drained water. The study's aim was to facilitate saline soil leaching in an arid environment, such as the Jordan Valley, and to decrease the electrical conductivity of soil salinity to less than $4.0\,dS\,m^{-1}$. Additionally, investigations were performed to analyze the leachability of major cations and anions in the soil profile. Finally, the study's other purpose was to ascertain sustainable water management (SWM) practices through preservation of fresh water by leaching saline soil using drained water.

2. Materials and Methods

2.1. Field Site and Experiment Preparations. This research was conducted in Ghor Al-Safi, an area in the southern part of the Jordan Valley approximately 20 km from the southern edge of the Dead Sea. The area is nearly flat with just a slight slope to the west. The climate is subtropical and is characterized by a hot, dry summer season (April to October) and a warm winter season (November to March). Rainfall is less than 80 mm/year occurring mainly in two or three main events during the winter season [9, 10]. The soil is characterized by low fertility, organic matter (1.01 to 4.71%), and CEC ($49\,cmol\,kg^{-1}$) [11, 12]. The natural vegetation in this area consists of scattered wild plants tolerant to high salinity, such as *Prosopis farcta*, *Proposi juliflord*, and *Ziziphus spina-christi* [13]. The dominant soil texture is silt with poor soil structure, and the average salinity is 98.1 dS/m based on a

saturated paste test. This land has been categorized as an extremely salty area [14]. A marly layer, 40 cm thick, was found at approximately 60 cm deep.

Soil reclamation experiments were conducted using three types of reclamation treatments. The chemical and physical soil properties were measured before and after the reclamation experiments to assess the efficiency of salt removal and to quantify amount of salt removed from the plant effective root zone using different treatments. The treatments were applied, namely, soil amended with gypsum (treatment A), untreated soil consistent with the area (treatment B), and untreated soil (treatment C). Treatment C was washed with agricultural drainage saline water (EC = $8.0\,dS\,m^{-1}$, Na^+ = 2743.3 mg/L, Ca^{+2} = 1802.7 mg/L, and Mg^{+2} = 1051.2 mg/L) at the start of the experiment, and then leaching continued with fresh water (EC = $1.1\,dS\,m^{-1}$, Na^+ = 73.3 mg/L, Ca^{+2} = 34.3 mg/L, and Mg^{+2} = 41.6 mg/L) similar to the other treatments. The treatment block consisted of three separate experimental units, each $50\,m^2$ ($5\,m \times 10\,m$) and confined by a trench 180 cm deep to collect drainage water, Figure 1. There were nine experimental units (3 treatments × three replicates). A continuous flooding method was used for soil leaching.

Gypsum was applied to treatment A at an application rate of $10.9\,t\,ha^{-1}$. The gypsum requirement is the amount of applied gypsum (application rate) required to reduce the exchangeable sodium percent (ESP) to 10% at the end of the experiment, given that the ESP was approximately 42% according to similar records for the study area [14]. The upper 30 cm of soil was well mixed and irrigated with fresh water once per week. Treatment B represented soil typically found in the study area. This treatment was continuously washed with fresh water during the experimental period. Conversely, treatment C was leached initially using saline water until the drainage water reached a salinity value of $20\,dS\,m^{-1}$. Afterwards, it was leached with fresh water until the drainage water reached an EC value of $4\,dS\,m^{-1}$.

2.2. Soil, Water Samplings, and Analysis. Soil samples were collected from three depths (0–50, 50–100, and 100–150 cm) for each treatment at the beginning and after at least 25 weeks of leaching treatments; the data is shown in Tables 1 and 2. From each treatment, 10 samples were taken from 3 depths (0–50, 50–100, and 100–150 cm) using a soil auger and then homogenized to represent each treatment, in total 30 samples. In summary, three homogenized samples were analyzed for each treatment. The leaching end point was decided when the drainage water's electrical conductivity (EC) reached a value $\leq 4\,dS\,m^{-1}$. The chemical and physical properties of the soil samples were measured first in the Chemistry Department and later in the Faculty of Agricultural Laboratories at Mu'tah University, Jordan. The soil soluble ions (Cl^-, SO_4^{-2}, HCO_3^-, Na^+, K^+, Ca^{+2}, and Mg^{+2}) were determined in the saturated soil paste according to US Salinity Laboratory Staff [15] and Rhoades [16]. The Cl^- and SO_4^{-2} ions were analyzed using gravimetric titration and turbidimetric method, respectively. Ca^{+2} and Mg^{+2} cations were

FIGURE 1: Layout of the experimental fields: treatment A (A1–A3), treatment B (B1–B3), and treatment C (C1–C3).

determined by atomic absorption spectrophotometry (AAS). Finally, Na$^+$ and K$^+$ were analyzed using flame emission photometry (FEP).

While exchangeable Na$^+$ and K$^+$ were extracted in ammonium acetate and Ca^{+2} and Mg^{+2} in sodium acetate according to Rowell [17], they were then analyzed using the same FEP and AAS techniques mentioned above, respectively. Furthermore, EC was measured in the saturated soil paste extract using an EC-meter [16]. Soil texture was determined using the hydrometer method according to Gee and Bauder [18]. OM was determined using the wet oxidation method according to Sparks et al. [19]. CEC was determined by displacement of cations from soil samples with 1 N ammonium acetate, and determination was done using Flame Photometer (Palemio and Rhoades 1977). Gypsum was analyzed in soil according to the standardized procedure described by Schlichting et al. [20]. The carbonate (HCO$_3$) content was determined using a volumetric titration method of soil paste extract (Rhoades et al. 1992).

Water samples were collected from fresh water used for irrigation and soil leaching and drainage water using polyethylene bottles. The water was collected through drainage canals around each treatment at a depth of 1.8 m during the leaching experiment until its EC value reached ≤4 dS m^{-1}. All of the drainage water was pumped out and its quantity measured. Furthermore, it was analyzed for EC and

major ionic composition. The following ions (Na$^+$, Ca^{+2}, Mg^{+2}, Cl$^-$, and SO$_4^{-2}$) were analyzed using Dionex Ion Chromatograph Model DX 100 (Dionex, USA).

2.3. Statistical Analysis. Design analysis of variance (ANOVA) was used to test the treatments effected by applying the following mathematical model [21]: $y_{ij} = \mu + \tau_i + \beta_j + \varepsilon ij$ (where y_{ij} represents the observation from ijth experimental unit (=plot), τ_i represents the effect of treatment i, such that the average of each treatment $\overline{T} = \mu + \tau_i$, εij are the residuals, the deviations of each observation from their expected values, β_j is the effect of jth replication, and μ is the overall mean). Fisher's least significant differences (LSD) were used as a method for comparing treatment means. The statistical analysis was done using Statistical Analysis System (SAS) software.

3. Results and Discussion

3.1. Physical and Chemical Soil Properties. The predominant soil texture in the experimental site was silty as recorded in Table 1. High silt content has an adverse impact on soil moisture movement due to its low hydraulic conductivity. The OM content varied from 0.7 to 3.0% and EC from 27 to 144 dS m^{-1} for the investigated soil, Table 1. Thus, the

TABLE 1: Chemical properties of the soil before starting the leaching experiment (minimum, maximum, average, and standard error).

Soil depth (cm)			0–50 cm				50–100 cm				100–150 cm		
Parameter	Min	Max	Average*	Std** error	Min	Max	Average	Std error	Min	Max	Average	Std error
EC (dS/m)	68	144	111	22.0	27	81	50.6	15.6	29	56	44	7.8
pH (unit)	6.9	8.2	7.3	0.4	7.1	8.2	7.6	0.3	7.2	8.3	7.6	0.3
Cl (meq/L)	614	1377	1098	222.9	230	595	403	105.4	319	542	408	64.8
SO_4 (meq/L)	29	307	76	85.9	21	60	40	11.3	19	72	33.5	15.8
HCO_3 (meq/L)	2.4	5.2	3.1	0.8	1.6	4.2	2.4	0.8	1.8	5	2.4	1.0
Na (meq/L)	330	670	480	98.4	144	320	240	50.9	108	253	209	42.9
K (meq/L)	17	42	32	7.3	10	20	13	3.0	8	1314	11	1.6
Ca (meq/L)	125	240	175	33.3	39	116	74.5	22.3	55	88	67	9.7
Mg (meq/L)	208	476	333	77.4	82	277	144	57.5	68	210	141.6	41.0
$CaCO_3$ (%)	37	47.6	40	3.2	35	46	40.7	3.2	35	45	38.6	2.9
CEC (meq/100 g)	14.6	18.8	17.5	1.2	10	15.7	13.1	1.6	10	14.7	11.3	1.4
Exch. Na (meq/100 g)	6.8	8.2	7.5	0.4	2.	6.4	4.1	1.3	2.6	5.6	3.8	0.9
Exch. K (meq/100 g)	2.8	3.8	3.0	0.3	2.5	3.0	2.7	0.1	1.2	3	1.8	0.5
Exch. Ca + Mg (meq/100 g)	5.0	6.8	6.7	0.6	5.5	6.3	6.3	0.3	6.0	6.1	5.7	0.1
ESP (%)	39	97	42.9	17.1	17	45	31	8.1	27	55	33	8.6
Organic matter (%)	1.1	2.4	1.8	0.4	1.0	3.0	1.8	0.6	0.7	2.1	1.1	0.4
Gypsum conc. (%)	1.5	6.2	4.1	1.4	1.8	4.5	2.8	0.8	1.7	5.0	3.3	1.0
Saturation percentage (%)	57	76	62	5.7	63	79	72	4.6	58	80	68	6.4

*Number of samples: 10 samples were taken from each depth (0–50, 50–100, and 100–150 cm) and homogenized to represent each treatment in each replicate, in total 30 samples.
**Std error: standard error.

TABLE 2: Exchangeable cation concentrations (meq/100 g soil), electrical conductivity (dS/cm), and pH of saturated soil paste extract after the leaching experiment.

Treatment	Exchangeable cations (meq/100 g)			EC ± Std error	pH ± Std error
	K ± Std error**	Na ± Std error	Ca & Mg ± Std error		
		Soil depth (0–50 cm)			
A	$1.68^a{}^* \pm 0.15$	$0.873^a \pm 0.02$	$14.96^a \pm 0.15$	$0.94^a \pm 0.08$	$7.8^a \pm 0.07$
B	$1.65^a \pm 0.03$	$0.983^a \pm 0.03$	$14.86^a \pm 0.01$	$1.05^a \pm 0.04$	$7.71^a \pm 0.07$
C	$1.747^a \pm 0.36$	$1.59^a \pm 0.46$	$14.16^a \pm 0.80$	$1.18^a \pm 0.04$	$7.83^a \pm 0.02$
		Soil depth (50–100 cm)			
A	$2.37^a \pm 0.08$	$1.48^a \pm 0.10$	$13.64^a \pm 0.18$	$1.76^a \pm 0.59$	$7.7^a \pm 0.13$
B	$2.46^a \pm 0.02$	$1.79^a \pm 0.17$	$13.48^a \pm 0.12$	$1.36^a \pm 0.33$	$7.81^a \pm 0.08$
C	$2.06^a \pm 0.28$	$2.02^a \pm 0.41$	$13.41^a \pm 0.68$	$1.82^a \pm 0.20$	$7.94^a \pm 0.01$
		Soil depth (100–150 cm)			
A	$2.44^a \pm 0.07$	$1.65^c \pm 0.05$	$13.4^a \pm 0.11$	$3.2^a \pm 0.24$	$7.64^b \pm 0.02$
B	$2.52^a \pm 0.03$	$2.38^b \pm 0.13$	$12.59^b \pm 0.15$	$2.01^b \pm 0.26$	$7.92^a \pm 0.03$
C	$2.51^a \pm 0.10$	$2.64^a \pm 0.07$	$12.33^c \pm 0.11$	$2.2^b \pm 0.12$	$7.93^a \pm 0.02$

*Superscript letters a, b, and c: statistical analysis significance at 95% confidence limit. Treatments followed by the same superscript letters were not significantly different at $P < 0.05$.
**Std error: standard error.

soil was classified as saline-sodic soil [22]. High soil salinity is a limiting factor for microbial soil activity; it is responsible for OM degradation; hence it accumulates under a saline environment. Moreover, the upper soil depth (0–50 cm) showed elevated ESP range (39–97%) because of evapotranspiration as a result of high temperatures in the Jordan Valley. As a

result of capillary phenomenon, more salt accumulated in the upper soil horizon [23]. While ESP ranged from 17 to 45% for the middle soil depth (50–100 cm), the deepest soil layers (100–150 cm) showed a high ESP range of 27–55% because they were in contact with the water table. Therefore, there were big variations in ESP values due to salt accumulation before the start of the leaching treatment; this is a normal heterogeneity of saline sodic soil matrix in such an arid environment.

Furthermore, the high concentration of cations and anions, particularly Na^+ and Cl^-, in the soil leads to high soil salinity. The Cl^- concentration ranged from 230 to 1377 meq/L within the surface and deep soil layers. The Na^+ showed a similar trend to Cl^- where its concentration ranged from 108 meq/L within the deep layers of the soil to 670 meq/L at the surface layer. The pH value ranged from 6.9 to 8.3. The alkaline pH was due to the presence of high $CaCO_3$ concentrations in the soil [24]. The soil $CaCO_3$ ranged from 35.0% up to 47.6%. Under alkaline soil, the dissolution of $CaCO_3$ is unlikely, resulting in low HCO_3 concentrations (2.4 to 5.0 meq/L) as in Table 1; therefore alkaline pH inhabits the Ca^{+2} solubility [25]. Thus, the major source of Ca^{+2} in the soil was gypsum, where gypsum concentration ranged from 1.5 to 6.2%. Gypsum was being utilized for reclamation of sodic soils as a source of Ca^{+2} to replace Na^+ (Bresler 1982). Therefore, an extra amount of gypsum was added to the soil since its original amount was not adequate to reclaim saline-sodic soil (Frenkel et al. 1989; Baumhardt et al. 1992). The cumulative water quantities were recorded during the leaching experiments which lasted for at least 25 weeks to reduce the soil salinity to $4.0\,dS\,m^{-1}$. These water quantities were 107, 101, and $102\,m^3$ per $50\,m^2$ for gypsum, fresh water, and saline water treatments, respectively (Figure 2). The estimated leached water amounts required for reclamation of the investigated saline-sodic soil were 21400, 20200, and $20400\,m^3$ per hectare, for gypsum, fresh water, and saline water treatments, respectively. These water amounts were considered high for Jordan which suffers from severe water shortage [12, 26]. Therefore, utilizing agricultural drainage water for leaching processes until reaching $20\,dS\,m^{-1}$ of drained water can save around 66% ($13200\,m^3$ per hectare) of fresh water requirements [27]. One drawback of this approach is that using brackish water consumes a larger amount of water compared to the amount needed for leaching with fresh water [28, 29]. Additionally, soil swelling might take place too which may lead to an increase in volume as the soil gets wet and shrinks as it dries out. This phenomenon might occur when a high salt content such as ESP > 15% in the soil is high [30], and an extensive degradation in soil structure might occur when irrigation waters have SAR > 13 [31]. A mass balance of Na^+ shows that not all of the exchangeable sodium can be removed during the reclamation stage, even though excess amounts of gypsum were applied [31].

3.2. Soil Leaching Experiment. Desalinization curves (Figure 3) showed that the three different leaching treatments were able to reduce soil salinity to less than $4\,dS\,m^{-1}$ at the end of the experiment. A sharp decline in soil salinity during the first few weeks of the leaching experiment was observed. Then a gradual decrease was observed with successive leaching until the end of the experiment. The gypsum treatment A was the fastest and most effective treatment as the EC declined to less than $20\,dS\,m^{-1}$ only in 7 weeks, while treatment B needed 14 weeks and treatment C needed

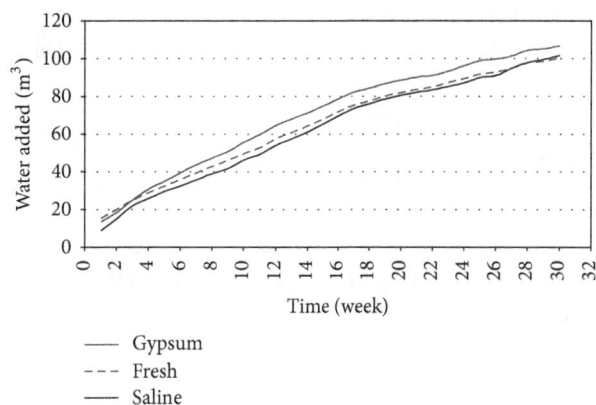

FIGURE 2: Cumulative water quantities (m^3) added during the leaching process (gypsum amended soil, fresh water, and saline water).

16 weeks, Figure 3. Similar behavior of soil salinity was also reported by Gharaibeh et al. [29]. Gharaibeh et al.'s research showed that leached salts were greater at the start of the experiment, but their efficiencies decreased consistently over time. The water quality used for leaching played an important role in the amount of water and time required for salt to be leached from treated soil. The lower the amount of salt in the leaching water, the greater the amount of salt removed. In this study, the phenomenon was also observed in fresh water as the time and amount of leached water were much lower than the saline water. By using different leaching water qualities, there was a difference in the salt removal rate between the three treatments. For example, treatment A showed a sharp decrease in salinity due to the use of gypsum as a rich source of calcium ion: the drop in EC was relatively fast when compared to the other treatments, as shown in Figure 3. This behavior relates to the flocculation of soil particles which is a distinctive occurrence when gypsum is applied. This will improve the soil's hydraulic conductivity and consequently help with the leaching of soluble salts (Oster 1982; Mahmoodabadi et al. 2012). Fresh water treatment B also showed a sharp decline in salinity when compared to the saline water treatment C. These results were in agreement with S. K. Gupta and I. C. Gupta [28]. The utilization of drainage water for leaching saline sodic soil can contribute effectively in sustainable water management of scarce fresh water resources in the Jordan Valley.

The relationships between the decline in electrical conductivity (EC) and the amount of drainage water (m^3) from the gypsum-amended soil (A), fresh water (B), and saline water (C) treatments over the experimental period are shown in Figures 3(a), 3(b), and 3(c). The EC of drainage water dramatically decreased in the first few weeks, and then a gradual decrease was observed until a steady state was reached at the end of the experiment. The gypsum-amended soil, treatment A, showed the fastest salt leaching process compared with both the fresh, B, and saline, C, water treatments. This may be the result of increased displacement of Na^+ due to an increased soil-solution of Ca^{+2} from amendments and subsequent leaching of the replaced Na^+ through

(a) Gypsum experiment: leaching water versus EC

(b) Fresh water experiment: leaching water versus EC

(c) Saline water experiment: leaching water versus EC

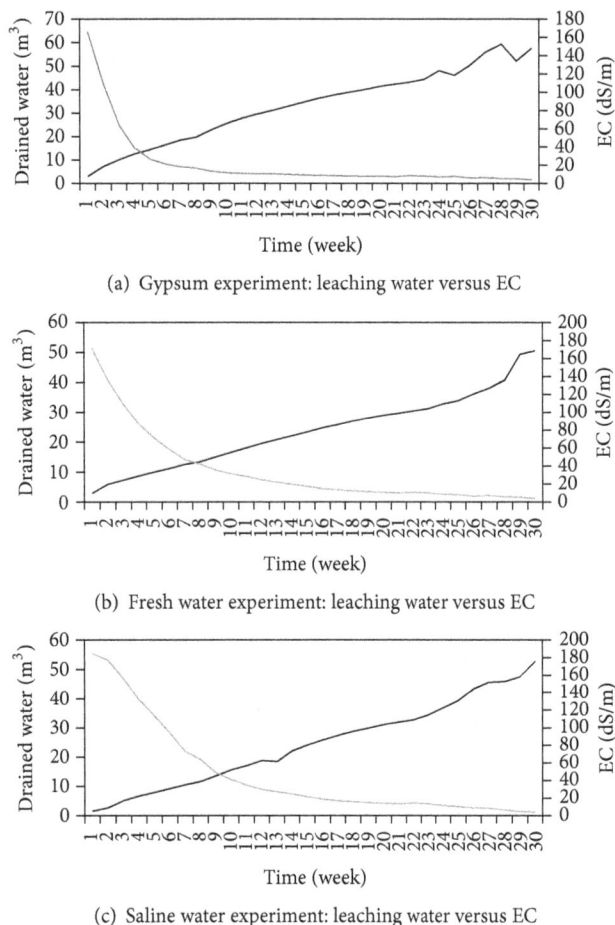

FIGURE 3: Electrical conductivity and quantities of drained water recorded during the experiment ((a) gypsum, (b) fresh water, and (c) saline water). The black line represents the leaching water quantity (m³), and the gray line is the EC.

drained water [29] (Mahmoodabadi et al. 2012). The results also showed that gypsum treatments required smaller quantities of water to remove Na⁺ from soil compared with fresh and saline water treatments. Similar trends were reported for saline and saline-alkaline soils [11, 29, 32, 33]. The leaching process at the beginning of the experiment showed the removal of salts from the easily percolated water in the macropores. However, in later stages of the leaching process, the drop of EC slowed as the leaching water diffused through micropores by capillary action. It is known that salinity affected soil undergoes variable equilibrium conditions through the leaching processes until complete equilibrium between soil solution and soil particles is achieved. At the end of the leaching experiment, the equilibrium process proceeded very slowly, consuming a large amount of water and time. It was found that a complete equilibrium need not be reached because partial equilibrium is sufficient to save water and time [34].

3.3. Changes in Drainage Water's Chemical Composition.
There was a sharp decline in cations (Na^+, Ca^{+2}, and Mg^{+2})

and Cl^- concentrations for drainage water during the early stages of the leaching experiments, Figure 4. This decrease coincided with a decrease in SAR and a sharp drop in ESP which was continuously represented in the soil leaching process (Figure 5). Furthermore, a similar trend was observed for Cl^- ion. The leaching process of ions from the soil was dependent upon ion radius, mobility, and adsorption affinity on soil paste. The observed reduction in SAR and ESP may have been caused by Na^+ reduction in the soil solution over time. The replacement of Na^+ and K^+ by Ca^{+2} in the exchangeable sites probably led to an initial increase in soluble Na^+ and K^+ which then decreased over time.

3.4. Physical and Chemical Soil Properties after the Leaching Experiment.
The chemical properties of the collected soil samples showed a decline in EC after the leaching experiment (see Table 2). PH values of the soil ranged between 7.64 and 7.94 for the three treatments at all depths. No sodicity development was observed in the leached soil during or after the leaching experiment. There was a sharp drop in EC values along the soil profile during the soil leaching for all treatment methods (A, B, and C). The EC values dropped from 111, 51, and $44 \, dS \, m^{-1}$ for the three soil depths (0–50, 50–100, and 100–150 cm) to 0.94–1.18, 1.36–1.82, and $2.01–3.2 \, dS \, m^{-1}$, respectively. Some lower EC values near the soil's surface for all treatments were noticed because of the excess leaching resulting from irrigation. Since salts are very mobile, soil EC values are being affected not only by soil amendment but also by the movement of water in the soil (Zhang et al. 2007). Moreover, since downward leaching of salt intensifies with irrigation or leaching, some of the salt should have been moved downward. Significant variations in cation distribution were found as a sharp decrease in exchangeable Na^+ and K^+ ions for all treatments at all depths. The results showed that the exchangeable Na^+ concentration dropped from 7.5, 4.1, and 3.8 meq/100 g to 1.59, 0.98, and 0.87 meq/100 g at 0–50 cm depth and to 2.02, 1.79, and 1.48 at 50–100 cm depth and 2.64, 2.38, and 1.65 at 100–150 cm depth for treatments A, B, and C, respectively, as displayed in Tables 1 and 2.

Further, application of gypsum provided a readily available source of Ca^{+2} ions in the soil solution. It replaced the exchangeable Na^+ ions which then came into the soil solution and subsequently increased the electrolyte concentration. This increase in electrolyte concentration in turn increased the soluble salt concentration of the percolating solution [35]. Therefore, Ca^{+2} and Mg^{+2} concentrations increased from 6.7, 6.3, and 5.7 meq/100 g to 14.2, 14.9, and 15.0 meq/100 g, to 13.4, 13.5, and 13.6, and to 12.3, 12.6, and 13.4 at 0–50, 50–100, and 100–150 cm depths for treatments A, B, and C, respectively. These results support the idea that Ca^{+2} ion substitutes Na^+ ion in the soil, which leads to leaching the Na^+ ion from the soil. Additionally, this can be related to the higher mobility of Na^+ and K^+ than Ca^{+2} and Mg^{+2} in the soil profile during the leaching process (Mahmoodabadi et al. 2012).

3.5. Statistical Analysis Significance.
The results of the statistical analysis were designated using three superscript letters a,

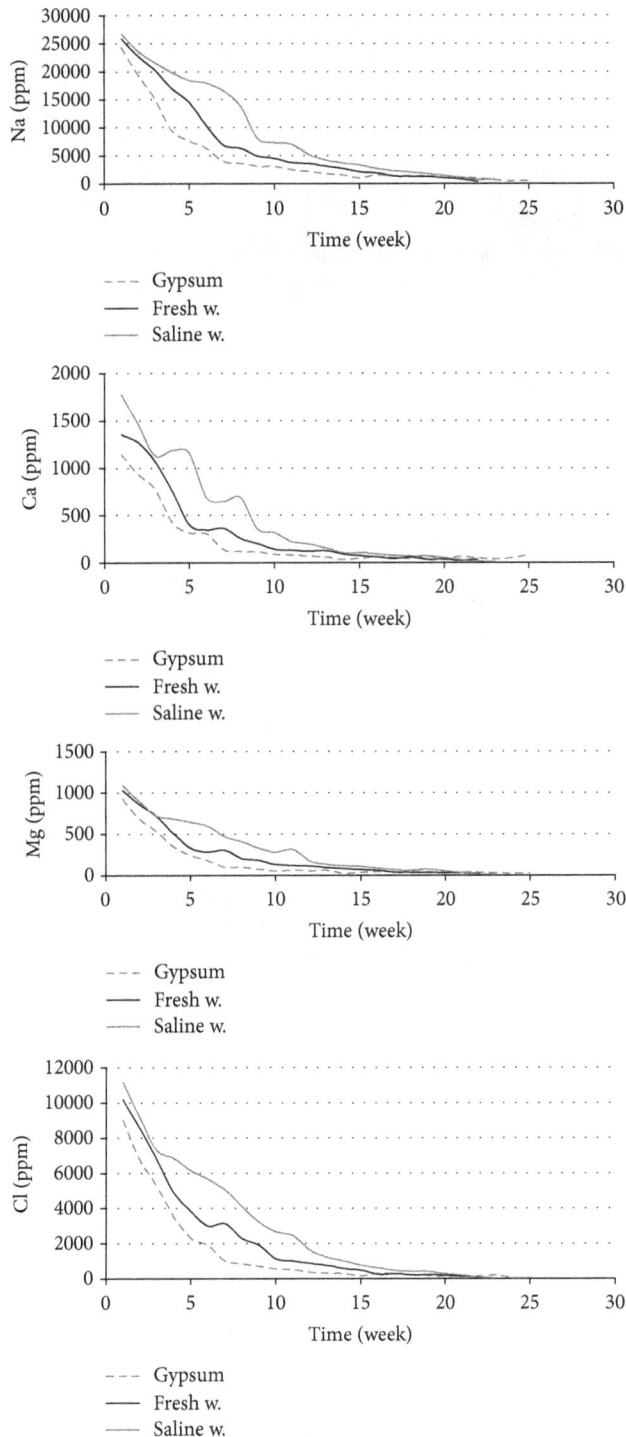

FIGURE 4: Decline in ionic composition (Na^+, Ca^{+2}, Mg^{+2}, and Cl^-) of drainage water for all treatments.

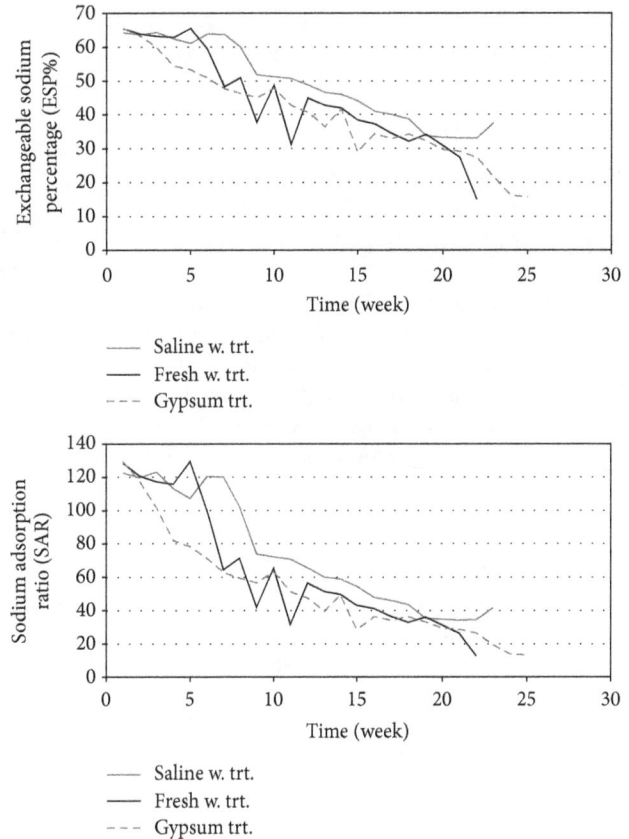

FIGURE 5: SAR and ESP in drainage water during the leaching experiment for gypsum, fresh water, and saline water treatments.

4. Conclusion

It can be concluded that utilization of agricultural drainage water in preliminary stages of soil leaching for saline-sodic soil can save more than 60% of fresh water requirements for soil reclamation. Likewise, adding gypsum to saline-sodic soil reduced the amount of water needed for leaching. The drop in EC with gypsum-amended soil was relatively faster than the other treatments because of flocculation of soil particles, which is a distinctive phenomenon with use of gypsum. A sharp decrease in exchangeable Na^+ and K^+ ions for all treatments and at all depths was also noticed with an increase in exchangeable Ca^{+2} and Mg^{+2}. Overall, the three investigated leaching treatment scenarios were able to reduce soil salinity to less than $4\,dS\,m^{-1}$ which conformed to the study's purpose.

However, the utilization of saline water resources, such as brackish and irrigation drainage water along with a gypsum application for leaching saline-sodic soils, can be an invaluable practice in the Jordan Valley, especially as fresh water becomes an increasingly valuable resource. Moreover, as gypsum is a relatively inexpensive material to purchase in large quantities, the research pointed out that using both saline water and gypsum might be a potential reclamation material for saline-sodic soils on a large scale in arid and semiarid environments. It is also recommended to investigate

b, or c, where the significance was tested at a 95% confidence limit (Table 2). The treatments followed by the same letter were not significantly different at $P < 0.05$. Significant differences were observed only at a soil depth 100–150 cm among the different treatments.

other soil reclaiming materials, such as phosphoric acid (H_3PO_4) with saline-sodic soils alongside saline water.

Competing Interests

The author declares that he has no competing interests.

Acknowledgments

This work was supported financially by the Higher Council for Science and Technology (HCST), Amman, Jordan. The researcher would like to extend his thanks to those who helped, encouraged, or participated directly or indirectly in this study. Additionally, the author would like to express deep gratitude to Mrs. Jennifer Benaggoun and Dr. Christopher Quinn for their editing skills.

References

[1] L. A. Richards, Ed., *Diagnosis and Improvements of Saline and Alkali Soils*, Handbook No 60 (USDA), United States Department of Agriculture (USDA), Washington, DC, USA, 1954.

[2] N. Al-Abed, J. Amayreh, A. Al-Afifi, and G. Al-Hiyari, "Bioremediation of a Jordanian saline soil: a laboratory study," *Communications in Soil Science and Plant Analysis*, vol. 35, no. 9-10, pp. 1457–1467, 2004.

[3] A. Nerd and D. Pasternak, "Growth, ion accumulation, and nitrogen fractioning in Atriplex barclayana grown at various salinities," *Journal of Range Management*, vol. 45, no. 2, pp. 164–166, 1992.

[4] F. I. Massoud, "Basic principles for prognosis and monitoring of salinity and sodicity," in *Proceedings of the International Conference on Managing Saline Water for Irrigation*, pp. 432–454, Texas Tech University, Lubbock, Tex, USA, August 1976.

[5] A. Jiries and K.-P. Seiler, "Water movement in typical soils in the Jordan Valley," *Jordan Metrological Journal and Agriculture*, vol. 22, no. 3, pp. 5–12, 1995.

[6] A. Khlaifat, M. Hogan, G. Phillips, K. Nawayseh, J. Amira, and E. Talafeha, "Long-term monitoring of the Dead Sea level and brine physico-chemical parameters "from 1987 to 2008"," *Journal of Marine Systems*, vol. 81, no. 3, pp. 207–212, 2010.

[7] F. Nasir, M. Batarseh, A. H. Abdel-Ghani, and A. Jiries, "Free amino acids content in some halophytes under salinity stress in arid environment, Jordan," *Clean—Soil, Air, Water*, vol. 38, no. 7, pp. 592–600, 2010.

[8] E. Amezketa, R. Aragüés, and R. Gazol, "Efficiency of sulfuric acid, mined gypsum, and two gypsum by-products in soil crusting prevention and sodic soil reclamation," *Agronomy Journal*, vol. 97, no. 3, pp. 983–989, 2005.

[9] O. E. Mohawesh, "Evaluation of evapotranspiration models for estimating daily reference evapotranspiration in arid and semiarid environments," *Plant, Soil and Environment*, vol. 57, no. 4, pp. 145–152, 2011.

[10] O. E. Mohawesh and S. A. Talozi, "Comparison of Hargreaves and FAO56 equations for estimating monthly evapotranspiration for semi-arid and arid environments," *Archives of Agronomy and Soil Science*, vol. 58, no. 3, pp. 321–334, 2012.

[11] M. Gharaibeh, N. Eltaif, and S. Shra'a, "Leaching curves of highly saline-sodic soil amended with phosphoric acid and phosphogypsum," in *Proceedings of the 2nd International Conference on Agricultural and Animal Science, International Proceedings of Chemical, Biological and Environmental Engineering (IPCBEE)*, vol. 22, IACSIT Press, Singapore, 2011.

[12] O. Mohawesh, "Assessment of pedotransfer functions (PTFs) in predicting soil hydraulic properties under arid and semi-arid environments," *Journal of Agricultural Sciences*, vol. 9, no. 4, pp. 475–492, 2013.

[13] MOA, "Ministry of Agriculture," Internal Report 2009, Ministry of Agriculture, Amman, Jordan, 2009.

[14] MWI, "Ministry of Water and Irrigation," Annual Report 1998, Laboratory Department at the Water Authority of Jordan, Dier-Ala Laboratories, Amman, Jordan, 1998.

[15] US Salinity Laboratory Staff, "Diagnosis and improvement of saline and alkali soils," in *USDA Agricultural Handbook No. 60*, US Government Printing Office, Washington, DC, USA, 1954.

[16] J. D. Rhoades, "Soluble salts," in *Methods of Soil Analysis, Part 2: Chemical and Microbiological Properties*, A. L. Page, R. H. Muler, and D. R. Keeney, Eds., no. 9, part 2, pp. 167–179, Soil Science Society of America, Madison, Wis, USA, 2nd edition, 1982.

[17] D. L. Rowell, *Soil Science: Methods and Applications*, Springer, Berlin, Germany, 1994.

[18] G. W. Gee and J. W. Bauder, "Particle-size analysis," in *Methods of Soil Analysis*, A. Klute, Ed., no. 9, part 1, pp. 383–411, Soil Science Society of America (SSSA), Madison, Wis, USA, 1986.

[19] D. Sparks, A. Page, P. Helmke, R. Loeppert, D. W. Nelson, and L. E. Sommers, "Total carbon, organic carbon, and organic matter," in *Methods of Soil Analysis Part 3—Chemical Methods*, SSSA Book Series, Soil Science Society of America, American Society of Agronomy, Madison, Wis, USA, 1996.

[20] E. Schlichting, H. P. Blume, and K. Stahr, *Boden Kundliches Prktikum*, Blackwell Wissenschaft, Wien, Austria, 1995.

[21] R. G. D. Steel and J. H. Torrie, *Principles and Procedures of Statistics*, McGraw-Hill, New York, NY, USA, 2nd edition, 1980.

[22] E. Bresler, B. L. McNeal, and D. L. Carter, *Saline and Sodaic Soil: Principles-Dynamic-Modeling*, Springer, Berlin, Germany, 1982.

[23] M. Qadir, A. Ghafoor, and G. Murtaza, "Amelioration strategies for saline soils: a review," *Land Degradation and Development*, vol. 11, no. 6, pp. 501–521, 2000.

[24] J. Bolton, "Changes in soil pH and exchangeable calcium in two liming experiments on contrasting soils over 12 years," *The Journal of Agricultural Science*, vol. 89, no. 1, pp. 81–86, 1977.

[25] A. Hardan, "Removal of salts from undisturbed saline-alkali soil columns by different leaching waters," in *Proceedings of the Man, Food, and Agriculture in the Middle East Symposium*, pp. 409–431, American University of Beirut, 1969.

[26] M. I. Batarseh, A. Rawajfeh, K. K. Ioannis, and K. H. Prodromos, "Treated municipal wastewater irrigation impact on olive trees (Olea Europaea L.) at Al-Tafilah, Jordan," *Water, Air, and Soil Pollution*, vol. 217, no. 1-4, pp. 185–196, 2011.

[27] A. Al Zubaidi and K. M. Hassan, "Leaching of some salt affected soils in Iraq using drainage water," *Iraqi Journal of Agriculture Sciences*, vol. 13, pp. 219–234, 1978.

[28] S. K. Gupta and I. C. Gupta, "Land development and leaching," in *Management of Saline Soils and Waters*, pp. 136–152, Mohan Primlani, New Dehli, India, 1987.

[29] M. A. Gharaibeh, N. I. Eltaif, and O. F. Shunnar, "Leaching and reclamation of calcareous Saline-sodic soil by moderately Saline and moderate-SAR water using Gypsum and Calcium Chloride," *Journal of Plant Nutrition and Soil Science*, vol. 172, no. 5, pp. 713–719, 2009.

[30] M. E. Sumner, "Sodic soils: new perspectives," *Australian Journal of Soil Research*, vol. 31, no. 6, pp. 683–750, 1993.

[31] I. Shainberg and J. Letey, "Response of soils to sodic and saline conditions," *Hilgardia*, vol. 52, no. 2, pp. 1–57, 1984.

[32] G. A. Al–nakshabandi, A. Alzubaidi, H. N. Ismail, F. Al–rayhani, and E. Al–hadithy, "Leaching of euphrates saline soil in lysimeters," *Journal of Soil Science*, vol. 22, no. 4, pp. 508–513, 1971.

[33] I. M. Habib, "Leaching of saline soils in monoliths of Iraq," *Egyptian Journal of Soil Science*, vol. 14, pp. 149–158, 1974.

[34] S. Muhammed, B. L. McNeal, C. A. Bower, and P. F. Pratt, "Modification of the high-salt water method for reclaiming sodic soils," *Soil Science*, vol. 108, no. 4, pp. 249–256, 1969.

[35] M. Qadir, D. Steffens, F. Yan, and S. Schubert, "Sodium removal from a calcareous saline-sodic soil through leaching and plant uptake during phytoremediation," *Land Degradation and Development*, vol. 14, no. 3, pp. 301–307, 2003.

Combining Geoelectrical Measurements and CO₂ Analyses to Monitor the Enhanced Bioremediation of Hydrocarbon-Contaminated Soils: A Field Implementation

Cécile Noel,[1,2] Jean-Christophe Gourry,[1] Jacques Deparis,[1] Michaela Blessing,[1] Ioannis Ignatiadis,[1] and Christophe Guimbaud[2]

[1]*BRGM, 3 avenue Claude Guillemin, 45060 Orléans, France*
[2]*LPC2E, CNRS, 3 avenue de la Recherche Scientifique, 45071 Orléans, France*

Correspondence should be addressed to Cécile Noel; cecile.noel@cnrs-orleans.fr

Academic Editor: Pantelis Soupios

Hydrocarbon-contaminated aquifers can be successfully remediated through enhanced biodegradation. However, *in situ* monitoring of the treatment by piezometers is expensive and invasive and might be insufficient as the information provided is restricted to vertical profiles at discrete locations. An alternative method was tested in order to improve the robustness of the monitoring. Geophysical methods, electrical resistivity (ER) and induced polarization (IP), were combined with gas analyses, CO₂ concentration, and its carbon isotopic ratio, to develop a less invasive methodology for monitoring enhanced biodegradation of hydrocarbons. The field implementation of this monitoring methodology, which lasted from February 2014 until June 2015, was carried out at a BTEX-polluted site under aerobic biotreatment. Geophysical monitoring shows a more conductive and chargeable area which corresponds to the contaminated zone. In this area, high CO₂ emissions have been measured with an isotopic signature demonstrating that the main source of CO₂ on this site is the biodegradation of hydrocarbon fuels. Besides, the evolution of geochemical and geophysical data over a year seems to show the seasonal variation of bacterial activity. Combining geophysics with gas analyses is thus promising to provide a new methodology for *in situ* monitoring.

1. Introduction

Petroleum hydrocarbon leaks and accidental spills happen commonly during the production, refining, transport, and storage of petroleum. Release of petroleum hydrocarbons into the environment causes damage to ecosystems [1] and to soil and water resources [2]. Increasing demand for drinking water and cropland with population growth requires effective remediation techniques to treat the contamination and to decrease hostile effects on health and environment. *In situ* remediation techniques such as enhanced bioremediation were shown to be effective in cleanup of the contamination [3, 4]. Enhanced bioremediation involves the addition of nutrients or electron acceptors to the subsurface environment to accelerate the natural biodegradation processes which degrade hydrocarbons [5]. Due to the potential of cost saving of *in situ* techniques compared to conventional *ex situ*

techniques, there is an economical interest for commercial providers to use enhanced bioremediation [6]. However, these processes remain partially unexploited, mainly because their *in situ* monitoring, before, during, and after soil treatment operations, is often expensive and technically challenging. Indeed, where significant subsurface heterogeneity exists, conventional intrusive groundwater sampling campaigns can be insufficient to obtain relevant information as they are restricted to costly monitoring piezometers at discrete locations. New monitoring tools are needed to overcome these limitations and make the enhanced bioremediation more reliable and robust, as well as economically competitive.

Previous studies suggest that geoelectrical techniques, especially electrical resistivity (ER) and induced polarization (IP), can be used to detect the presence of LNAPLs (Light Nonaqueous Phase Liquids) [7–10] as well as monitoring the effects of their biodegradation [11–14]. Biodegradation processes modify ground electrical properties because they change biophysicochemical conditions in the subsurface: (i) bacteria modify local redox conditions, inducing changes in self-potential (SP) [15]; (ii) microbial activity can produce organic acids and/or carbonic acid that affect the pore water conductivity, modifying both the in-phase and the quadrature conductivity [16]; (iii) during microbial growth and formation of biofilms, biomass can clog pores and potentially change the porosity and hydraulic conductivity, increasing the storage of electrical charges [17, 18]. Thus, ER and IP are expected to be effective nonintrusive tools to monitor enhanced bioremediation.

Some studies have implemented geoelectrical methods to characterize hydrocarbon-contaminated sites under biodegradation [19–22], but only few apply these methods for long-term monitoring or to prove the efficiency of enhanced bioremediation [23]. These previous field studies suggest that ER and IP are highly sensitive to the biophysicochemical processes associated with biodegradation. Nevertheless, the interpretation of geophysical data remains challenging partly because several factors may contribute to the observed electrical response (presence of metallic particles or clays, e.g.) and/or influence the electrical response in function of time (variation of water saturation and temperature). That is why geoelectrical methods are often used in conjunction with geochemical measurements (temperature, pH, redox potential, water conductivity, and dissolved oxygen content) to detect changes in the chemical and physical properties of the soil and groundwater. With the aim of minimizing the use of piezometers, microbial activity is followed by studying gas emissions at the ground surface. Indeed, aerobic degradation of hydrocarbons results in the production of CO_2 [24]. Due to its limited solubility (in alkalinity saturated groundwaters), CO_2 gas will tend to migrate toward the ground surface and surface CO_2 fluxes can be directly linked to the biodegradation intensity. Meanwhile, this CO_2 production occurs concurrently with natural root and microbial soil respiration. Thus, tools capable of quantifying CO_2 sources are also needed. Carbon stable isotopic analyses (ratio of $^{13}C/^{12}C$) use the fact that petroleum hydrocarbon isotopic signature is distinguishable (range from −18 to −34‰ versus Vienna Pee Dee Belemnite or VPDB [25]) from other aquifer components: for example, carbonate minerals have a signature between +2 and −12‰ versus VPDB [26] and the C_3 plants (plants that use the Calvin or C_3 cycle of carbon fixation) produce organic matter with isotopic signature near −25 ± 5‰ versus VPDB [25, 27, 28]. Moreover, bacteria can induce a carbon isotopic fractionation as they preferentially use the lighter isotope ^{12}C (due to the lower energy required to break intramolecular bonds) [29]. Some studies linked the CO_2 production with contaminant degradation rate in laboratory [30–32] and at contaminated field sites [33]. Some researchers have examined the isotopic signature of CO_2 and dissolved inorganic carbon from aerobic biodegradation of petroleum hydrocarbons to support their analysis [34–38]. Results reflect the difficulty in definitively attributing CO_2 produced from biodegradation of contaminants versus that from possible interferences of background CO_2 under field

conditions. The other natural sources of CO_2, plant root respiration in shallow sediments and carbonate dissolution in subsurface sediments, must be taken into account in field experiments [25].

Integrating both geoelectrical methods and gas analyses could help to develop a less invasive technique with a little geochemical sampling in piezometers in order to monitor enhanced biodegradation of hydrocarbons. These tools were previously tested at laboratory scale in sand-columns with toluene as model pollutant and an exogenous bacterial strain [39]. The promising results from the column study allowed implementing a monitoring strategy at the field scale. The site was a gasoline station where gasoline and diesel fuels leaked eighteen years ago. A reactive barrier supplied the necessary oxygen to the aquifer in order to stimulate the aerobic bacterial processes since April 2014. Geoelectrical measurements and ER and IP surveys, as well as CO_2 analyses, had been performed from February 2014 until June 2015 on this site.

2. Materials and Methods

The experimentation had been implemented at a gasoline station over a 16-month period, from February 2014 to June 2015. After a preliminary characterization of the site (baseline, February 2014), and the start of the oxidative reactive barrier in April 2014, the long-term monitoring had been carried out.

The site had been equipped to monitor time-lapse changes in electrical properties of the subsurface with electrical resistivity (ER) and time domain induced polarisation (TDIP) systematic 2D tomographies. This monitoring had been combined with regular CO_2 fluxes measurements and its carbon isotopic ratio at the ground surface.

The indirect monitoring was completed with several analyses in piezometers to assess the noninvasive survey: (i) measurements of pollutant concentration and carbon isotopic signature of selected aromatic hydrocarbons and (ii) monitoring of groundwater physicochemical properties with probes.

2.1. Site Description. The site studied was a gasoline station where gasoline and diesel fuels leaked in 1997 (Figure 1). It is still in activity but the former tank installation (source of contamination) was dismantled. Several piezometers had been drilled for previous studies on this site. These studies showed that

(i) the ground was mainly composed of silts and clays: a first layer (thickness ranging from 0.5 to 3.5 m) of backfills made of brown silts, which often contained gravels, followed by a succession of loamy strata, with varying degrees of sands, and even sandy clays with limestone or millstone blocks (until 5 to 7.5 m depth); and then argillaceous limestones are interbedded with thin layers of sandy clays;

(ii) the ground had a low permeability (between 10^{-6} and 10^{-7} m/s);

(iii) the direction of flow was northwest;

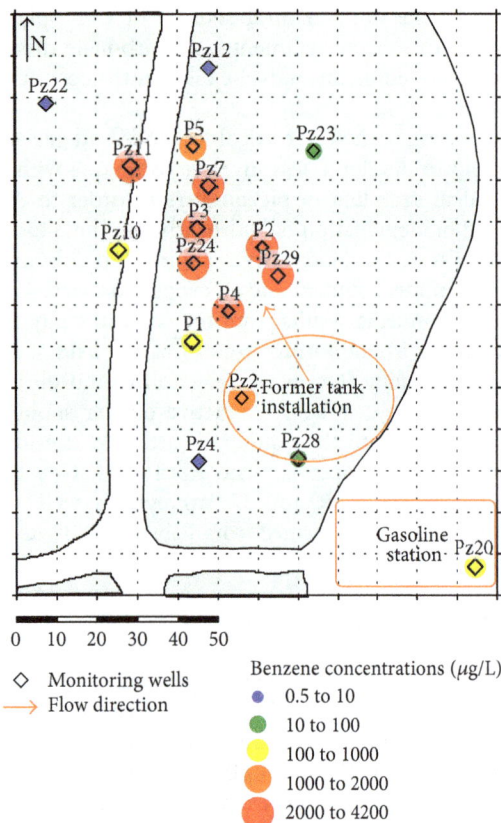

FIGURE 1: Map of the site showing gasoline station, former tank installation, monitoring piezometers, and the benzene concentrations measured in water in September 2013.

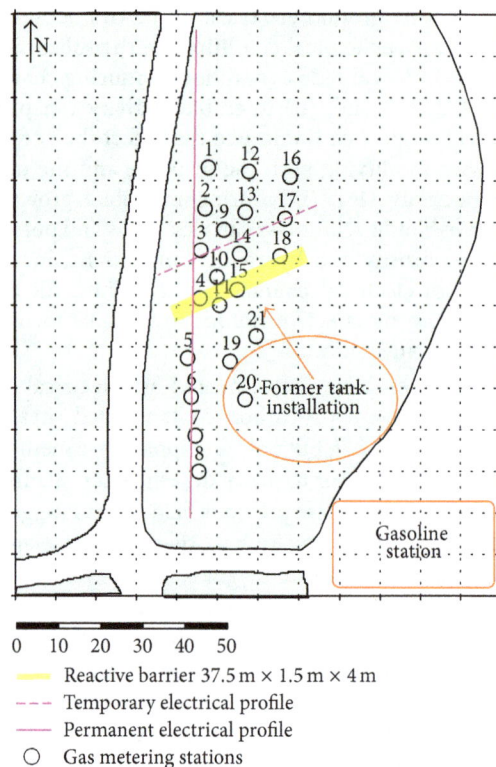

FIGURE 2: Map of the site showing the reactive barrier and the monitoring disposal: 2 electrical profiles and 21 gas metering stations.

(iv) the water table depth ranged between 2.5 and 4.5 m and depends on seasons (winter, summer);

(v) there was a presence of hydrocarbons in the area of the former tank installation and the fraction of benzene, toluene, ethylbenzene, and xylenes (BTEX) represented a consequent part of the pollutants (Figure 1);

(vi) there were indications of a natural attenuation (hydrocarbon degradation) which prove the presence of a bacterial flora, such as absence of dissolved oxygen, that matched the BTEX plume in water [40].

An important limiting factor in bioremediation of soils contaminated with hydrocarbons is the lack of oxygen to support microbial activities [41, 42]. Injection of aqueous H_2O_2 was the selected method to overcome this limitation and to stimulate aerobic metabolic processes.

A permeable reactive barrier of 4 m depth × 35 m long × 1.5 m wide was implemented to stop the plume migration (Figure 2). The site was equipped with 3 pumping wells just upstream the barrier, 3 injection wells in the barrier, and 2 injection wells just downstream the barrier. Pumped water was filtered through active charcoal. After filtration, diluted H_2O_2 was added to the treated water and oxygenated water was reinjected in the barrier, through the 3 injection wells, slotted from −2 to −4 m in order to deliver oxygen in

the upper part of the water table. The permeability of this barrier (10^{-3} m/s) allowed an homogeneous distribution of the treating fluid. Finally, the last part of the oxygenated water was injected in 2 boreholes downstream the barrier, slotted from −6 to −8 m, to bring oxygen in the deepest part of the water table. The biostimulation was started in April 2014.

2.2. Electrical Geophysical Measurements. Geophysical methods are a standard tool for obtaining information on volumetric distributions of subsurface physical properties of rocks and fluids. Among several electrical methods which measure electrical properties of the ground, two of them were used: the electrical resistivity (ER) and the time domain induced polarization (TDIP).

In a typical ER measurement, four electrodes are used. ER instrumentation measures potential differences between pairs of electrodes, where the potentials result from a current applied by the instrumentation between two other electrodes. By making measurements with current and potential electrodes at many different locations, one can collect sufficient data which allow the construction of an electrical resistivity section (a 2D slice of the earth).

Whereas ER is sensitive to pore fluid resistivity (and consequently to total dissolved solids or ionic strength), the IP method measures the capacitive behaviour of the subsurface, which strongly depends on the properties of the mineral surface. Different studies have shown the sensitivity of IP to mineral precipitation and changes occurring at pore-grain interfaces, where biological reactions occur [12]. TDIP measurements use the same instrumentation as ER measurements. Here, the injected current is alternately ON and OFF. The decrease of the induced voltage is measured to calculate the chargeability, which is the ground capacity to store charges. Hereafter, in this paper, the normalized chargeability (defined as the chargeability divided by the resistivity magnitude) will be calculated to free the chargeability from variation of resistivity [43].

Two time domain ER and IP profiles were carried out at the study site. Positioning of the profiles was performed according to the contaminant plume and barrier location: both profiles went across the plume, one was perpendicular to the barrier, and the other was parallel, 8 m downstream the barrier (Figure 2).

The perpendicular profile was buried permanently at 50 cm of depth. Apparent resistivity and chargeability measurements were obtained using a system with 60 stainless steel electrodes: 30 electrodes, with a constant spacing of 4 m, were dedicated to current injection, and the other 30, with a spacing of 4 m as well, were used for potential measurement to avoid electromagnetic (EM) coupling effect. The two sets of electrodes are disposed staggered, so that each electrode is laid out at a constant interval of 2 m. To achieve a good lateral resolution and to avoid EM coupling effect, the dipole-dipole array was applied to obtain the apparent resistivity and chargeability pseudosections. A Syscal R1+ (from IRIS Instruments©, Orleans, France) has been used in this study. Measurements were performed continuously every two days. The second profile, parallel to the barrier, was a temporary profile. It consists in 48 electrodes separated by 1 m. Measurements were performed every two months with an Elrec Pro (IRIS Instruments, Orleans, France).

After performing time-lapse ER and IP measurements, the electrical data were analysed in three stages:

(i) Initial anomalous data were filtered by a process of elimination:

   (a) current $I < 100$ mA;
   (b) potential $V < 0.1$ mV;
   (c) apparent chargeability $m < 0$;
   (d) error $Q > 5\%$ (resistivity measurements) and $Q > 2\%$ (chargeability measurements);

(ii) The quadrupoles in common to all the data sets were selected;

(iii) Data processing was performed with RES2DINV software [44], in time-lapse mode. The process is based on several iterations comparing the measured data with a model calculated by the software for each time series data set. This process enables obtaining a 2D distribution of resistivity and chargeability sections at every time. The following settings were used to define the type of constrains applied during the time-lapse inversion process:

   (a) The relative importance given to minimize the difference between models at different times is controlled by the time-lapse damping factor. A value of 0 to 5 can be generally used: if a value of 0 is used, the inversion of the different time series data sets would be carried out independently; if a value of 1 is used, equal weight would be given to reducing the difference between the models at different times; a larger value of the damping factor forces the different time models to be more similar. After some previous tests, a value of 3 was selected.

   (b) Time-difference roughness filter was set for "smooth changes" between time models.

   (c) The maximum number of iterations was 5.

2.3. CO₂ Analyses: Emissions Fluxes and Carbon Isotopic Ratio. The geoelectrical measurements were combined with CO_2 analyses. The measurements of CO_2 effluxes at the ground surface demonstrate the occurrence and rate of aerobic hydrocarbon biodegradation [33]. However, to determine the accuracy of the method, contaminated-derived soil respiration must be distinguished from chemistry of mineral-carbonates and from the natural soil respiration, resulting from biodegradation of dead organic matter and plant root respiration.

CO_2 produced by fossil hydrocarbon degradation may be distinguished from that produced by other processes based on the carbon isotopic compositions, characteristic of the source material, and/or fractionation accompanying microbial metabolism. Indeed, carbon isotopic signature of CO_2 reaches those of contaminant and it can be accompanied by significant carbon isotope fractionation [37].

Stable isotopic ratio represents the abundance of the rare isotope with respect to the abundant isotope ($^{13}C/^{12}C$). Due to slight analytical variations, it is common to compare isotopic ratios measured in an unknown sample to those in a standard material; this results in the delta notation:

$$\delta^{13}C\,(‰) = \left[\frac{\left(^{13}C/^{12}C\right)_{\text{sample}}}{\left(^{13}C/^{12}C\right)_{\text{VPDB}}} - 1 \right] * 1000 \qquad (1)$$

with $\left(^{13}C/^{12}C\right)_{\text{VPDB}} = 0.011237$ [45].

CO_2 concentration and $\delta^{13}C$ isotopic ratio in produced CO_2 are measured by a high-resolution laser infrared spectrometer, called SPIRIT (Spectrometer Infrared *In Situ* Tropospheric) and developed by the LPC2E (CNRS, Orleans, France). This spectrometer is an adaptation of the original SPIRIT, described in Guimbaud et al. (2011) [46] and used for greenhouse gas emissions at the air-land interface [47, 48], by the implementation of a quantum cascade laser using the $^{12}CO_2$ and $^{13}CO_2$ ro-vibrational lines. The uncertainty for $\delta^{13}C$ determination on field deployment is on the order of 0.4‰. A complete description of the adapted SPIRIT for $^{13}C/^{12}C$ isotopic ratio quantification is available in Guimbaud et al. [49]. The spectrometer is directly connected with a closed accumulation chamber set on a permanent PVC cylinder collar (diameter = 30 cm) sunk into the soil to measure CO_2 emission fluxes and $\delta^{13}(CO_2)$. The ground

TABLE 1: List of probes in the monitoring piezometers (function and depth).

Monitoring piezometer	Probes	Depth (m)
Pz4	O_2	6.15
	Water level	4.15
	Conductivity	4
Pz7	O_2	3.68
	Conductivity	4.13
Pz2	Temperature A	5.29
	Temperature B	7.36
	Temperature C	9.3
	Water level	6.3
P2	Temperature	5.09
Pz24	O_2	4.98
	Conductivity	5.14
P4	O_2	5.09
	Conductivity	4.98
	Water level	5.06

surface inside the PVC collar was cleared of vegetation (grass, roots) over 15 cm depth. We had 21 measuring points on the site, upstream, above, and downstream the reactive barrier (Figure 2). Measurements were performed every two months since February 2014.

2.4. Geochemical Monitoring in Piezometers. To assess the indirect monitoring of the enhanced bioremediation, chemical analyses were performed every two months in water samples from monitoring piezometers. BTEX concentration and carbon isotopic ratios of these BTEX were measured. Moreover, some piezometer probes had measured continuously the water table level, temperature, conductivity, and oxygen rate in monitoring piezometers (Table 1).

3. Results

3.1. Baseline Characterization. Baseline characterization consisted of collecting datasets prior to the oxygen injection (February-March 2014).

3.1.1. Geophysical Data. Figure 3 shows the results for the permanent profile in March 2014. This profile went through the barrier (metric point 51 m) and was within close proximity to several piezometers (from north to south: Pz12, P5, Pz7, P3, Pz24, P1, and Pz4) whose positions are plotted on Figure 3. The temporary profile was located around 8 m downstream the barrier and it overlapped with two piezometers (from west to east: Pz24 and P2, Figure 4).

The electrical resistivity tomography from the permanent profile showed resistivities between 40 and 80 $\Omega \cdot$m, over the 6 first meters, which correspond to the sandy-loamy layer of the aquifer. From 6-meter depth, resistivity was much smaller (<20 $\Omega \cdot$m) and this may be associated with a sandy clay level. The tomography from the temporary profile displayed the same resistivity range since the pilot site is quite homogeneous. Nevertheless, areas less resistant (<30 $\Omega \cdot$m)

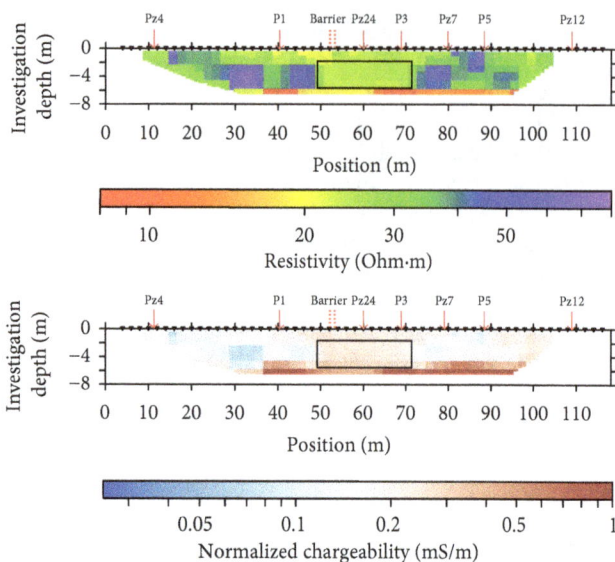

FIGURE 3: Resistivity and normalized chargeability sections across the plume for the permanent electrical profile, on 6th March 2014. The black rectangle highlights a zone of lower resistivity and higher normalized chargeability.

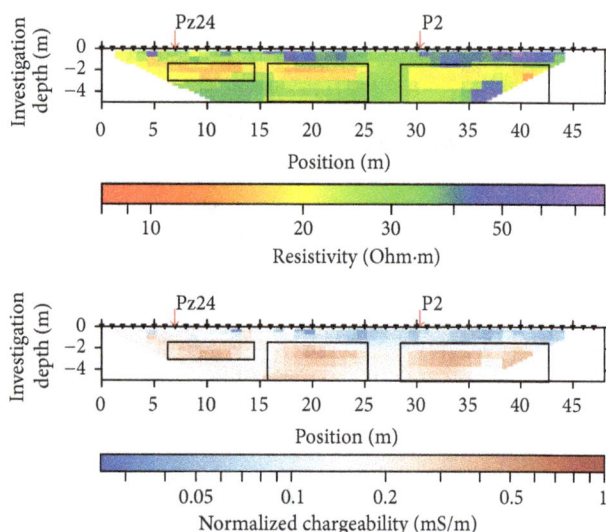

FIGURE 4: Resistivity and normalized chargeability sections across the plume for the temporary electrical profile, on 6th March 2014. The black rectangles highlight zones of lower resistivity and higher normalized chargeability.

could be identified between 1.5 and 6 m depth, on both profiles:

(i) Between the metric points 50 and 72 m for the permanent profile.

(ii) Between the metric points 7 and 14 m; 16 and 25 m; and 29 and 42 m, for the temporary profile.

Regarding the normalized chargeability tomographies (chargeability divided by resistivity), the values were in the range of 0.025 to 1 mS/m, and the zones of higher normalized

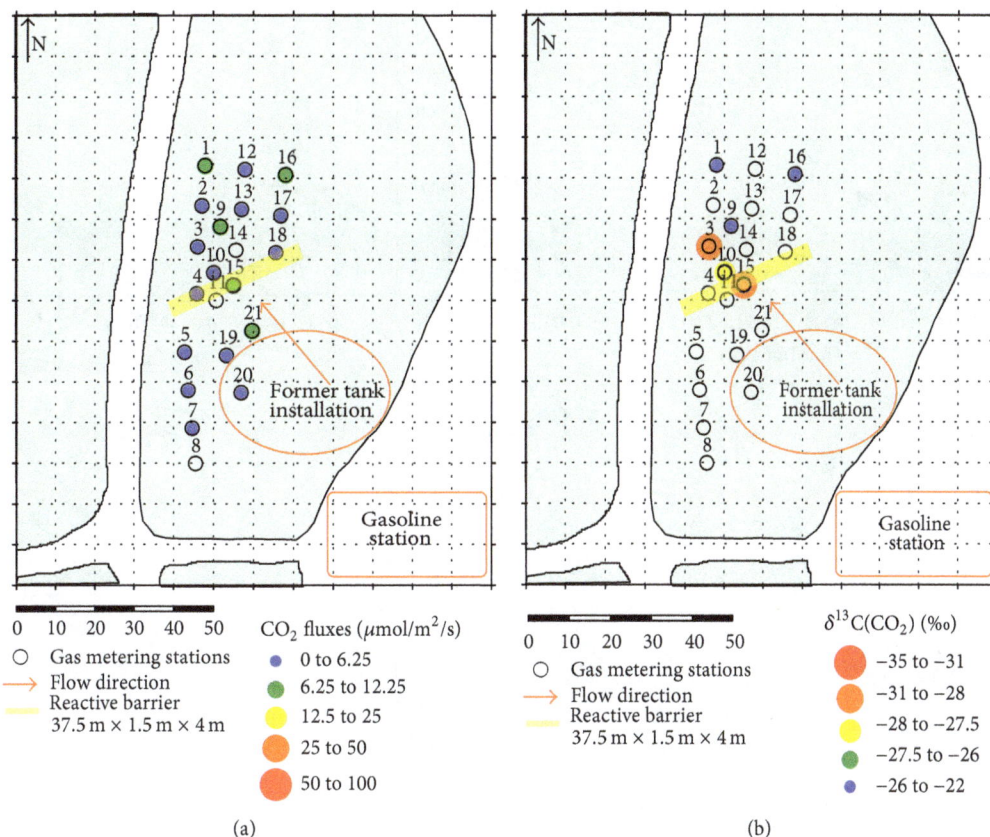

FIGURE 5: Maps of measured CO_2 emissions (a) and its carbon isotopic ratio versus VPDB (b), in February 2014.

chargeability (>0.2 mS/m) matched the lower resistivity zones. These normalized chargeability anomalies correspond to a polarization effect. On the permanent profile, the normalized chargeability was quite high beyond 6 m depth. This confirms the presence of clays, very chargeable, at these depths.

According to the geochemical data (Figure 1), the localization of these anomalies (higher normalized chargeability and lower resistivity) matched the hydrocarbon polluted area, downstream the former tank installation, at the top of the water table, where natural biodegradation produces conductive byproducts and where chargeable bacteria are numerous (proximity of an organic carbon source).

3.1.2. SPIRIT Data. Figure 5 displays the maps of measured CO_2 emissions and its carbon isotopic ratio in February 2014. There were quite high CO_2 emissions on this site, around $3.8 \, \mu mol \cdot m^{-2} \cdot s^{-1}$, despite the fact that, during cold periods (December–March), gas fluxes are presumably very low [50, 51]. These results were consistent with those found by Sihota et al. (2011) [33] where the average CO_2 efflux associated with contaminant degradation in the hydrocarbon-contaminated zone was estimated at $2.6 \, \mu mol \cdot m^{-2} \cdot s^{-1}$. Moreover, the measured $\delta^{13}C(CO_2)$, ranging between $-22.2 \pm 0.4‰$ and $-28.8 \pm 0.4‰$ versus VPDB (averaging $-25.3 \pm 3‰$), were very negative compared to $\delta^{13}C$ of atmospheric CO_2 (around $-11‰$ versus VPDB, one meter above the ground) and

distinguishable from $\delta^{13}C$ of CO_2 from vegetation respiration (around $-28‰$ versus VPDB [25, 27, 28]). This was an indication of natural biodegradation of hydrocarbons. Indeed, during biodegradation, CO_2 has an isotopic signature near those of the carbon source, which is quite negative in the case of fossil organic carbon (range from -23 to $-29‰$ versus VPDB for the BTEX of the site [40]).

3.2. Cleanup Progress after One Year of Biostimulation. Figure 6 presents the BTEX (benzene, toluene, ethylbenzene, and xylenes) concentrations over time for 4 piezometers, between February 2014 (two months before barrier activation) and April 2015 (one year after barrier activation). There was an increase in the pollutant concentrations due to the remobilization of contaminants during the barrier construction. The discharge of BTEX may have had significantly slowed down the oxygen diffusion downstream the barrier: it had been quickly consumed and the amount of H_2O_2 could not be increased for avoiding bacteria stress. And, one year after the beginning of biostimulation, there was just a little decrease in benzene and toluene concentrations.

3.3. One Year of Electrical Monitoring. Before barrier activation, it was possible to localize a zone of natural biodegradation on the permanent and the temporary profile (Figures 3 and 4). Figure 7 shows the evolution of the electrical data over a year of enhanced biodegradation, for the permanent profile.

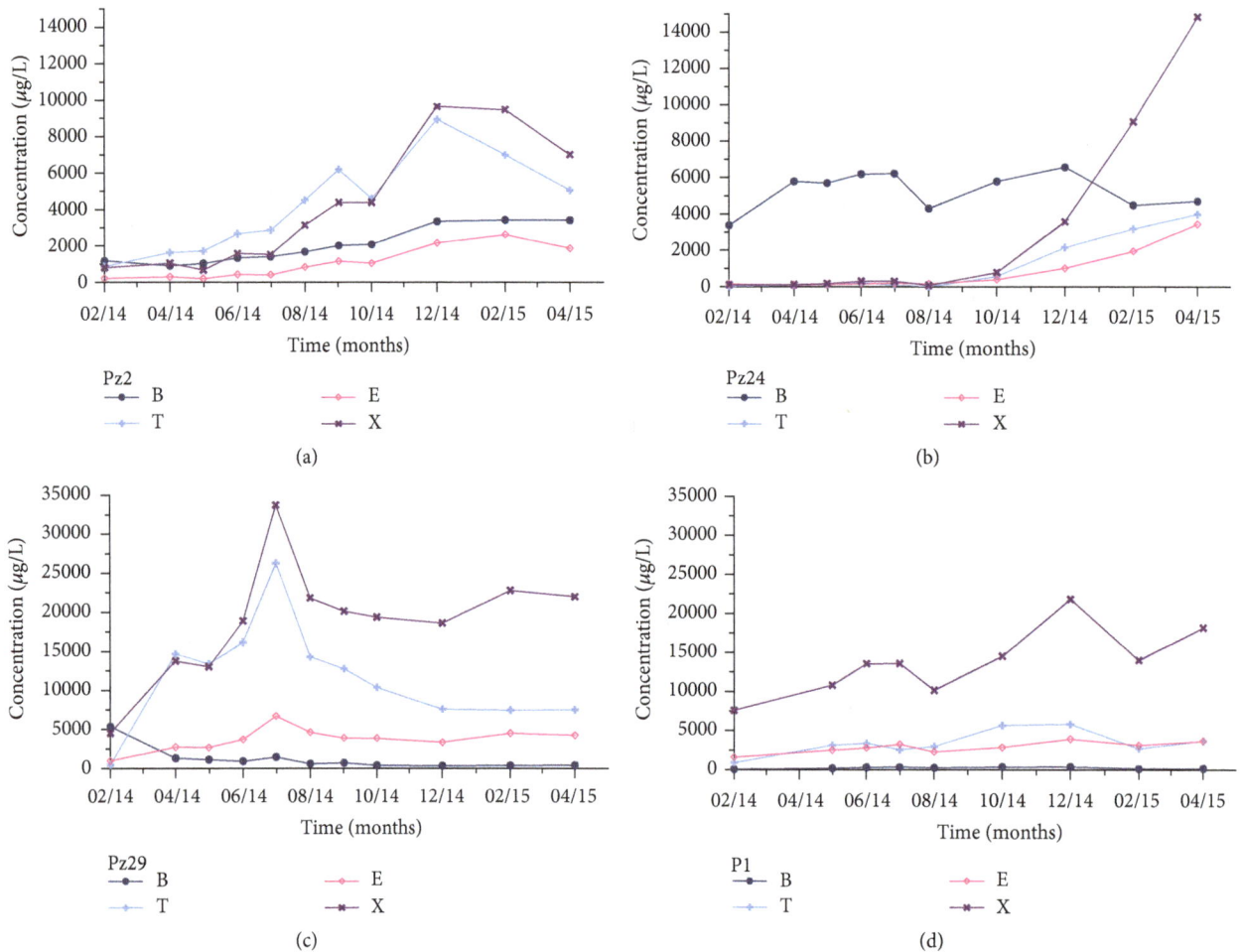

Figure 6: Evolution of BTEX (benzene, toluene, ethylbenzene, and xylenes) concentrations in water over time, measured in four monitoring piezometers: Pz2 (a), Pz24 (b), Pz29 (c), and P1 (d), between February 2014 and April 2015.

The resistivity model sections were obtained after time-lapse inversion of 12 datasets, spaced about a month, acquired between March 2014 and April 2015. The convergence of iterative inversion is excellent (<1% for resistivity and <2 for normalized chargeability) and it is comparable with the noise level of the resistivity meter (1%, according to Iris Instruments).

In order to compare the electrical data taking into account the impact of temperature and water saturation on electrical resistivity and chargeability, the inversed data were considered only in the saturated zone and adjusted at 25°C thanks to a linear law [52, 53]:

$$\frac{\rho_{25}}{\rho_T} = \alpha * (T - 25) + 1, \qquad (2)$$

where ρ_T is resistivity at temperature T (°C), ρ_{25} is resistivity at 25°C, and α (°C^{-1}) is a temperature compensation factor. The temperature compensation factor α was calculated based on Pz4 (located in a nonpolluted area) temperature data and found to be equal to 0.014, which means that resistivity decreases by 1.4% when temperature increases by 1°C.

Figure 7 shows the anomaly of lower resistivity and higher normalized chargeability on all the tomographies,

from March 2014 to April 2015. It was possible to identify two periods of decreasing resistivity and increasing normalized chargeability: a first period from March 2014 to October 2014, and a second period from March 2015 to April 2015. In order to quantify the evolution of the differences, the percentage difference of resistivity and normalized chargeability from March 2014 were calculated (Figure 8). A decrease in resistivity up to 15% is noticed in the area of interest (between metric points 50 and 70 m) at the end of the summer of 2014. However, such a seasonal variation cannot be seen on water conductivity (Figure 9(a)). Regarding normalized chargeability, the tendency is less clear and there are increases outside the area of interest as well.

3.4. One Year of CO_2 (Fluxes and Carbon Isotope Ratio) Monitoring.

Gas monitoring was carried out every one-two months, between February 2014 and June 2015. The time allowed for measurements was very short (2 days per campaign). As a consequence, priority was given to flux measurements and then to isotopic ratio measurements of metering points with the highest emissions (more accurate).

CO_2 fluxes ranged between 2.5 and 100 μmol·m^{-2}·s^{-1} (Figure 10(a)). As there was no vegetation inside the PVC

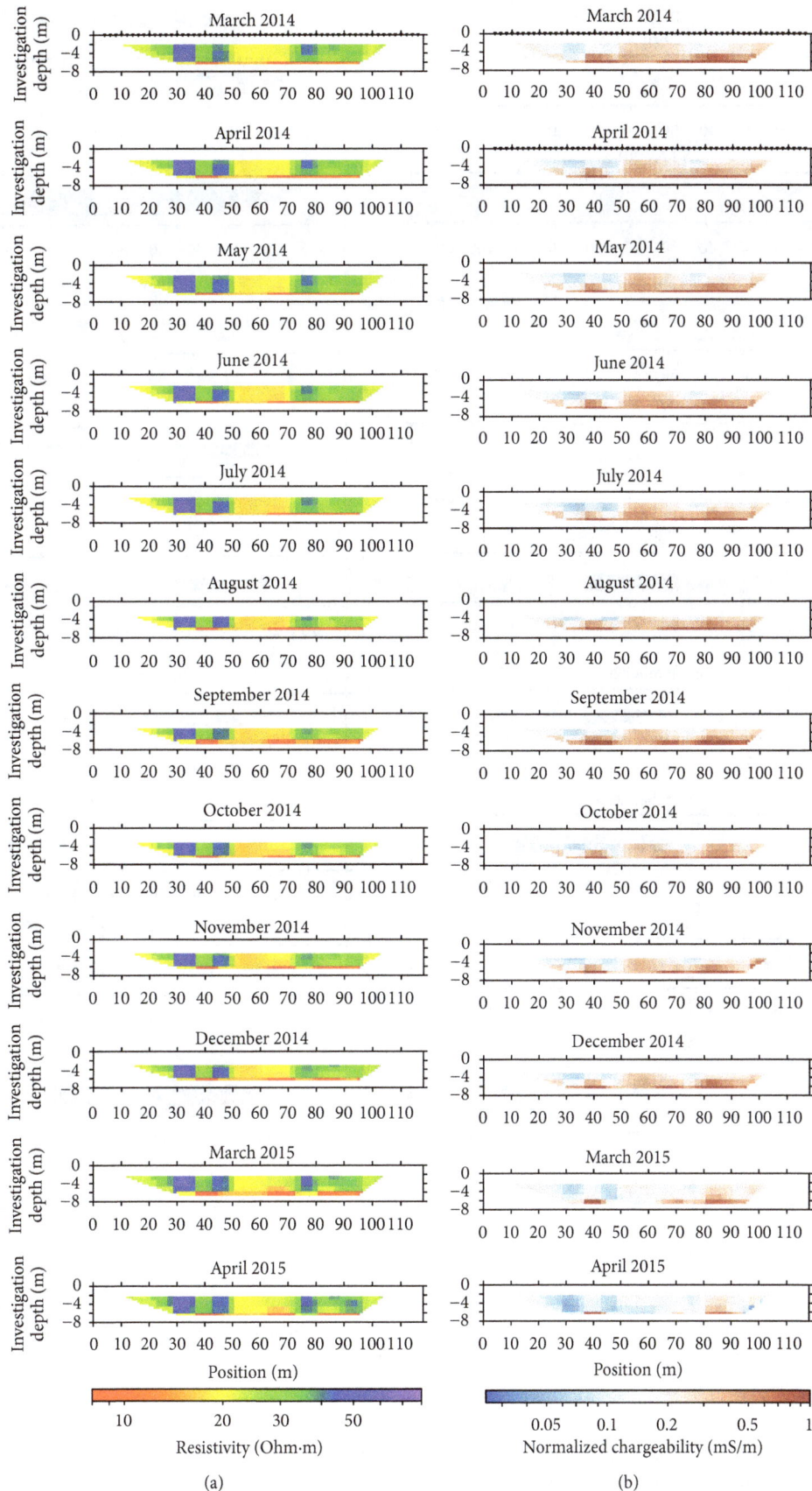

FIGURE 7: Resistivity (a) and normalized chargeability (b) sections, adjusted at 25°C in the saturated zone, for the permanent electrical profile, between March 2014 and April 2015. The inversion results were obtained after 5 iterations, with 0.58 and 1.82% of error for resistivity and chargeability, respectively.

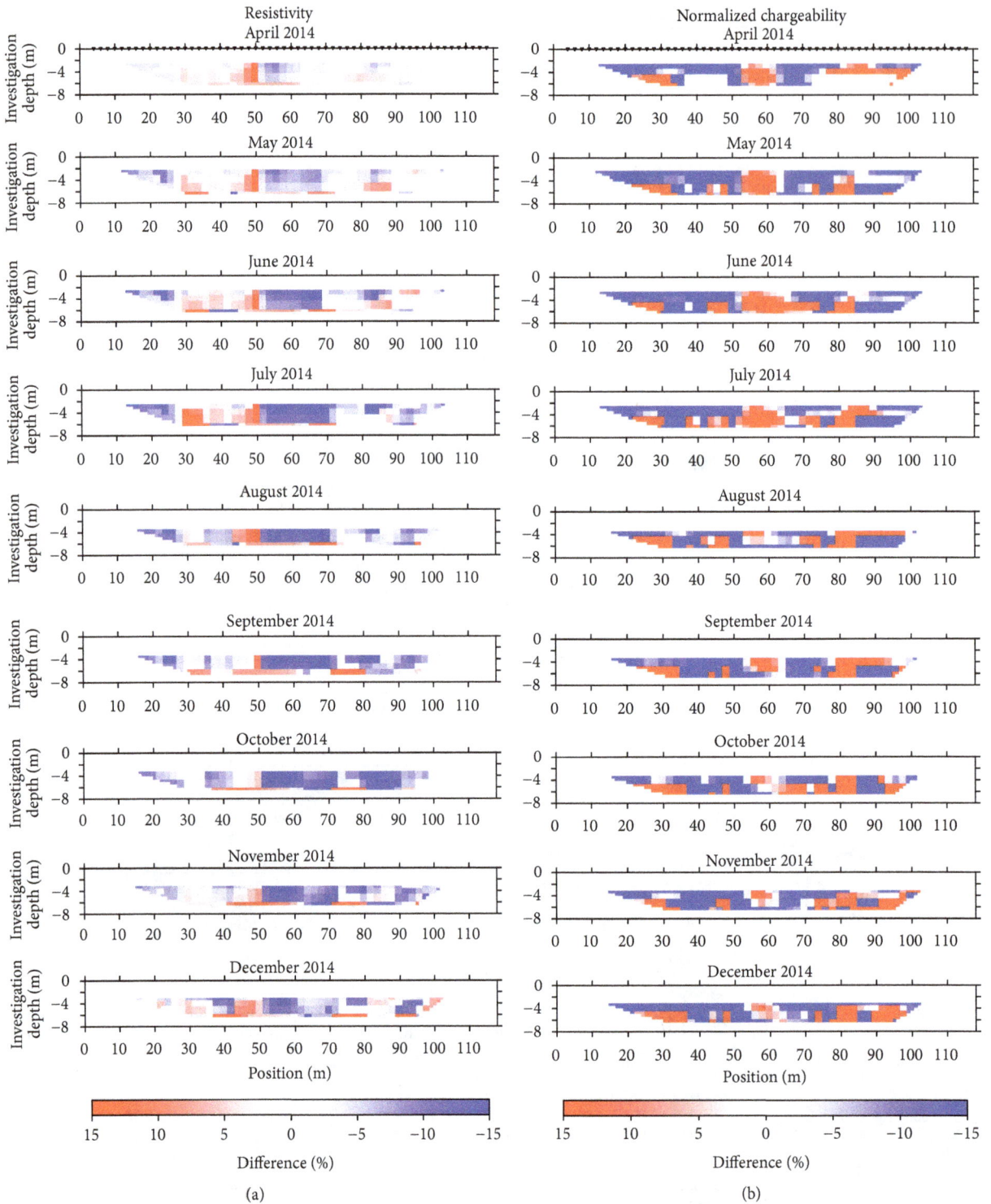

FIGURE 8: Percentage difference (P) over March 2014 of resistivity ρ (a) and normalized chargeability nm (b) sections, for the permanent electrical profile, with $P_\rho = (\rho_{ti} - \rho_{t0})/\rho_{t0} * 100$ and $P_{nm} = (nm_{ti} - nm_{t0})/nm_{t0} * 100$.

collar used to put the SPIRIT chamber, the CO_2 released at the ground surface should come from degradation of soil organic matter and mainly from hydrocarbon biodegradation. The lower emissions were recorded in February 2014. Then, they gradually increased up to August 2014, and decreased during

winter 2014-2015. However, meteorological conditions affect gas emissions: CO_2 fluxes were lower when temperatures were low and ground was wet (high rainfalls) (Figure 10) [50]. In order to throw off this effect, we chose a reference measuring point, no. 7, located in the nonpolluted area, where

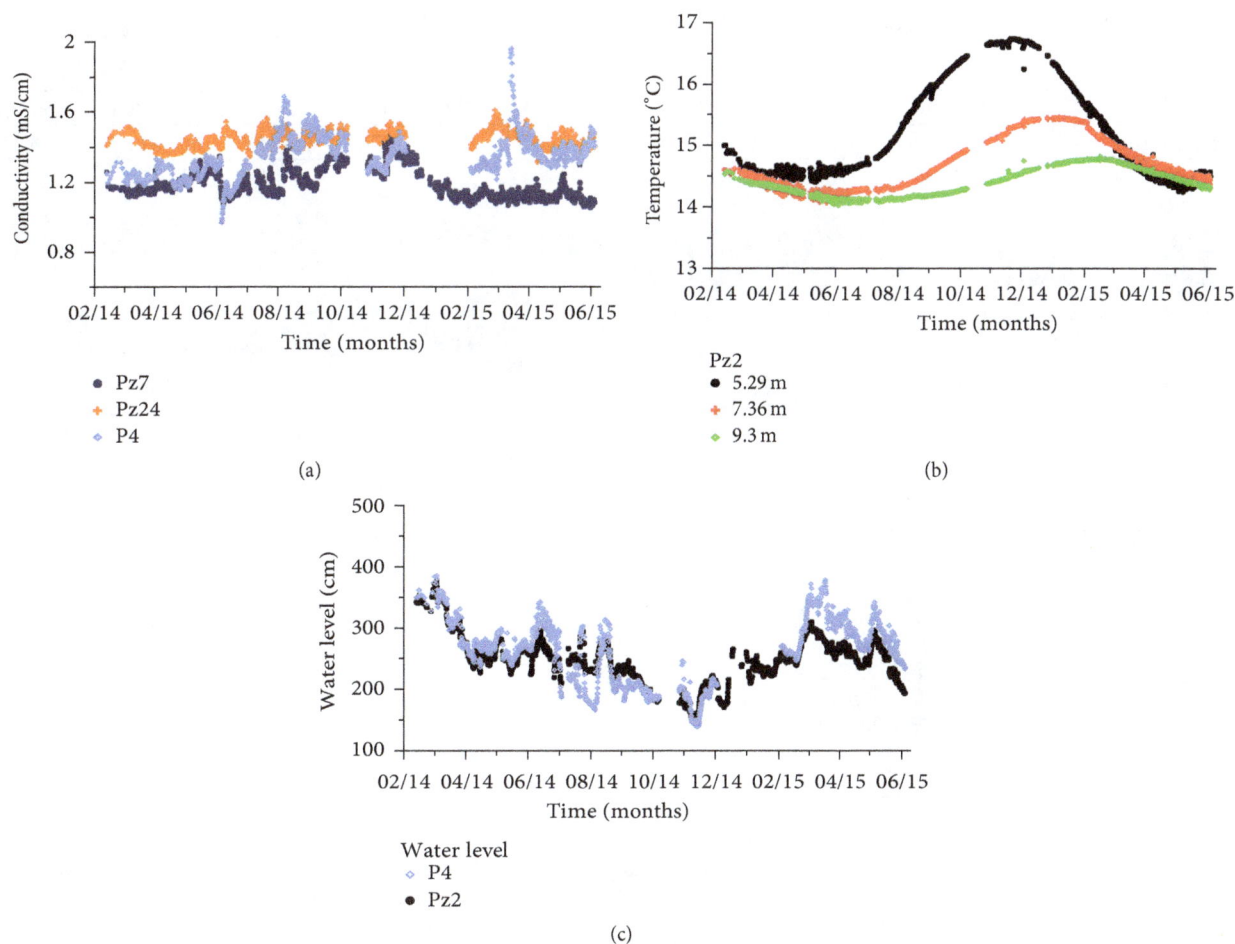

FIGURE 9: Water conductivity (a) and temperature (b) and ground water level (c) evolution over time.

changes in fluxes are supposed to be only due to temperature and moisture conditions. The difference of flux with season at this point was subtracted from the values obtained at the other measuring points to keep only the flux variations due to bacterial activity (degradation of natural organic matter and hydrocarbons) (Table 2). A seasonal variation of the CO_2 emissions is still monitored (Figure 11): there were more emissions during summer and fall, when groundwater level is low (Figure 9(c)), water temperature is warmer (Figures 9(b) and 10(b)), and ground is less moist (Figure 10(c)).

Moreover, CO_2 emissions had been measured with an isotopic signature typical of hydrocarbon biodegradation. Indeed, $\delta^{13}C(CO_2)$ were measured between −27 and −30‰ versus VPDB since February 2014 (Figure 12(a)), in accordance with the BTEX $\delta^{13}C$ signature measured in water (around −25 to −26‰ versus VPDB) (Figure 12(b)). Besides, CO_2 released at the surface originates from a more or less degraded BTEX located below the gas measuring station. Indeed, in December 2014, average evolution of $\delta^{13}C$(benzene) increased from (−28.1 ± 0.5)‰ to (−26.7 ± 1.2)‰ and $\delta^{13}C$(toluene) increased from (−27.4 ± 0.4)‰ to (−24.8 ± 1.9)‰, from upstream to downstream the pollution plume. At the same time, $\delta^{13}C(CO_2)$ increased, from upstream to downstream as well, from (−30.5 ± 0.7)‰ to

(−28.3 ± 1.6)‰ in July 2014 and from (−28.1 ± 1.1)‰ to (−27.0 ± 1.2)‰ in October 2014 [49]. Although these $\delta^{13}C(CO_2)$ values were only a few per mil less than the $\delta^{13}C(CO_2)$ from microbial degradation of plant material, $\delta^{13}C(CO_2)$ showed a light isotopic fractionation (1 to 3‰) through the stream of the pollution plume:

(i) $\delta^{13}C(CO_2)$ was lower than $\delta^{13}C$(hydrocarbon source).

(ii) The pollutant was enriched in ^{13}C when it was degraded, when residual BTEX concentrations were lower than 2000 μg·L^{-1}.

(iii) The CO_2 tended to be depleted in ^{13}C, when fluxes were higher than 20 μmol·m^{-2}·s^{-1}.

It means that molecules with light isotopes (^{12}C) were preferred by bacterial metabolism. This is a typical isotopic fractionation due to biodegradation. This is in agreement with one example of a successful field application of isotopic assessment of biodegradation in an aviation gasoline contaminated aquifer where aerobic biodegradation of the gasoline constituents produced significantly enriched CO_2 [54]: the $\delta^{13}C(CO_2)$ values were ranging from −22.0 to −26.3‰, soil organic matter for the site ranged from −13.5 to −26.1‰, and the aviation gasoline $\delta^{13}C$ values were between

TABLE 2: Corrections made to CO_2 measurements from April 2014 to April 2015 (February 2014 is the baseline, uncorrected) from gas measuring point number 7.

	Apr. 2014	May 2014	June 2014	July 2014	Aug. 2014	Oct. 2014	Dec. 2014	Feb. 2015	Apr. 2015
Corrections (μmol·m^{-2}·s^{-1})	−0.6	−1.4	−3.05	−7.87	−9.9	−7.65	+1.22	+1.2	−3.27

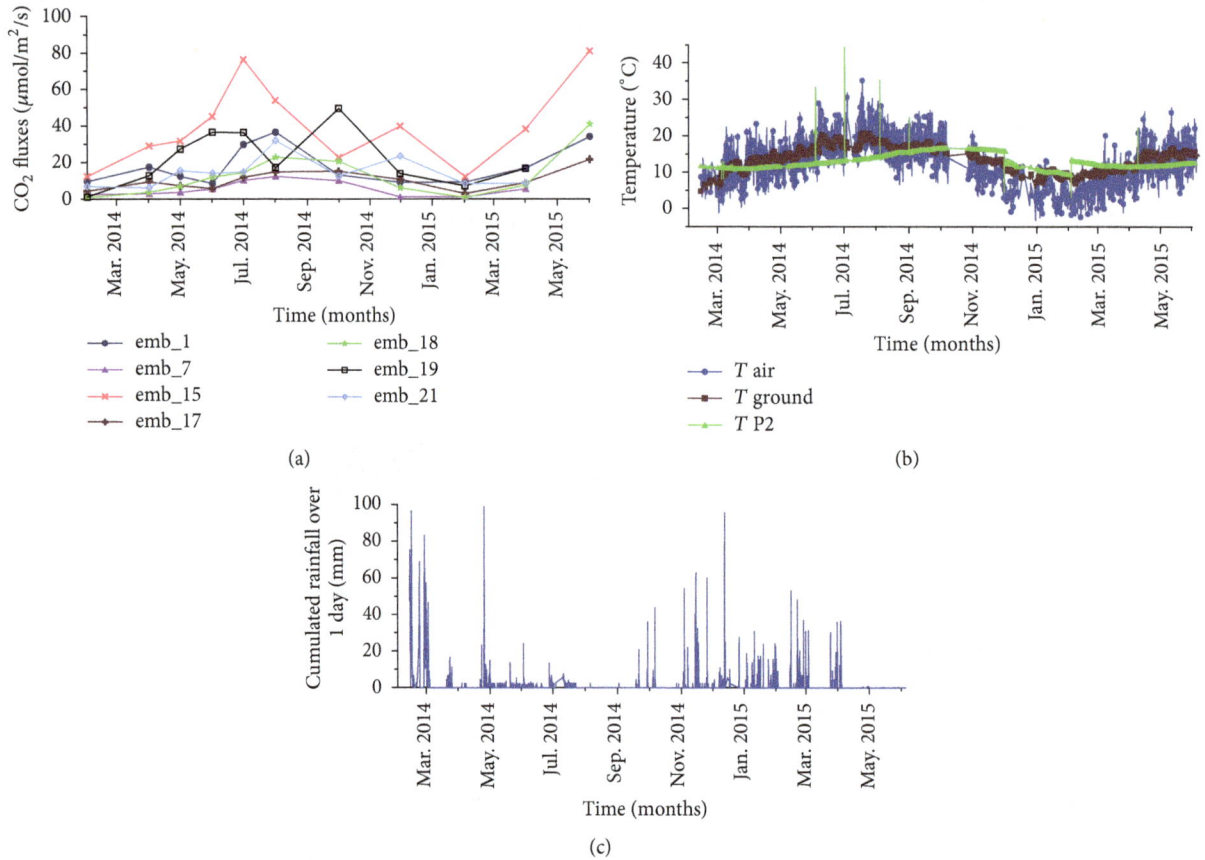

FIGURE 10: Measured CO_2 emissions (a); air, ground, and water temperatures (b); and rainfall (c) over time.

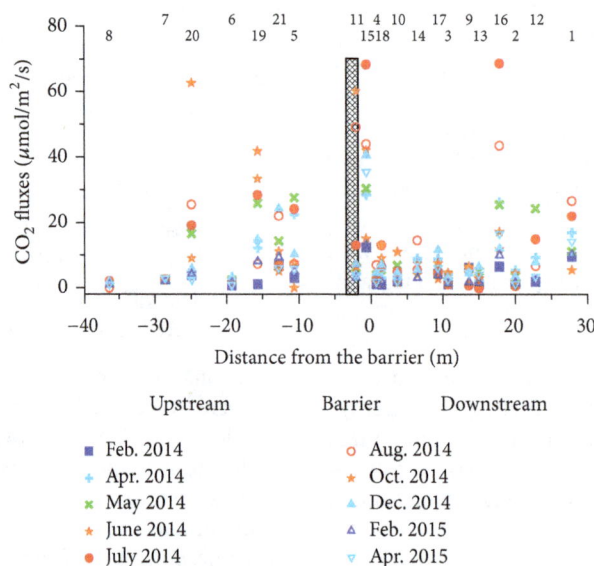

FIGURE 11: Measured CO_2 emissions (corrected from meteorological conditions) as a function of gas metering station location, for one year of monitoring.

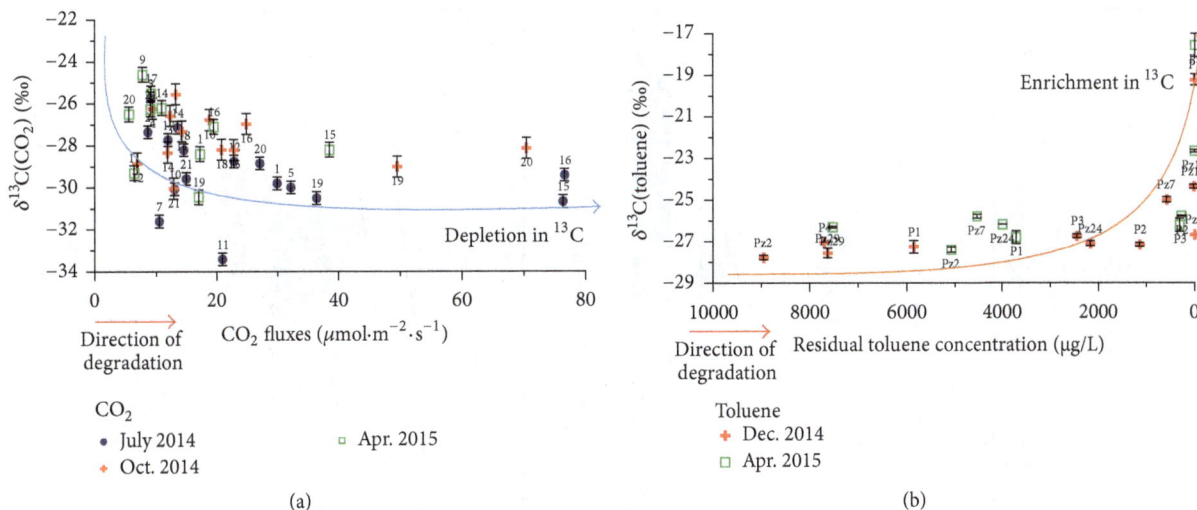

FIGURE 12: Carbon isotopic ratio versus VPDB of CO_2 (a) in function of CO_2 fluxes at the ground surface and carbon isotopic ratio versus VPDB of residual toluene (b) as a function of toluene concentrations in water.

−20.5 and −27.3‰. Conrad et al. (1999) [54] concluded that the difference of $\delta^{13}C$ was discernible. Another study from Aelion et al. (1997) [55] was carried out at a gasoline contaminated groundwater site under aerobic degradation. Although some of the $\delta^{13}C(CO_2)$ values from the contaminated site (between −35.9 and −22.0‰) overlapped with background $\delta^{13}C(CO_2)$ (between −22.9 and −23.9‰) from uncontaminated areas, on average the stable isotopic values were distinguishable between the contaminated and background sites.

4. Integration of Results and Conclusion

At the pilot site, the combination of geoelectrical measurements (ER and TDIP) with CO_2 analyses (fluxes and carbon isotopic ratios) was tested as monitoring tools to follow an enhanced biodegradation of hydrocarbons.

The results from the baseline characterisation in February-March 2014 had shown a more conductive and chargeable zone, around 3 meters in depth, which matches the polluted zone defined by geochemical borehole analyses, at the top of the water table (Figures 1, 3, 4, and 13). In this area, it can be assumed that there are numerous chargeable bacteria and production of conductive metabolites due to the pollutant degradation. This is consistent with the age of the gasoline leakage (almost twenty years ago), sufficient to allow the development of natural bacterial flora able to degrade hydrocarbons [56, 57] and to transform an electrical resistive substance in a more conductive one [12]. Moreover, CO_2 emissions were strong in this area and had an isotopic signature typical of hydrocarbon biodegradation. However, the overlap between $\delta^{13}C(CO_2)$ values from uncontaminated areas (plant root respiration and organic matter degradation) and from contaminated areas (with petroleum hydrocarbon degradation in addition) suggests that the combined measurement of $\delta^{13}C(CO_2)$ and $\delta^{13}C(BTEX)$ may be needed to confirm that the CO_2 is indeed from the contaminant degradation and not from microbial metabolism of natural

plant materials. The northern zone of the site must be considered as a potential contaminated zone as well, accepting that the CO_2 emissions (measuring points 1, 12, and 16) are a sign of biodegradation and not the result of the presence of gas preferential pathways in this area (Figure 13).

The treatment for enhanced biodegradation was started in April 2014, with the activation of the reactive barrier that brought oxygen to the water table to stimulate aerobic biometabolism. The regular monitoring over one year showed that the area of biodegradation was still detected by both indirect methods (Figure 13). Moreover, a seasonal variation (after correction of effects associated with groundwater and soil temperature and soil moisture) for both electrical and gas monitoring was highlighted. Indeed, we observed highest conductivities of the ground at the end of summer and during fall, seasons when the piezometer probes measured the warmer groundwater temperatures (Figure 9(b)). The inverted resistivity sections were adjusted from temperature effect [52, 53], with a correction limited at groundwater level. In piezometer Pz2, at 5.29 m depth, the highest variation of temperature over a year is of the order of 2°C (Figure 9(b)), which corresponds to a variation of resistivity (1.4% per °C) lower than 3%. Nonetheless, a significant decrease in resistivity up to 15% was observed in the area of biodegradation at the end of the summer of 2014 and such a seasonal variation cannot be seen on water conductivity (Figure 9(a)). However, there is also a seasonal variation of the bacterial activity: when groundwaters are warmer, bacterial activity is more intensive and degrades more hydrocarbons and should produce more conductive metabolites [58–62]. Similarly, gas analyses, mainly CO_2 fluxes measurements, showed a seasonal variation. However, this should be mainly due to variation of soil respiration processes; otherwise we would have observed a seasonal variation in BTEX isotopic signatures. On the basis of this analysis, it is possible to make an assertion that seasonal variation of biodegradation processes was detected by our nonintrusive monitoring tools, especially by geoelectrical measurements.

(a)

(b)

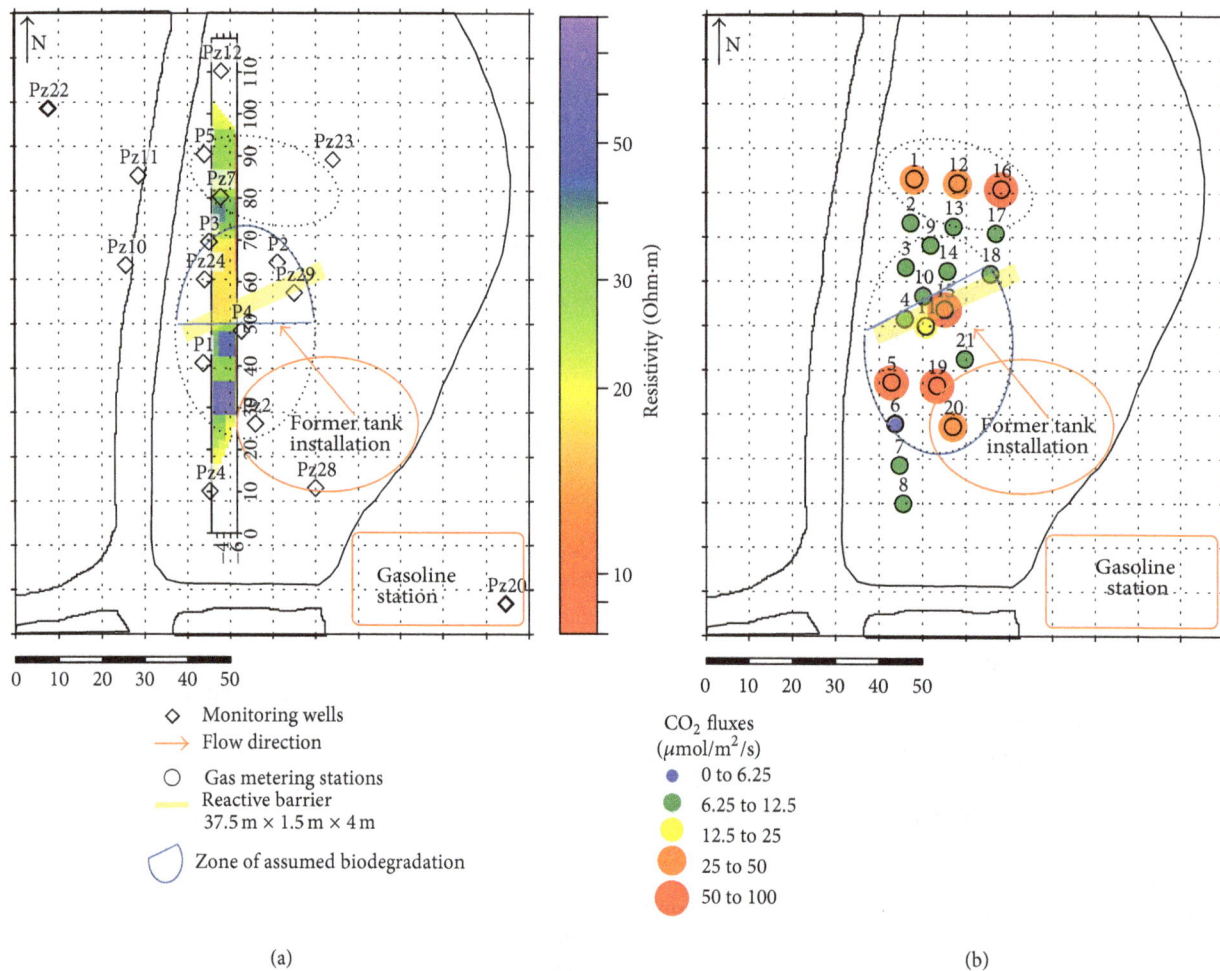

FIGURE 13: Maps of electrical resistivity measured on the permanent profile (a) and of measured CO_2 emissions (b), in July 2014. The elliptical shape highlights the area of biodegradation with production of conductive metabolites and CO_2.

However, the permeability of the ground is low and the groundwater table velocity is around 17 m/year. The oxygen from the barrier diffused slowly and its effect was not detected by both direct and indirect monitoring. The seasonal variation of bacterial activity was not yet "erased" by the continuous oxygen injection. Nonetheless, our first results over a year of monitoring show the interest of using geophysical methods and gas analyses to monitor and evaluate *in situ* remediation.

By applying the technology described here, fewer piezometers will be required for the monitoring and the understanding of bioremediation processes, leading to significant cost saving due to fewer monitoring requirements (piezometers, samples, and lab analyses) and more optimized remedial applications based on rapid identification of missed target zones.

Acknowledgments

The authors acknowledge the French Research Agency (Project BIOPHY, ANR-10-ECOT-014) and LABEX VOLTAIRE (Excellence Laboratory for Atmosphere, Resources and Environmental Interactions Study, University of Orleans, France) (ANR-10-LABX-100-01) for financial support. They also thank all the cooperation partners of the BIOPHY project: BRGM, LPC2E, Serpol, and Total. The LPC2E technical research team is gratefully acknowledged for their contribution to the development of the SPIRIT instrument. Technical support from BRGM is greatly appreciated as well. And they thank P. Gaudry, S. Williams, and G. Belot (LPC2E master's students) for their help during *in situ* measurement campaigns, as well as E. Verardo (BRGM Ph.D. student) for her contribution.

References

[1] N. Das and P. Chandran, "Microbial degradation of petroleum hydrocarbon contaminants: an overview," *Biotechnology Research International*, vol. 2011, Article ID 941810, 13 pages, 2011.

[2] F. Nadim, G. E. Hoag, S. Liu, R. J. Carley, and P. Zack, "Detection and remediation of soil and aquifer systems contaminated with petroleum products: an overview," *Journal of Petroleum Science and Engineering*, vol. 26, no. 1–4, pp. 169–178, 2000.

[3] K.-F. Chen, C.-M. Kao, C.-W. Chen, R. Y. Surampalli, and M.-S. Lee, "Control of petroleum-hydrocarbon contaminated groundwater by intrinsic and enhanced bioremediation," *Journal of Environmental Sciences*, vol. 22, no. 6, pp. 864–871, 2010.

[4] M. Farhadian, C. Vachelard, D. Duchez, and C. Larroche, "In situ bioremediation of monoaromatic pollutants in groundwater: a review," *Bioresource Technology*, vol. 99, no. 13, pp. 5296–5308, 2008.

[5] J. V. Weiss and I. M. Cozzarelli, "Biodegradation in contaminated aquifers: incorporating microbial/molecular methods," *Ground Water*, vol. 46, no. 2, pp. 305–322, 2008.

[6] M. Majone, R. Verdini, F. Aulenta et al., "In situ groundwater and sediment bioremediation: barriers and perspectives at European contaminated sites," *New Biotechnology*, vol. 32, no. 1, pp. 133–146, 2015.

[7] A. Flores Orozco, A. Kemna, C. Oberdörster et al., "Delineation of subsurface hydrocarbon contamination at a former hydrogenation plant using spectral induced polarization imaging," *Journal of Contaminant Hydrology*, vol. 136-137, pp. 131–144, 2012.

[8] J. A. Sogade, F. Scira-Scappuzzo, Y. Vichabian et al., "Induced-polarization detection and mapping of contaminant plumes," *Geophysics*, vol. 71, no. 3, pp. B75–B84, 2006.

[9] H. Vanhala, "Mapping oil-contaminated sand and till with the Spectral Induced Polarization (SIP) method," *Geophysical Prospecting*, vol. 45, no. 2, pp. 303–326, 1997.

[10] G. R. Olhoeft, "Direct detection of hydrocarbon and organic chemicals with ground penetrating radar and complex resistivity," in *Proceedings of the NWWA/API Conference on Petroleum Hydrocarbons and Organic Chemicals in Ground Water-Prevention, Detection and Restoration*, Houston, Tex, USA, November 1986.

[11] F. M. Mewafy, D. D. Werkema, E. A. Atekwana et al., "Evidence that bio-metallic mineral precipitation enhances the complex conductivity response at a hydrocarbon contaminated site," *Journal of Applied Geophysics*, vol. 98, pp. 113–123, 2013.

[12] E. A. Atekwana and E. A. Atekwana, "Geophysical signatures of microbial activity at hydrocarbon contaminated sites: a review," *Surveys in Geophysics*, vol. 31, no. 2, pp. 247–283, 2010.

[13] E. A. Atekwana and L. D. Slater, "Biogeophysics: a new frontier in Earth science research," *Reviews of Geophysics*, vol. 47, no. 4, Article ID RG4004, 2009.

[14] W. A. Sauck, "A model for the resistivity structure of LNAPL plumes and their environs in sandy sediments," *Journal of Applied Geophysics*, vol. 44, no. 2-3, pp. 151–165, 2000.

[15] V. Naudet and A. Revil, "A sandbox experiment to investigate bacteria-mediated redox processes on self-potential signals," *Geophysical Research Letters*, vol. 32, no. 11, Article ID L11405, 2005.

[16] N. Schwartz, T. Shalem, and A. Furman, "The effect of organic acid on the spectral-induced polarization response of soil," *Geophysical Journal International*, vol. 197, no. 1, pp. 269–276, 2014.

[17] R. Albrecht, J. C. Gourry, M.-O. Simonnot, and C. Leyval, "Complex conductivity response to microbial growth and biofilm formation on phenanthrene spiked medium," *Journal of Applied Geophysics*, vol. 75, no. 3, pp. 558–564, 2011.

[18] C. A. Davis, E. Atekwana, E. Atekwana, L. D. Slater, S. Rossbach, and M. R. Mormile, "Microbial growth and biofilm formation in geologic media is detected with complex conductivity measurements," *Geophysical Research Letters*, vol. 33, no. 18, Article ID L18403, 2006.

[19] A. Arato, M. Wehrer, B. Biró, and A. Godio, "Integration of geophysical, geochemical and microbiological data for a comprehensive small-scale characterizationof an aged LNAPL-contaminated site," *Environmental Science and Pollution Research*, vol. 21, no. 15, pp. 8948–8963, 2014.

[20] G. Cassiani, A. Binley, A. Kemna et al., "Noninvasive characterization of the Trecate (Italy) crude-oil contaminated site: links between contamination and geophysical signals," *Environmental Science and Pollution Research*, vol. 21, no. 15, pp. 8914–8931, 2014.

[21] R. M. Rosales, P. Martínez-Pagan, A. Faz, and J. Moreno-Cornejo, "Environmental monitoring using Electrical Resistivity Tomography (ERT) in the subsoil of three former petrol stations in SE of Spain," *Water, Air, and Soil Pollution*, vol. 223, no. 7, pp. 3757–3773, 2012.

[22] V. Che-Alota, E. A. Atekwana, E. A. Atekwana, W. A. Sauck, and D. D. Werkema Jr., "Temporal geophysical signatures from contaminant-mass remediation," *Geophysics*, vol. 74, no. 4, pp. B113–B123, 2009.

[23] K. H. Williams, A. Kemna, M. J. Wilkins et al., "Geophysical monitoring of coupled microbial and geochemical processes during stimulated subsurface bioremediation," *Environmental Science & Technology*, vol. 43, no. 17, pp. 6717–6723, 2009.

[24] K. Kaufmann, M. Christophersen, A. Buttler, H. Harms, and P. Höhener, "Microbial community response to petroleum hydrocarbon contamination in the unsaturated zone at the experimental field site Værløse, Denmark," *FEMS Microbiology Ecology*, vol. 48, no. 3, pp. 387–399, 2004.

[25] C. M. Aelion, P. Höhener, D. Hunkeler, and R. Aravena, *Environmental Isotopes in Biodegradation and Bioremediation*, CRC Press, Taylor & Francis, 2010.

[26] T. W. Boutton, "Stable carbon isotope ratios of natural materials: II. Atmospheric, terrestrial, marine, and freshwater environments," in *Carbon Isotope Techniques*, vol. 1, chapter 11, pp. 173–185, Academic Press, 1991.

[27] J. H. Troughton and K. A. Card, "Temperature effects on the carbon-isotope ratio of C3, C4 and crassulacean-acid-metabolism (CAM) plants," *Planta*, vol. 123, no. 2, pp. 185–190, 1975.

[28] C. Körner, G. D. Farquhar, and Z. Roksandic, "A global survey of carbon isotope discrimination in plants from high altitude," *Oecologia*, vol. 74, no. 4, pp. 623–632, 1988.

[29] R. U. Meckenstock, B. Morasch, C. Griebler, and H. H. Richnow, "Stable isotope fractionation analysis as a tool to monitor biodegradation in contaminated acquifers," *Journal of Contaminant Hydrology*, vol. 75, no. 3-4, pp. 215–255, 2004.

[30] S.-J. Kim, D. H. Choi, D. S. Sim, and Y.-S. Oh, "Evaluation of bioremediation effectiveness on crude oil-contaminated sand," *Chemosphere*, vol. 59, no. 6, pp. 845–852, 2005.

[31] O. Schoefs, M. Perrier, and R. Samson, "Estimation of contaminant depletion in unsaturated soils using a reduced-order biodegradation model and carbon dioxide measurement," *Applied Microbiology and Biotechnology*, vol. 64, no. 1, pp. 53–61, 2004.

[32] N. E.-D. Sharabi and R. Bartha, "Testing of some assumptions about biodegradability in soil as measured by carbon dioxide evolution," *Applied and Environmental Microbiology*, vol. 59, no. 4, pp. 1201–1205, 1993.

[33] N. J. Sihota, O. Singurindy, and K. U. Mayer, "CO$_2$-Efflux measurements for evaluating source zone natural attenuation rates in a petroleum hydrocarbon contaminated aquifer," *Environmental Science and Technology*, vol. 45, no. 2, pp. 482–488, 2011.

[34] N. J. Sihota and K. U. Mayer, "Characterizing vadose zone hydrocarbon biodegradation using carbon dioxide effluxes, isotopes, and reactive transport modeling," *Vadose Zone Journal*, vol. 11, no. 4, 2012.

[35] P. K. Aggarwal, M. E. Fuller, M. M. Gurgas, J. F. Manning, and M. A. Dillon, "Use of stable oxygen and carbon isotope analyses for monitoring the pathways and rates of intrinsic and enhanced in situ biodegradation," *Environmental Science & Technology*, vol. 31, no. 2, pp. 590–596, 1997.

[36] M. E. Conrad, P. F. Daley, M. L. Fischer, B. B. Buchanan, T. Leighton, and M. Kashgarian, "Combined 14C and δ13C monitoring of in situ biodegradation of petroleum hydrocarbons," *Environmental Science and Technology*, vol. 31, no. 5, pp. 1463–1469, 1997.

[37] P. K. Aggarwal and R. E. Hinchee, "Monitoring in situ biodegradation of hydrocarbons by using stable carbon isotopes," *Environmental Science & Technology*, vol. 25, no. 6, pp. 1178–1180, 1991.

[38] K. H. Suchomel, D. K. Kreamer, and A. Long, "Production and transport of carbon dioxide in a contaminated vadose zone: a stable and radioactive carbon isotope study," *Environmental Science & Technology*, vol. 24, no. 12, pp. 1824–1831, 1990.

[39] C. Noel, J. C. Gourry, J. Deparis, I. Ignatiadis, F. Battaglia-Brunet, and C. Guimbaud, "Suitable real time monitoring of the aerobic biodegradation of toluene in contaminated sand by Spectral Induced Polarization measurements and CO$_2$ analyses," *Near Surface Geophysics*, In press.

[40] E. Verardo, A. Saada, V. Guerin et al., "Cas d'étude de gestion de site par atténuation naturelle: site 3—hydrocarbures pétroliers," Project ATTENA—Phase 2, Final Report, ADEME, 2013.

[41] R. Boopathy, "Factors limiting bioremediation technologies," *Bioresource Technology*, vol. 74, no. 1, pp. 63–67, 2000.

[42] R. M. Atlas, "Microbial hydrocarbon degradation—bioremediation of oil spills," *Journal of Chemical Technology & Biotechnology*, vol. 52, no. 2, pp. 149–156, 1991.

[43] L. D. Slater and D. Lesmes, "IP interpretation in environmental investigations," *Geophysics*, vol. 67, no. 1, pp. 77–88, 2002.

[44] M. H. Loke, J. E. Chambers, and R. D. Ogilvy, "Inversion of 2D spectral induced polarization imaging data," *Geophysical Prospecting*, vol. 54, no. 3, pp. 287–301, 2006.

[45] H. Craig, "Isotopic standards for carbon and oxygen and correction factors for mass-spectrometric analysis of carbon dioxide," *Geochimica et Cosmochimica Acta*, vol. 12, no. 1-2, pp. 133–149, 1957.

[46] C. Guimbaud, V. Catoire, S. Gogo et al., "A portable infrared laser spectrometer for field measurements of trace gases," *Measurement Science and Technology*, vol. 22, pp. 1–17, 2011.

[47] S. Gogo, C. Guimbaud, F. Laggoun-Défarge, V. Catoire, and C. Robert, "In situ quantification of CH4 bubbling events from a peat soil using a new infrared laser spectrometer," *Journal of Soils and Sediments*, vol. 11, no. 4, pp. 545–551, 2011.

[48] A. Grossel, B. Nicoullaud, H. Bourennane et al., "Simulating the spatial variability of nitrous oxide emission from cropped soils at the within-field scale using the NOE model," *Ecological Modelling*, vol. 288, pp. 155–165, 2014.

[49] C. Guimbaud, C. Noel, M. Chartier et al., "A quantum cascade laser infrared spectrometer for CO$_2$ stable isotope analysis: field implementation at a hydrocarbon contaminated site under bioremediation," *Journal of Environmental Sciences*, vol. 40, pp. 60–74, 2016.

[50] J. W. Raich and A. Tufekcioglu, "Vegetation and soil respiration: correlations and controls," *Biogeochemistry*, vol. 48, no. 1, pp. 71–90, 2000.

[51] J. W. Raich and C. S. Potter, "Global patterns of carbon dioxide emissions from soils," *Global Biogeochemical Cycles*, vol. 9, no. 1, pp. 23–36, 1995.

[52] K. Hayley, L. R. Bentley, M. Gharibi, and M. Nightingale, "Low temperature dependence of electrical resistivity: implications for near surface geophysical monitoring," *Geophysical Research Letters*, vol. 34, no. 18, 2007.

[53] M. Hayashi, "Temperature-electrical conductivity relation of water for environmental monitoring and geophysical data inversion," *Environmental Monitoring and Assessment*, vol. 96, no. 1–3, pp. 119–128, 2004.

[54] M. E. Conrad, A. S. Templeton, P. F. Daley, and L. Alvarez-Cohen, "Isotopic evidence for biological controls on migration of petroleum hydrocarbons," *Organic Geochemistry*, vol. 30, no. 8, pp. 843–859, 1999.

[55] C. M. Aelion, B. C. Kirtland, and P. A. Stone, "Radiocarbon assessment of aerobic petroleum bioremediation in the vadose zone and groundwater at an AS/SVE site," *Environmental Science & Technology*, vol. 31, no. 12, pp. 3363–3370, 1997.

[56] C.-H. Chaîneau, J.-L. Morel, and J. Oudot, "Microbial degradation in soil microcosms of fuel oil hydrocarbons from drilling cuttings," *Environmental Science & Technology*, vol. 29, no. 6, pp. 1615–1621, 1995.

[57] J. R. Van der Meer, W. M. de Vos, S. Harayama, and A. J. B. Zehnder, "Molecular mechanisms of genetic adaptation to xenobiotic compounds," *Microbiological Reviews*, vol. 56, no. 4, pp. 677–694, 1992.

[58] J. G. Leahy and R. R. Colwell, "Microbial degradation of hydrocarbons in the environment," *Microbiological Reviews*, vol. 54, no. 3, pp. 305–315, 1990.

[59] F. Garnier, *Contribution à L'évaluation Biogéochimique des Impacts Liés à L'exploitation Géothermique des Aquifères Superficiels: Expérimentations et Simulations à L'échelle D'un Pilote et D'installations Réelles*, Université d'orléans, 2012.

[60] M. Lenczewski, P. Jardine, L. McKay, and A. Layton, "Natural attenuation of trichloroethylene in fractured shale bedrock," *Journal of Contaminant Hydrology*, vol. 64, no. 3-4, pp. 151–168, 2003.

[61] B. A. Bekins, I. M. Cozzarelli, E. M. Godsy, E. Warren, H. I. Essaid, and M. E. Tuccillo, "Progression of natural attenuation processes at a crude oil spill site: II. Controls on spatial distribution of microbial populations," *Journal of Contaminant Hydrology*, vol. 53, no. 3-4, pp. 387–406, 2001.

[62] G. Lu, T. P. Clement, C. Zheng, and T. H. Wiedemeier, "Natural attenuation of BTEX compounds: model development and field-scale application," *Ground Water*, vol. 37, no. 5, pp. 707–717, 1999.

3

Diurnal Variation of Soil Heat Flux at an Antarctic Local Area during Warmer Months

Marco Alves and Jacyra Soares

Department of Atmospheric Science, IAG, University of São Paulo, 05508-090 São Paulo, SP, Brazil

Correspondence should be addressed to Jacyra Soares; jacyra@usp.br

Academic Editor: Marco Trevisan

Soil heat flux (G) is one term in the energy balance equation, and it can be particularly important in regions with arid, bare, or thinly vegetated soil surfaces. However, in remote areas such as the Antarctic, this measurement is not routinely performed. The analysis of observational data collected by the ETA Project at the Brazilian Antarctic Station from December 2013 to March 2014 showed that, for the total daily energy flux, the surface soil flux heats the deeper soil layers during December and January and G acts as a heat source to the outer soil layers during February and March. With regard to daytime energy flux, G acts as a source of heat to the deeper layers. During the night-time, the soil is a heat source to the shallower soil layers and represents at least 29% of the net night-time radiation. A relatively simple method—the objective hysteresis method (OHM)—was successfully applied to determine the surface soil heat flux using net radiation observations. *A priori*, the OHM coefficients obtained in this study may only be used for short-time parameterizations and for filling data gaps at this specific site.

1. Introduction

Energy from the Sun plays an important role in climatic systems as a whole and a specific role in the Earth's radiation balance. Net radiation and soil heat flux provide the energy for sensible and latent atmospheric turbulent heat fluxes near the surface.

Surface soil heat flux (G), defined as the heat exchange between different soil depths, with each layer possessing different temperature values [1], is a particularly important component of surface energy in regions with arid, bare, or thinly vegetated soil surfaces due to its capacity to work seasonally as a heat source (winter season) or a heat sink (summer season) [2–4]. Therefore, accuracy in the estimation of G is important to atmospheric systems [5]. However, such estimates are not readily available, particularly in locations such as the Antarctic, which possesses extreme climatic conditions for measurements *in situ* and heterogeneity in surface properties.

Bare soils generally have low albedo, whereas ice-covered soils show high albedo [6, 7]. Consequently, these soil types will have different soil temperatures even when in proximity to one another. For this reason, changes in Antarctic soil coverage can lead to local variation in temperature, directly affecting the Antarctic ecosystem [8].

Previous research has also suggested that soil temperature changes can have a marked impact on affecting the biological processes [9–15].

Prosek et al. [16] investigating the components of energy balance in a vegetated oasis at the Polish Station (King George Island, Antarctic region) have found that the boundary atmosphere and the soil substrate represent the basic components of the ecotopes of the Antarctic vegetation oasis.

Sturm et al. [17] discussed the warming in Alaska, of $0.5°C$ per decade in the last 30 years, and the positive feedback involving biological processes in the winter soil, which have contributed to the conversion of tundra to shrubland.

The direct estimation of soil heat flux by remotely sensed data is not feasible [18]. However, empirical relationships between G and the net radiation (R_n) can be used to determine G [19].

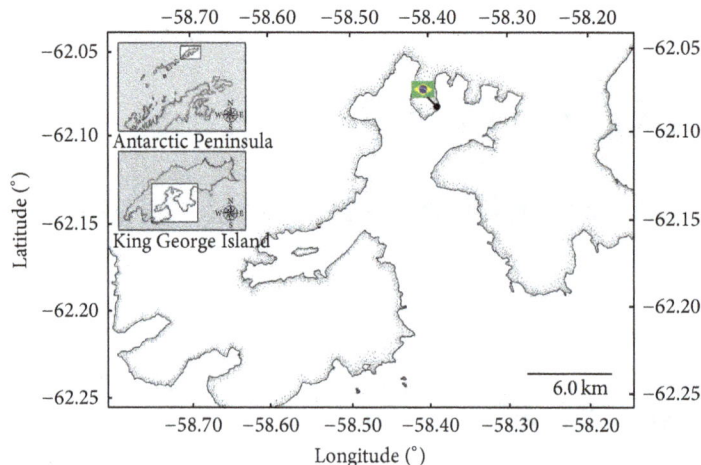

FIGURE 1: Geographical location of the EACF. Adapted from [32].

The objective hysteresis method (OHM) was proposed by Camuffo and Bernardi [20] to estimate G using values of R_n. The OHM model was initially developed for urban areas in an attempt to estimate urban heat storage as a residual term with different sources/sinks of energy such as buildings, vegetation, and the ground responsible for the exchange of sensible and latent heat fluxes [21–23]. However, there is nothing in the formulation of the OHM that limits its application only to urban areas.

This study uses a method of hysteresis to estimate soil heat flux and compare these estimates to direct measurements performed at the Brazilian Antarctic Station "Comandante Ferraz" (EACF) from December 2013 to March 2014. The data used in this study was collected by the ETA Project (*Estudo da Turbulência na Região Antártica*). To our knowledge, this is the first application of this method to the EACF region. Most of the previously published values are for urban areas and they are not valid for the studied region and time period.

2. Materials and Methods

The measurements were performed at the EACF (62°05′07″S, 58°23′33″W, 20 m above mean sea level) at a micrometeorological tower, located on King George Island, which is part of the South Shetland Islands of the Antarctic Peninsula (Figure 1).

The micrometeorological tower is surrounded by surface of different characteristics. Figure 2 presents an overview of the investigated area and shows the soil cover commonly observed, varying from bare soil to snow-covered soil.

During warmer months, the surface is covered mostly by bare soil, with presence of diverse sizes of rocks and gravels (<1 m in width). The occurrence of snow due to atmospheric systems, in these months, is unavailable. Nearby the tower there is a lake (<10 m), commonly frozen during the year, except on some days of summer.

The data used in this study were collected *in situ* from December 2013 to March 2014 (the warmer months). The net radiation data were obtained using a NR Lite2 Net Radiometer installed in a micrometeorological tower (3.4 m in height, Figures 3(a) and 3(c)) and the surface soil heat flux was measured using a Hukseflux HFP01 at a depth of 0.05 m (Figure 3(b)). The air temperature was measured by temperature sensor Model CS215, mounted with a 6-plate radiation shield, installed at 2.2 m height. The plate louvered construction allows air to pass freely through the shield, serving to keep the probe at or near ambient temperature. The shield's white colour reflects solar radiation. The soil temperature was measured using a probe 107L at 0.05 m depth. All sensors used in this work are from Campbell Scientific Inc.

The data were stored at 5 min (average) intervals by a CR5000 data logger (Campbell Scientific Inc.) using the local time (LT = −04 UTC) as the standard time (Figure 3). The data logger was connected to a laptop that automatically transmitted the data, every 30 min, to the Air-Sea Interaction Laboratory at IAG, USP. All data were reviewed and questionable data were removed considering the values located out of the 2-standard-deviation interval centred on the average value of the investigated variable.

The sign convention used for this study is that R_n and G are positive when energy is moving up in agreement with the vertical coordinate z.

2.1. The Objective Hysteresis Method and Statistical Evaluation of the OHM Application. The estimation of G using the objective hysteresis method (OHM) was based on the expression given by [20]

$$G = a_1 R_n + a_1 \frac{\partial R_n}{\partial t} + a_3, \tag{1}$$

where a_1, a_2, and a_3 are empirical coefficients related to the response of the surface properties due to solar energy

(a)

(b)

(c)

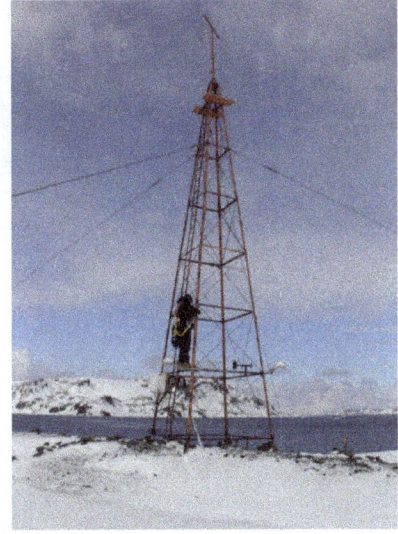

(d)

FIGURE 2: (a) Overview of the investigated area (photography copyright Renato Torlay). Micrometeorological tower region surrounded by (b) bare soil, (c) bare and snow-covered soil, and (d) snow-covered soil.

and $\partial R_n/\partial t$ is the temporal variation of R_n at the surface, discretized in time as $\partial R_n/\partial t = 0.5(R_n^{t+1} - R_n^{t-1})$. The a_1 and a_3 coefficients describe the best-fit straight line for the data over the entire day and are therefore constants. The slope of the best-fit straight line is represented by a_1 and the intercept with the ordinate is represented by a_3. The a_2 coefficient indicates the departure of actual values from the best-fit straight line.

The coefficients play different roles in the equation: a_1 is dimensionless and always positive and indicates the intensity of the relation between R_n and G; a_2 (s) shows the magnitude of the hysteresis, indicating the direction and the degree of the phase relationship between R_n and G. In summary, a_1 and a_2 are coefficients related to the mean values resulting from the soil characteristics (including the presence of water) and to the magnitude of R_n with its temporal variation [20].

The coefficient a_3 (Wm^{-2}) may be negative or positive depending on the local atmospheric conditions, and it represents the spontaneous heat flux between the soil and the adjacent atmosphere, when R_n and $\partial R_n/\partial t$ approach zero. In other words, this term represents the average heat flux released from the soil during the transitional periods of the day [20, 22, 24].

Statistical tests were performed to evaluate the application of the OHM model to the EACF region, including a linear regression with slope and intercept, coefficient of determination (R^2), root mean square error (RMSE), and the mean absolute error (MAE).

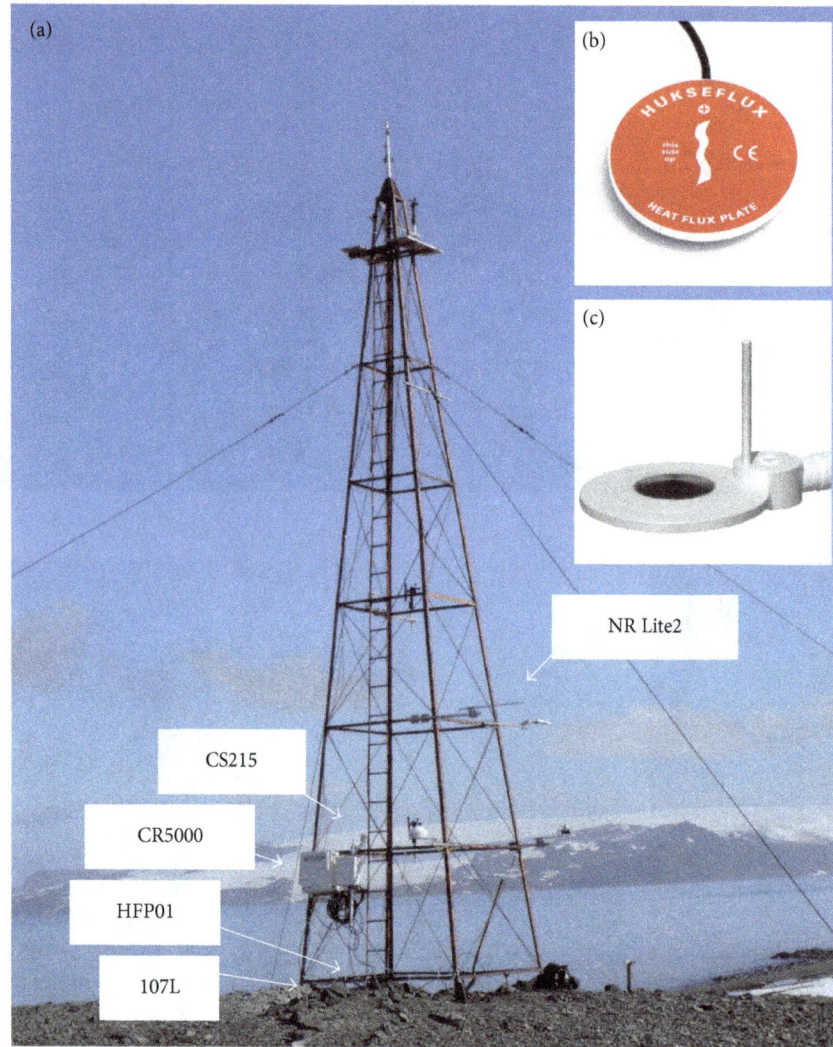

FIGURE 3: (a) Micrometeorological tower at the EACF, (b) soil heat flux instrument (HFP01), and (c) net radiation instrument (NR Lite2).

R^2 represents the percentage of the data that is close to the line of the best fit; these variances can be understood by the regression model [25]. The coefficient is given by

$$R^2 = \frac{\sum_{i=1}^{N} \left(G_{i,\text{obs}} - \overline{G_{\text{obs}}}\right)\left(G_{i,\text{mod}} - \overline{G_{\text{mod}}}\right)}{\left[\sum_{i=1}^{N} \left(G_{i,\text{obs}} - \overline{G_{\text{obs}}}\right)^2\right]^{0.5}\left[\sum_{i=1}^{N} \left(G_{i,\text{mod}} - \overline{G_{\text{mod}}}\right)^2\right]^{0.5}}, \quad (2)$$

where N is the total number of observations, $G_{i,\text{obs}}$ is the "i" observation value, and $G_{i,\text{mod}}$ is the "i" modelled result. The overbar denotes the time average for the period of evaluation. R^2 ranges from 0 to 1, with values near 1 indicating a good fit of the modelled results.

The performance of a model can also be quantified using an error value with the same units as the variable. The RMSE and MAE represent such quantifications [25–27]:

$$\text{RMSE} = \sqrt{\frac{1}{N}\sum_{i=1}^{N}\left(G_{i,\text{obs}} - G_{i,\text{mod}}\right)^2},$$

$$\text{MAE} = \frac{1}{N}\sum_{i=1}^{N}\left|G_{i,\text{obs}} - G_{i,\text{mod}}\right|. \quad (3)$$

In general, the RMSE is greater than or equal to the MAE for the range of most values because the MAE is less sensitive to extreme values than the RMSE.

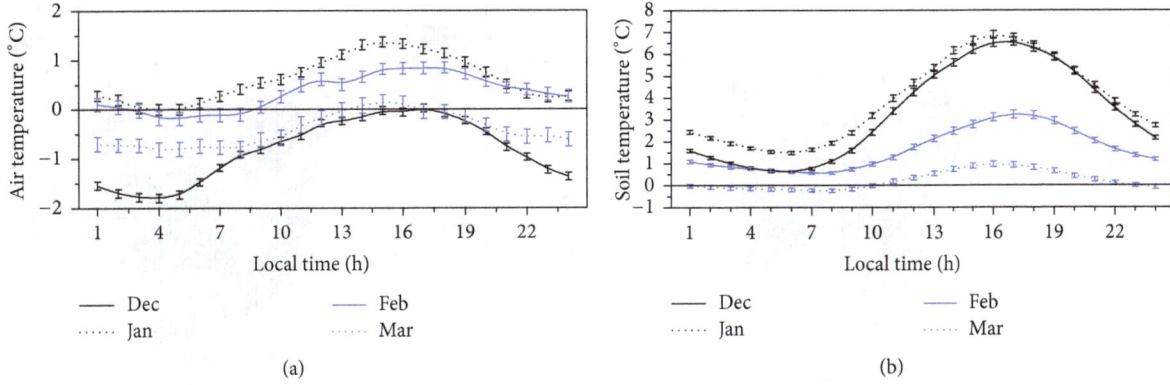

FIGURE 4: Diurnal variation of hourly average of (a) air temperature and (b) soil temperature for December 2013 to March 2014. The vertical bars indicate the standard error.

TABLE 1: Summary of daily energy fluxes (R_n and G), partitioned into daytime, night-time, and daily fluxes.

Month	Average daylight time (h)	Daytime energy flux			Night-time energy flux			Daily total energy flux		
		R_n	G	G/R_n	R_n	G	G/R_n	R_n	G	G/R_n
		(MJ m^{-2} d^{-1})		(%)	(MJ m^{-2} d^{-1})		(%)	(MJ m^{-2} d^{-1})		(%)
December 2013	19.1	−8.72	−0.66	7.6	0.54	0.31	57.4	−8.18	−0.35	4.3
January 2014	18.1	−8.39	−0.40	4.8	0.72	0.36	50.0	−7.67	−0.04	0.5
February 2014	15.4	−4.09	−0.17	4.2	0.54	0.20	37.0	−3.55	0.04	−1.1
March 2014	12.5	−2.19	−0.03	1.4	0.88	0.26	29.5	−1.31	0.23	−17.6

3. Results and Discussion

The soil temperature, during the investigated period, was always higher than the air temperature, as observed in cold regions [28]. January presented higher air and soil temperatures with a diurnal amplitude around, respectively, 1.4°C and 5.4°C (Figure 4). Among the investigated months, the diurnal variation of the soil temperature presented larger amplitudes than the air temperature, with maximum amplitude in December in both cases. Unfortunately, there is no data available during the investigated months directly related to the snow presence (surface albedo and emissivity, latent heat, etc.) but, in general, large soil temperature amplitude is characteristic of bare soil. During February and March most of the air temperature values were below zero and the diurnal amplitudes of the soil temperatures are comparatively smaller which could indicate the snow presence during these months. Previous studies have indicated that a characteristic of snow cover would be a long period of relatively stable soil temperatures with smaller amplitude of the temperature signal [13, 29].

December and January have more hours of daylight (Table 1) and consequently larger amount of net radiation compared to the other investigated months (Figure 5).

Through all investigated months there was a delay between G and R_n during the two periods of the day corresponding to the transitional day/night/day periods when the signal of R_n and G are inverted, with the soil acting as a source

TABLE 2: Monthly coefficients of the OHM applied to the EACF region.

Month	Coefficient		
	a_1	a_2 (s)	a_3 (W m^{-2})
December 2013	0.17	−0.09	11.8
January 2014	0.13	−0.03	11.5
February 2014	0.11	−0.10	5.6
March 2014	0.10	−0.068	4.3

of heat to the shallower soil layers and R_n as a source of heat to the soil (Figure 5).

In general, during the daytime, the energy provided by R_n is shared between G and the turbulent fluxes, but during the night-time, the turbulent fluxes are less important and G represents a comparatively larger portion of the net radiation. Therefore, at night, the soil played an important role as an energy source to the outer soil layers accounting for approximately 31% of R_n in March and up to approximately 55% of R_n in December (Table 1, Figure 6). High values of G/R_n occurred near the transitional periods, when R_n was low but, near noon, G was approximately 10% of R_n (Figure 6) as observed by several authors in different locations [30, 31].

During the daytime of the studied period, part of the net energy was stored in the soil, with maximum storage of approximately 7.6% observed in December (Table 1). During January, the quantity of net radiation energy flux incident

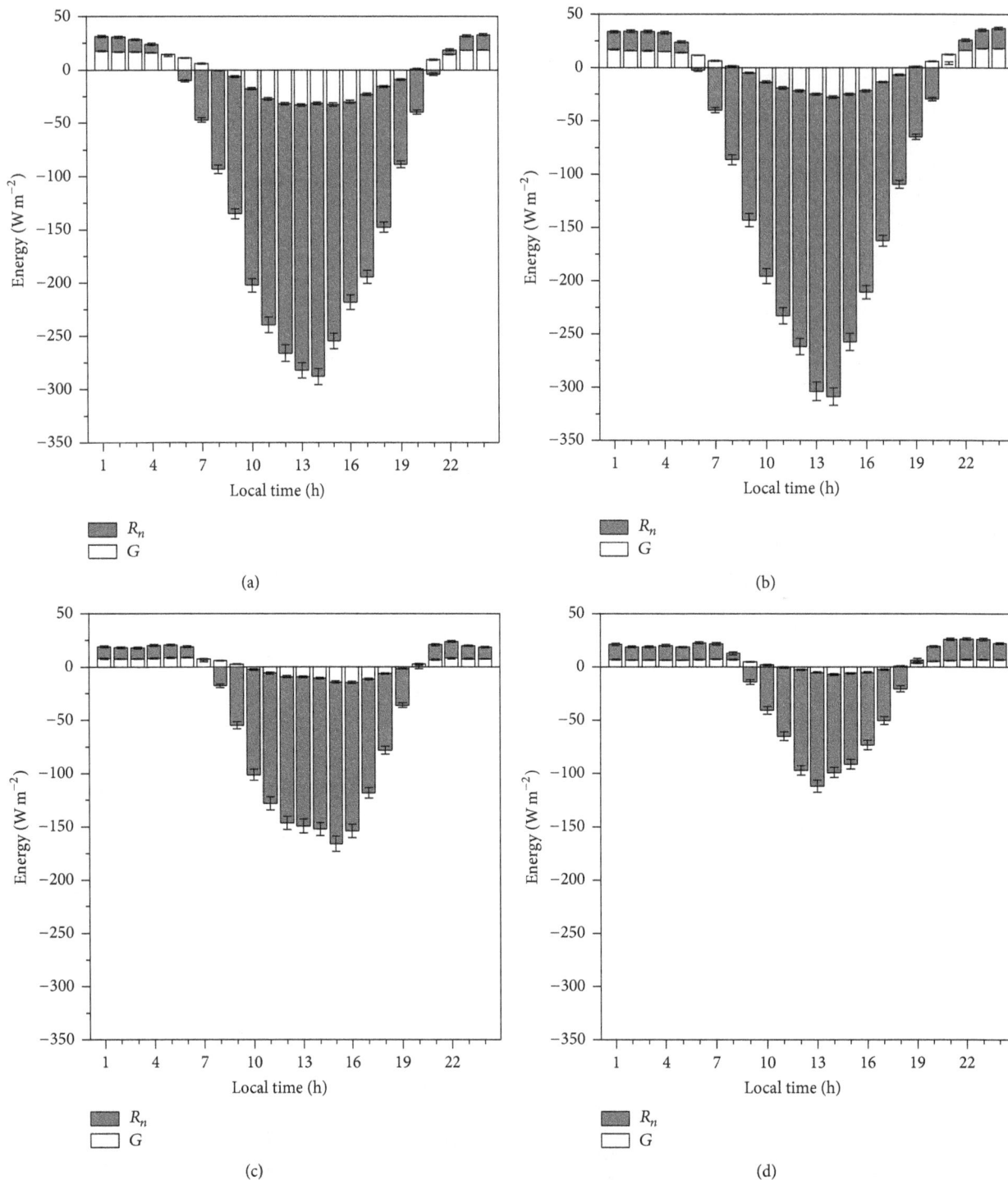

FIGURE 5: Diurnal variation of hourly average net radiation (R_n) and soil heat flux (G) for (a) Dec 2013, (b) Jan 2014, (c) Feb 2014, and (d) Mar 2014. The vertical bars indicate the standard error.

on the surface was more than twice the quantity of incident energy during February, but the proportion of energy stored in the soil was not so different.

In the EACF region, the daily ratio of G/R_n varied from 4.3% in December to −17.6% in March (Table 1) with the soil acting as a heat source to the deeper soil layers during December and January (positive values of G/R_n) and as a source of heat to the shallower soil layers (negative values of G/R_n) during February and March (Table 1).

The best-fit coefficients for the investigated months are shown in Table 2 and, a priori, these values are valid for this particular site during this period of study because

TABLE 3: Statistical evaluation of observed and modelled soil heat flux for the investigated months.

Month	Slope	Intercept (W m^{-2})	R^2	RMSE (W m^{-2})	MAE (W m^{-2})
December 2013	0.97	0.29	0.992	1.85	1.32
January 2014	1.03	−0.33	0.991	1.64	1.33
February 2014	1.01	−0.10	0.994	0.64	0.50
March 2014	0.99	0.03	0.990	0.51	0.35

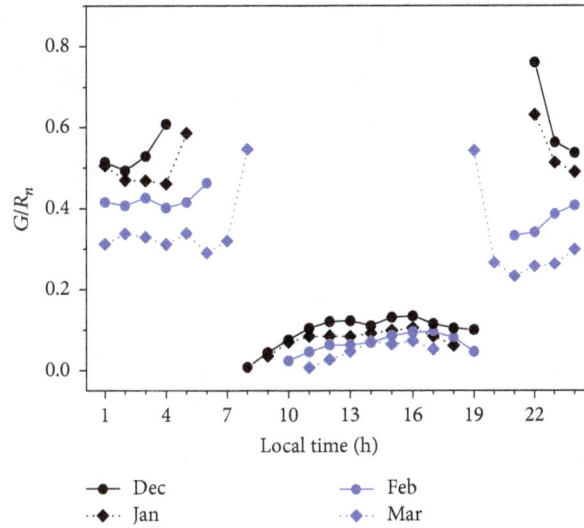

FIGURE 6: Diurnal variation of hourly average fraction of soil heat flux (G) as a percentage of net radiation (R_n) for December 2013 to March 2014. Data for the transitional periods were removed from the figure.

these coefficients are related to the presence of precipitation, atmospheric system, soil state, and soil characteristics. Using these coefficients (Table 2), the OHM was able to simulate the diurnal variation of G during the investigated months, as displayed in Figure 7.

Statistical tests were performed to evaluate the OHM application to the investigated region and all months showed a high degree of statistical agreement between observed and modelled values with slopes near 1, intercept values of approximately zero, and R^2 values greater than 0.99. In addition, the RMSE and MAE values were less than 1.85 Wm^{-2} and 1.33 Wm^{-2}, respectively (Table 3).

From the hysteresis graph of diurnal variation in the observed and modelled values of G against the R_n values, it can be seen that the eccentricities of the ellipses from December 2013 and January 2014 were larger than those from February and March 2014 (Figure 7) and during the transitional period, G and R_n showed opposite signals (I in Figure 7), with the soil releasing more heat to the shallower layers at the beginning of the day than at the end of the day (Figures 5 and 7).

4. Conclusions

Despite its importance, measurement of soil heat flux is not performed routinely, particularly in remote areas such as the region investigated here. This study applied an indirect

method (OHM) proposed by [20] to estimate soil heat flux using values of observed net radiation.

This study used 5-minute averages of surface soil heat flux and net radiation observed at the Brazilian Station Comandante Ferraz from December 2013 to March 2014. The observed daily total energy flux indicated that, during December and January, G was a source of heat to the soil deeper layers and R_n was a heat source to the soil. However, during February and March, G and R_n have inverse directions with G releasing heat to the outer soil layers. During the daytime of the investigated months, G and R_n heated the soil. During the night-time, G represents at least 29% of R_n. Therefore, G cannot be ignored in the energy balance equation of the EACF region.

The OHM was able to estimate the surface soil heat flux for each studied month, with all correlation coefficients exceeding 99%.

Future analyses will involve the application of the OHM to all months of the year and investigate the validity of the expressions obtained here for the same months of different years. It is important to note that the coefficients obtained in this study can only be used for short-time parameterizations at a specific site and for filling data gaps.

Competing Interests

The authors declare that there are no competing interests regarding the publication of this paper.

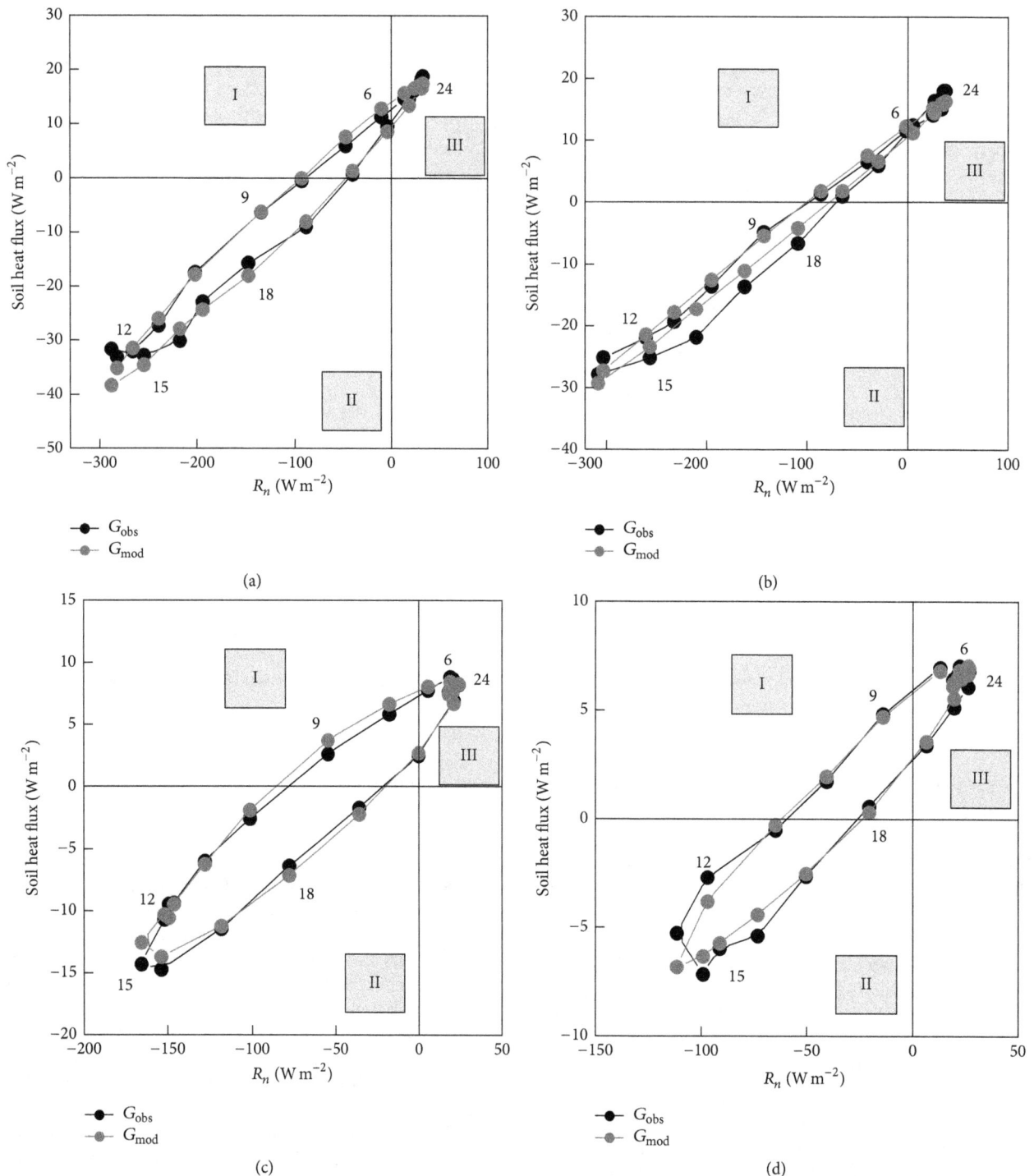

FIGURE 7: Hysteresis loop relations between observed (obs, black colour) and modelled (mod, grey colour) soil heat flux (G) and net radiation (R_n) in the EACF: (a) December 2013, (b) January 2014, (c) February 2014, and (d) March 2014. The numbers represent the local time (hour). I, II, and III indicate the period of the day: transitional, daytime, and night-time periods, respectively.

Acknowledgments

The first author acknowledges a scholarship from CAPES. All authors would like to thank the grants of "INCT-APA" (CNPq 574018/2008-5 and FAPERJ E-16/170.023/2008). The second author also thanks CNPq (305357/2012-3 and 407137/2013-0).

References

[1] C. M. Bhumralkar, "Numerical experiments on the computation of ground surface temperature in an atmospheric general circulation model," *Journal of Applied Meteorology*, vol. 14, no. 7, pp. 1246–1258, 1975.

[2] K. Anandakumar, R. Venkatesan, and T. V. Prabha, "Soil thermal properties at Kalpakkam in Coastal South India," *Proceedings of the Indian Academy of Sciences, Earth and Planetary Sciences*, vol. 110, no. 3, pp. 239–245, 2001.

[3] T. Choi, B. Y. Lee, S.-J. Kim, Y. J. Yoon, and H.-C. Lee, "Net radiation and turbulent energy exchanges over a non-glaciated coastal area on King George Island during four summer seasons," *Antarctic Science*, vol. 20, no. 1, pp. 99–112, 2008.

[4] S. E. Sofyan, E. Hu, and A. Kotousov, "A new approach to modelling of a horizontal geo-heat exchanger with an internal source term," *Applied Energy*, vol. 164, pp. 963–971, 2016.

[5] P. Viterbo, A. Beljaars, J.-F. Mahfouf, and J. Teixeira, "The representation of soil moisture freezing and its impact on the stable boundary layer," *Quarterly Journal of the Royal Meteorological Society*, vol. 125, no. 559, pp. 2401–2426, 1999.

[6] R. J. L. MacCulloch, *The microclimatology of Antarctic soils [Master of Sciences in Earth Sciences]*, University of Waikato, Hamilton, New Zealand, 1996.

[7] D. I. Campbell, R. J. MacCulloch, and I. B. Campbell, "Thermal regimes of some soils in the McMurdo Sound region, Antarctica," in *Ecosystem Processes in Antarctic Ice-Free Landscapes: Proceedings of an International Workshop on Polar Desert Ecosystems Christchurch, New Zealand 1–4 July, 1996*, W. B. Lyons, C. Howard-Williams, and I. Hawes, Eds., A.A. Balkema, Rotterdam, The Netherlands, 1997.

[8] A. J. Oliphant, R. C. A. Hindmarsh, N. J. Cullen, and W. Lawson, "Microclimate and mass fluxes of debris-laden ice surfaces in Taylor Valley, Antarctica," *Antarctic Science*, vol. 27, no. 1, pp. 85–100, 2015.

[9] F. S. Chapin III, "Direct and indirect effects of temperature on arctic plants," *Polar Biology*, vol. 2, no. 1, pp. 47–52, 1983.

[10] M. C. Davey, J. Pickup, and W. Block, "Temperature variation and its biological significance in fellfield habitats on a maritime Antarctic island," *Antarctic Science*, vol. 4, no. 4, pp. 383–388, 1992.

[11] L. S. Peck, P. Convey, and D. K. A. Barnes, "Environmental constraints on life histories in Antarctic ecosystems: tempos, timings and predictability," *Biological Reviews of the Cambridge Philosophical Society*, vol. 81, no. 1, pp. 75–109, 2006.

[12] N. S. Haussmann, J. C. Boelhouwers, and M. A. Mcgeoch, "Fine scale variability in soil frost dynamics surrounding cushions of the dominant vascular plant species (*Azorella selago*) on sub-Antarctic Marion Island," *Geografiska Annaler Series A: Physical Geography*, vol. 91, no. 4, pp. 257–268, 2009.

[13] N. Cutler, "Vegetation-environment interactions in a sub-arctic primary succession," *Polar Biology*, vol. 34, no. 5, pp. 693–706, 2011.

[14] S. Bokhorst, A. H. L. Huiskes, R. Aerts et al., "Variable temperature effects of Open Top Chambers at polar and alpine sites explained by irradiance and snow depth," *Global Change Biology*, vol. 19, no. 1, pp. 64–74, 2013.

[15] W. Kellmann-Sopyła and I. Giełwanowska, "Germination capacity of five polar *Caryophyllaceae* and *Poaceae* species under different temperature conditions," *Polar Biology*, vol. 38, no. 10, pp. 1753–1765, 2015.

[16] P. Prosek, M. Janouch, and K. Láska, "Components of the energy balance of the ground surface and their effect on the thermics of the substrata of the vegetation oasis at Henryk Arctowski Station, King George Island, South Shetland Islands," *Polar Record*, vol. 36, no. 196, pp. 3–18, 2000.

[17] M. Sturm, J. Schimel, G. Michaelson et al., "Winter biological processes could help convert arctic tundra to shrubland," *BioScience*, vol. 55, no. 1, pp. 17–26, 2005.

[18] C. S. T. Daughtry, W. P. Kustas, M. S. Moran et al., "Spectral estimates of net radiation and soil heat flux," *Remote Sensing of Environment*, vol. 32, no. 2-3, pp. 111–124, 1990.

[19] C. Liebethal and T. Foken, "Evaluation of six parameterization approaches for the ground heat flux," *Theoretical and Applied Climatology*, vol. 88, no. 1-2, pp. 43–56, 2007.

[20] D. Camuffo and A. Bernardi, "An observational study of heat fluxes and their relationships with net radiation," *Boundary-Layer Meteorology*, vol. 23, no. 3, pp. 359–368, 1982.

[21] C. S. B. Grimmond, H. A. Cleugh, and T. R. Oke, "An objective urban heat storage model and its comparison with other schemes," *Atmospheric Environment—Part B: Urban Atmosphere*, vol. 25, no. 3, pp. 311–326, 1991.

[22] C. S. B. Grimmond and T. R. Oke, "Heat storage in urban areas: local-scale observations and evaluation of a simple model," *Journal of Applied Meteorology*, vol. 38, no. 7, pp. 922–940, 1999.

[23] S. K. Meyn and T. R. Oke, "Heat fluxes through roofs and their relevance to estimates of urban heat storage," *Energy and Buildings*, vol. 41, no. 7, pp. 745–752, 2009.

[24] M. J. Ferreira, A. P. De Oliveira, and J. Soares, "Diurnal variation in stored energy flux in São Paulo city, Brazil," *Urban Climate*, vol. 5, pp. 36–51, 2013.

[25] D. R. Legates and G. J. McCabe Jr., "Evaluating the use of 'goodness-of-fit' measures in hydrologic and hydroclimatic model validation," *Water Resources Research*, vol. 35, no. 1, pp. 233–241, 1999.

[26] C. J. Willmott, "Some comments on the evaluation of model performance," *Bulletin of the American Meteorological Society*, vol. 63, pp. 1309–1313, 1982.

[27] J. Soares, A. P. Oliveira, M. Z. Božnar, P. Mlakar, J. F. Escobedo, and A. J. Machado, "Modeling hourly diffuse solar-radiation in the city of São Paulo using a neural-network technique," *Applied Energy*, vol. 79, no. 2, pp. 201–214, 2004.

[28] T. Zhang, "Influence of the seasonal snow cover on the ground thermal regime: an overview," *Reviews of Geophysics*, vol. 43, no. 4, Article ID RG4002, 2005.

[29] G. Kudo, "Effects of snow-free period on the phenology of alpine plants inhabiting snow patches," *Arctic & Alpine Research*, vol. 23, no. 4, pp. 436–443, 1991.

[30] B. E. Clothier, K. L. Clawson, P. J. Pinter Jr., M. S. Moran, R. J. Reginato, and R. D. Jackson, "Estimation of soil heat flux from net radiation during the growth of alfalfa," *Agricultural and Forest Meteorology*, vol. 37, no. 4, pp. 319–329, 1986.

[31] D. D. Baldocchi, S. B. Verma, and N. J. Rosenberg, "Water use efficiency in a soybean field: influence of plant water stress," *Agricultural and Forest Meteorology*, vol. 34, no. 1, pp. 53–65, 1985.

[32] R. B. Moura, *Estudo taxonômico dos holothuroidea (echinodermata) das ilhas shetland do sul e do estreito de bransfield, antártica [M.S. thesis]*, Museu Nacional, Universidade Federal do Rio de Janeiro, Rio de Janeiro, Brazil, 2009.

4

Oil and Gas Production Wastewater: Soil Contamination and Pollution Prevention

John Pichtel

Natural Resources and Environmental Management, Ball State University, Muncie, IN 47306, USA

Correspondence should be addressed to John Pichtel; jpichtel@bsu.edu

Academic Editor: Ezio Ranieri

During oil and natural gas production, so-called "produced water" comprises the largest byproduct stream. In addition, many oil and gas operations are augmented via injection of hydraulic fracturing (HF) fluids into the formation. Both produced water and HF fluids may contain hundreds of individual chemicals, some known to be detrimental to public health and the environment. Oil and gas production wastewater may serve a range of beneficial purposes, particularly in arid regions, if managed correctly. Numerous treatment technologies have been developed that allow for injection, discharge to the land surface, or beneficial reuse. Although many papers have addressed the effects of oil and gas production wastewater (OGPW) on groundwater and surface water quality, significantly less information is available on the effects of these fluids on the soil resource. This review paper compiles fundamental information on numerous chemicals used and produced during oil and gas development and their effects on the soil environment. Additionally, pollution prevention technologies relating to OGPW are presented. An understanding of the effects of OGPW on soil chemical, physical, and biological properties can provide a foundation for effective remediation of OGPW-affected soils; additionally, sustainable reuse of oil and gas water for irrigation and industrial purposes may be enhanced.

1. Introduction

Production of conventional oil and gas and coal bed methane is often accompanied by production of large volumes of produced water. The United States generates an estimated 21 billion barrels of produced water every year [1].

In certain geologic strata substantial volumes of oil and natural gas are present, yet they experience poor recovery rates due to low permeability of local strata. This is especially true for shales, tight sands, oil sands, and coal beds [2]. In hydraulic fracturing (HF) ("fracking"), a specially tailored mixture of fluids is pumped into recovery wells under high pressure to fracture low permeability formations and enhance gas and oil production [3–5]. Extraction of hydrocarbon resources using HF is commonly referred to as "unconventional production." Unconventional wells include those drilled horizontally, allowing the borehole to bend 90 degrees and penetrate the target formation laterally up to thousands of meters [6]. Within the past two decades the combination of HF with horizontal drilling has opened immense new oil and gas reserves worldwide which were previously considered inaccessible or unprofitable [7, 8] and brought large-scale drilling to new regions [3, 9].

Hydraulic fracturing is performed at depths between 5,000 and 10,000 feet and requires 2,500,000–4,200,000 gallons of water per well [10]. Fracturing operations inject highly pressurized fluids, that is, between 2,000 and 12,000 psi, at an average flow rate of 2000 gpm (47 bbl/min) [11]. The water is mixed with 0.5–2.0% (by volume) of selected chemical additives to increase water flow and improve deposition efficiency. Approximately 1,000 chemicals are known to be used in the HF process [7, 11].

Following initial injection into the well to generate fractures, a portion of the injected water returns to the surface immediately and is termed "flowback" [3]. The remaining fluids either permeate into the formation or return to the surface over the life of the producing well and are termed "produced water." Both types of wastewater may contain HF fluids, naturally occurring salts, radioactive materials, heavy metals, and other compounds from the formation such as polycyclic aromatic hydrocarbons, alkenes, alkanes, and other volatile and semivolatile organics [12–18]. In this paper, oil and gas flowback water, produced water, and hydraulic fracturing fluids will collectively be termed oil and gas production wastewater (OGPW).

As HF operations are expanding, the volumes of wastewater being generated are increasing exponentially [19]. Wastewater from drilling activities is typically managed via disposal in injection wells or evaporation ponds, application to fields, spreading on roads, and/or treatment and reuse for future oil and gas operations [8, 13, 20, 21].

Hydraulic fracturing components may pose a threat to public health and the environment as some are known to be acutely toxic, some are carcinogenic, and others are believed to be endocrine-disruptors [12, 22–24]. Other chemicals remain proprietary information [25] whose effects on public health and the environment are unavailable. Recent work found that 67%, 37%, and 18% of assessed wells were fractured with $\geq$ 1, 5, or 10 proprietary chemicals, respectively [12, 26].

Contamination of soil can occur through spills of fluids during drilling and fracturing processes and during transport by truck or through wastewater pipelines and failure of well casings and equipment failures and corrosion of pipes and tanks. In some regions OGPW is transferred to wastewater treatment plants [20]; however, facilities may be unable to remove several anthropogenic or naturally occurring compounds [27–29]. This can result in their discharge, following treatment, to surface water and ultimately to soil [30, 31].

Between 2009 and 2013 a total of 1933 spills were documented in Colorado [32]. In 2013, spills were reported at 1% of Colorado wells (550 of 51,000 active wells). An analysis of permitted Pennsylvania wells shows a spill rate of 2% (103 of 5,580 active wells) [26]. A total of 24 states with active shale reservoirs currently report spills; however, reporting limits and information required vary markedly. Only five states require maintenance of public records for spills and violations [12, 26, 33]. In light of the limited mandatory reporting, it is possible that the degree of oil and gas operations on water quality is underestimated [26, 33]. For example, an analysis in Pennsylvania found that only 59% of documented spills were reported [12, 26]. Elevated concentrations of benzene, toluene, ethylbenzene, and xylenes (BTEX) have been detected in groundwater near surface spills [34, 35]; soils have been affected by excess salinity and sodicity [36, 37].

A great number of papers have described the effects of OGPW on groundwater [14, 38–40] and surface water [40–42]; however, studies of the effects of OGPW on the soil resource are scant, and reclamation of OGPW-affected soils has received minimal attention in the scientific literature. In order to understand the potential effects of OGPW on soil chemical, physical, and biological properties as well as potential effects on plant growth, it is necessary to identify the chemicals used in HF and those produced from active wells, as well as their behavior in soil. The objectives of this paper are to cite common chemicals that are used for, and produced during, oil and gas development and compile essential information on their effects on the soil environment. Additionally, both remediation of OGPW-affected soils and pollution prevention technologies will be presented.

2. Hydraulic Fracturing Fluids

Oil and gas production chemicals can be pure compounds or mixtures containing active ingredients dissolved in a solvent or cosolvent and used to serve numerous processes (Table 1) [53].

In response to concerns about the potential public health and environmental impacts associated with HF, key reagents have, in recent years, been compiled and made publicly available. Regulatory agencies in many states have established reporting requirements for unconventional production; however, not all such requirements are mandatory [54, 55]. Many oil and gas producers choose to publish lists of HF chemicals on company websites or in the FracFocus Chemical Disclosure Registry [56].

The broad categories of HF fluids in routine use consist of [57] (1) viscosified water-based fluids; (2) nonviscosified water-based fluids; (3) gelled oil-based fluids; (4) acid-based fluids; and (5) foam fluids. For many hydrocarbon reservoirs, water-based fluids are most suitable due to the historic ease with which large volumes of mix water can be acquired.

Hydraulic fracturing fluids contain approximately 98 to 99.5% water plus a specially prepared mixture that helps optimize the fracturing process [3, 56]. Typical additives include proppants (propping agents), gelling and foaming components, friction reducers, cross-linkers, breakers, pH adjusters, biocides, corrosion inhibitors, scale inhibitors, iron controlling compounds, clay stabilizers, and surfactants (Table 1) [46, 58]. Not all these additives are used in every fracturing project, and sometimes one class of additives can serve multiple purposes; that is, a surfactant can be used as a cross-linker and gelling agent in certain situations [46]. Chemicals are added throughout the entire production process including drilling and fracturing and through closure to serve numerous functions [7, 13]. Some common HF additives are listed in Table 2.

A comprehensive study of the properties of HF chemicals was carried out by Stringfellow et al. [46] and includes commonly used compounds for each class of agents as well as data for toxicity and biodegradability.

2.1. Proppants. Propping agents are employed to "prop open" the fracture once pumps are turned off and fractures begin to close. The ideal propping agent is resistant to crushing and to corrosion, of low density, and is readily available and inexpensive [59]. Common propping agents are silica sand, resin-coated sand (RCS), and ceramic proppants (e.g., sintered bauxite, intermediate-strength proppant [ISP], and lightweight proppant [LWP]) [60]. Generally, sand is used to prop open fractures in shallow formations. RCS is stronger than sand and is used where more compressive strength is required to minimize proppant crushing. Ceramic proppants are used to stimulate deep (>8,000 ft) wells where significant in situ stresses impart large forces on the propping agent [61].

2.2. Gelling Agents. Gellants increase the viscosity of HF fluids. Greater viscosities increase fracture width so higher concentrations of proppant can be injected, fluid loss is

TABLE 1: Common classes of hydraulic fracturing compounds and their uses.

Chemical category	Application in hydraulic fracturing	Example compounds
Proppants	Hold fissures open and allow gas to flow out of the formation	Sand, sintered bauxite, zirconium oxide, ceramic beads, and graphite
Gellants	Increase viscosity and suspend sand during proppant transport	Propylene glycol, guar gum, ethylene glycol, and petroleum distillate
Foamers	Increase carrying capacity while transporting proppants and decrease overall volume of fluid needed	2-Butoxyethanol, diethylene glycol
Cross-linkers	Thicken fluids to increase viscosity and proppant transport into fractures	Potassium hydroxide, ethylene glycol, borate salts, and petroleum distillates
Breakers	Reduce the viscosity of the fluid so proppant will flow into fractures; added near the end of hydraulic fracturing to enhance flowback	Ammonium persulfate, magnesium peroxide
Acids	Clean up cement and drilling mud before fracturing fluid is injected and clear the path through the formation. Used later to dissolve minerals and clays to reduce clogging, allowing gas to flow to the surface	Hydrochloric acid
pH control	Maintains pH at various stages to ensure maximum effectiveness of various additives	Sodium hydroxide, acetic acid
Biocides	Kill bacteria that produce gases (particularly H_2S) which could contaminate methane gas, corrode pipes and fittings, and break down gellants	Glutaraldehyde, 2-bromo-2-nitro-1,2-propanediol
Corrosion inhibitors	Reduce damage to steel from acidic HF fluids	Ethoxylated octylphenol and nonylphenol, isopropanol
Scale inhibitors	Prevent buildup of mineral scale that can block fluid and gas passage through the pipes. Prevent steel materials from being damaged by acidic fracking fluids	Acrylamide, sodium polycarboxylate, methanol, and ammonium bisulfate
Iron control	Prevents carbonate and sulfate compounds from precipitating to form plugs in shale formation	Ammonium chloride, ethylene glycol, and polyacrylate
Clay stabilizers	Block clays from swelling to block the open channels created in the mining operation	Tetramethyl ammonium chloride, sodium chloride
Defoamers	Reduce foaming after it is no longer needed; lowers surface tension; and allows trapped gas to escape	2-Ethylhexanol, oleic acid, and oxalic acid
Friction reducers	To make water slick and minimize the friction created under high pressure and to increase the rate and efficiency of moving the HF fluid	Acrylamide, ethylene glycol, petroleum distillate, methanol, sodium acrylate-acrylamide copolymer, polyacrylamide (PAM), and petroleum distillates
Surfactants	Reduce surface tension and improve fluid passage through pipes in either direction	Methanol, ethanol, isopropanol, naphthalene, 1,2,4-trimethylbenzene, and 2-butoxyethanol

[3, 12, 43–45].

reduced, proppant transport is improved, and friction pressure is reduced [62].

Gelling agents primarily consist of guar and derivatives (e.g., hydroxypropyl guar, carboxymethyl guar, and carboxymethyl hydroxypropyl guar), celluloses, acids, and alcohols [62, 63]. Gellants can be linear or cross-linked to increase fluid viscosity. Agents are selected based on site-specific conditions in the well including temperature and salinity [5, 46].

Diesel fuel is sometimes used to form a viscous HF gel when combined with guar concentrate. The US Environmental Protection Agency (US EPA) has attempted to regulate the use of diesel fuel in HF; however, it is still used in place of water as it can carry more guar concentrate per unit volume [43].

2.3. Friction Reducers. Friction reducers are sometimes used as an alternative to gelling agents. The most commonly used friction reducer is 2-propenamide (polyacrylamide,

$[C_3H_5NO]_n$) [46]. Friction reducers are water-soluble, nonvolatile, and nontoxic.

2.4. Cross-linkers. Cross-linkers bind gel molecules and thereby increase viscosity and proppant transport. Cross-linkers frequently used in HF include borate salts; titanium, zirconium, and aluminum compounds; monoethanolamine; and monoethylamine [5, 47, 56]. Ammonium chloride, ethylene glycol, and potassium hydroxide are also used. Concentrations of cross-linkers in HF fluid are relatively low and range from 0.5 to 250 mg/L [47, 64–67]. Borate ions are the most commonly used cross-linking agents for guar polymer applications [68]. Borax (sodium tetraborate decahydrate) and boric acid plus caustic soda and cross-linking agents (0.024–0.09% w/w) have been used as sources of borate ions to cross-link guar [68].

2.5. Breakers. The viscous HF fluid, whether cross-linked or linear, must be degraded in order to achieve high conductivity

TABLE 2: Chemicals and chemical mixtures identified as being commonly used in hydraulic fracturing based on available sources.

Chemical name	CAS number	Formula
Acetaldehyde	75-07-0	C_2H_4O
Acetic acid	64-19-7	$C_2H_4O_2$
Acetone	67-64-1	C_3H_6O
Adipic acid	124-04-9	$C_6H_{10}O_4$
Alkyl benzyl dimethyl ammonium chloride	68424-85-1	Various
Ammonium chloride	12125-02-9	ClH_4N
Ammonium persulfate	7727-54-0	$(NH_4)_2S_2O_8$
Ammonium sulfate	7783-20-2	$(NH_4)_2SO_4$
Borate salts	Various	Various
Boric acid sodium salt	1333-73-9	Na_3BO_3
Calcium chloride	10043-52-4	$CaCl_2$
Calcium peroxide	1305-79-9	CaO_2
Carbon dioxide	124-38-9	CO_2
Carboxymethyl guar	39346-76-4	Various
Carboxymethyl hydroxyethyl cellulose	9004-30-2	Various
Carboxymethyl hydroxypropyl guar	68130-15-4	Various
Choline chloride	67-48-1	$C_5H_{14}ClNO$
Citric acid	77-92-9	$C_6H_8O_7$
Copolymer of acrylamide and sodium acrylate	25987-30-8	Various
Copper compounds	Various	Various
Didecyl dimethyl ammonium chloride	7173-51-5	$C_{22}H_{48}ClN$
Diesel fuel	Various	Various
Diethanolamine	111-42-2	$C_4H_{11}NO_2$
Dimethyl dihydrogenated tallow ammonium chloride	Various	Various
Ester salt	Various	Various
Ethanol	64-17-5	C_2H_6O
Ethyl methyl derivatives	Various	Various
Ethylene glycol	107-21-1	$C_2H_6O_2$
Formic acid	64-18-6	CH_2O_2
Fumaric acid	110-17-8	$C_4H_4O_4$
Glutaraldehyde	111-30-8	C_5H_8O
Glycol ethers	Various	Various
Guar gum	9000-30-0	Various
Isopropanol	67-63-0	C_3H_8O
Magnesium oxide	1309-48-4	MgO
Magnesium peroxide	14452-57-4	MgO_2
Methanol	67-56-1	CH_4O
Monoethanolamine	141-43-5	C_2H_7NO
Monoethylamine	75-04-7	C_2H_7N
N,n-Dimethyl formamide	68-12-2	C_3H_7NO
Naphthalene	91-20-3	$C_{10}H_8$
Nitrogen	7727-37-9	N_2
Petroleum distillate	64741-85-1	Various
Phosphonic acid salt	Various	Various
Polyacrylamide	9003-05-8	$(C_3H_5NO)_n$
Polyglycol ether	Various	Various
Potassium carbonate	584-08-7	K_2CO_3

TABLE 2: Continued.

Chemical name	CAS number	Formula
Potassium chloride	7447-40-7	KCl
Potassium hydroxide	1310-58-3	KOH
Potassium metaborate	13709-94-9	BKO_2
Potassium persulfate	7727-21-1	$K_2O_8S_2$
Propargyl alcohol	107-19-7	C_3H_4O
Pyridinium	16969-45-2	C_5H_6N
Quaternary ammonium chloride	61789-71-1	Various
Sodium carbonate	497-19-8	Na_2CO_3
Sodium chloride	7647-14-5	$NaCl$
Sodium erythorbate	6381-77-7	$C_6H_7NaO_6$
Sodium hydroxide	1310-73-2	$NaOH$
Sodium lauryl sulfate	151-21-3	$C_{12}H_{25}NaO_4S$
Sodium persulfate	7775-27-1	$Na_2O_8S_2$
Sodium polycarboxylate	Various	Various
Sodium tetraborate decahydrate	1303-96-4	$B_4O_7 \cdot 2Na \cdot 10H_2O$
Tetrakis hydroxymethyl phosphonium sulfate	55566-30-8	$(C_4H_{12}O_4P)_2O_4S$
Tetramethyl ammonium chloride	75-57-0	$C_4H_{12}ClN$
Thioglycolic acid	68-11-1	$C_2H_4O_2S$
Thiourea	62-56-6	CH_4N_2S
Tributyl tetradecyl phosphonium chloride	81741-28-8	$C_{26}H_{56}PCl$
Triethanolamine zirconate	101033-44-7	$C_{24}H_{56}N_4O_{12}Zr$
Zirconium hydroxy lactate sodium complex	113184-20-6	$C_{12}H_{19}NaO_{16}Zr$
Zirconium nitrate	13746-89-9	$Zr(NO_3)_4$
Zirconium sulfate	14644-61-2	$Zr(SO_4)_2$
1-Bromo-3-chloro-5,5-dimethylhydantoin	16079-88-2	$C_5H_6BrClN_2O_2$
2,2-Dibromo-3-nitrilopropionamide	10222-01-2	$C_3H_2Br_2N_2O$
2-Bromo-3-nitrilopropionamide	1113-55-9	$C_3H_3BrN_2O$
2-Butoxyethanol	111-76-2	$C_6H_{14}O_2$

[5, 44, 46–48].

in the proppant pack. Likewise, the filter cake formed on the face of the rock, which may restrict the flow of oil and gas and reduce well productivity, must be degraded. Breakers reverse cross-linking and cleave polymers into low molecular weight fragments thus reducing viscosity of gelled fluids [46, 69–71].

The general types of breakers are oxidizers, acids, and enzymes [62]. Oxidizers are the most commonly used class of breakers, in particular ammonium, potassium, and sodium salts of peroxydisulfate (persulfate) [68]. Enzymes may be used depending on fracturing conditions, particularly pH and temperature.

2.6. Acids and Bases. Acids and bases are added to HF fluids to adjust pH, which improves the effectiveness of almost all HF compounds, particularly cross-linked polymers. The use of acids also clears debris in the wellbore and provides an open channel for other HF fluids by dissolving carbonate minerals [56]. Lastly, pH adjustment prevents unwanted microbial activity in the wellbore.

Typical pH adjusters include inorganic acids such as hydrochloric and sulfuric acids, as well as organics such as acetic acid and fumaric acid. Common bases include potassium hydroxide, sodium hydroxide, sodium carbonate, and potassium carbonate [46].

2.7. Biocides. Biocides are used to control microbial growth in the boreholes and well areas, as such growths degrade HF chemicals and accelerate corrosion of well tubing, casings, and equipment [47, 70]. Biocides used for HF include quaternary ammonium compounds (QACs), glutaraldehyde, tetrakis hydroxymethyl phosphonium sulfate (THPS), tributyl tetradecyl phosphonium chloride (TTPC), and brominated compounds including 2,2-dibromo-3-nitrilopropionamide (DBNPA) [47, 56, 71]. QACs are extensively used as bioactive agents; the most commonly used ones are dialkonium and benzalkonium chlorides. Ammonium chloride is also used [46, 72].

2.8. Corrosion Inhibitors. Corrosion inhibitors form a protective layer on metal well components, thus preventing corrosion by acids, salts, and corrosive gases [73–75]. Common corrosion inhibitors include acetaldehyde, acetone, ethyl methyl derivatives, formic acid, and isopropanol [46, 76].

2.9. Scale Inhibitors. Scale inhibitors protect piping in the wells and prevent formation plugging. These inhibitors consist of polycarboxylates and acrylate polymers [46].

2.10. Iron Control Substances. Precipitates of ferric iron (Fe^{3+}) block paths within pipes and rock formations, which impact productivity [77, 78]. Ferric iron also inadvertently acts as a cross-linker in HF fluids containing gelling agents, thereby altering fluid viscosity [77]. Iron precipitation is prevented using citric acid, acetic acid, thioglycolic acid, and sodium erythorbate [79]. Iron controlling agents act as chelating agents, forming complexes with ferrous iron (Fe^{2+}) to prevent oxidation and subsequent precipitation as Fe^{3+} [70, 80].

2.11. Clay Stabilizers. In order to prevent clay swelling around shale formations, clay stabilizers are injected with HF fluids. These work via ion exchange, replacing cations such as Na^+ in the clay with other, often divalent cations that undergo less hydration and have a lesser tendency to swell the clay [81]. Commonly used clay stabilizers are choline chloride, potassium chloride, and tetramethyl ammonium chloride [46]. There has been some shift towards choline chloride use, which is nontoxic and readily biodegradable.

2.12. Surfactants. Surfactants are used to achieve optimal viscosity of HF fluids, reduce surface tension, and assist in fluid recovery after fracturing [5, 46, 70]. Surfactants can be used in place of cross-linkers and gelling agents in high temperature or high pressure formations. Surfactant formulations used in HF vary greatly, but common compounds include sodium lauryl sulfate and dimethyl dehydrogenated tallow ammonium chloride [46, 76].

The large quantity and diversity of compounds used in HF additives underscore the complexity of understanding their fates in the event of release, whether accidental or managed, to soil. Furthermore, the compounds described for each agent are only the known compounds; hazards relating to proprietary compounds remain unknown.

3. Flowback and Produced Water

As oil and gas production proceed, formation water eventually reaches the production well, and water begins to appear alongside the hydrocarbons. This produced water is a mixture of injected water, formation water, HF chemicals, and hydrocarbons [82–85].

Produced water has a complex composition but its constituents can be broadly classified into organic and inorganic compounds. These include dissolved and dispersed oil components, grease, heavy metals, radionuclides, HF chemicals, dissolved formation minerals, salts, dissolved gases (including CO_2 and H_2S), scale products, waxes, microorganisms, and dissolved oxygen [49, 50, 53, 83, 86, 87]. The composition will vary widely as a function of geologic formation, lifetime of the reservoir, and type of hydrocarbon produced [83].

A generalized chemical composition of produced water appears in Table 3.

3.1. Production Chemicals. Production chemicals, that is, HF fluids, enter produced water in traces and sometimes significant quantities [88] and vary from platform to platform. Active ingredients partition themselves into all phases present depending on their relative solubilities in oil, gas, or water.

3.2. Dissolved Minerals. Flowback water tends to have extremely high concentrations of total dissolved solids (TDS); this is due to dissolution of constituents from the formation following injection of HF fluids [89, 90]. High salinity may also originate from release of in situ brines (formation water) [90–93]. Levels of TDS can be 5–10 times the concentration in seawater [90]. Na^+ and Cl^- are responsible for salinity and range from a few mg/L to 300,000 mg/L [94]. For comparison, seawater and salt lakes are defined as having an upper limit of 50,000 mg/L [95]. Ions such as Cl^-, SO_4^{2-}, CO_3^{2-}, HCO_3^{2-}, Na^+, K^+, Ca^{2+}, Ba^{2+}, Mg^{2+}, Fe^{2+}, and Sr^{2+} affect conductivity and scale-forming potential. High levels of organic carbon also occur in substantial levels in flowback fluids and produced water [89, 96].

Fluid chemical composition is dependent, in part, upon its interaction time with the shale play. It has been found that TDS levels in produced water and late flowback can increase four-fold over that of early flowback. Similarly, total suspended solids (TSS) concentrations increase over 100-fold between early and late flowback. Concentrations of inorganic ions in produced water from Marcellus shale (PA) wells increased over the course of oil production [89, 96–98], rising significantly during the initial days after fracturing and then increasing more slowly as the well aged [76, 89, 96].

TABLE 3: Composition of oilfield produced water.

Parameter	Range	Metal	Range (mg/L)
Density (kg/m^3)	1014–1140	Ca	13–29,222
Conductivity (μS/cm)	4200–58,600	Na	132–97,000
Surface tension (dyn/cm)	43–78	K	24–4,300
Turbidity (NTU)	182	Mg	8–6,000
pH	4.3–10	Fe	<0.1–100
TOC (mg/L)	0–1,500	Al	310–410
TDS	267,588	B	5–95
TSS (mg/L)	1.2–10,623	Ba	1.3–650
Dissolved oxygen (mg/L)	8.2	Cd	<0.005
Total oil (mg/L)	2–565	Cu	<0.02–1.5
Volatiles (BTEX; mg/L)	0.39–35	Cr	0.02–1.1
TPH (mg/L)	>20	Li	3–50
Chloride (mg/L)	80–200,000	Mn	<0.004–175
Bicarbonate (mg/L)	77–3,990	Pb	0.002–8.8
Sulfate (mg/L)	<2–1,650	Sr	0.02–2,204
Sulfite (mg/L)	10	Ti	
NH$_3$-N (mg/L)	10–300	Zn	<0.01–0.7
Phenol (mg/L)		As	0.01–35
Volatile fatty acids (mg/L)	0.009–23	Hg	<0.005–0.3
	2–4,900	Ag	<0.005–0.3
		Be	<0.001–0.15
		Ni	<0.001–0.004
			<0.001–1.7

[49–51].

3.3. Metals.
Oilfield produced water contains heavy metals such as mercury and lead, as well as metalloids such as arsenic, in varied concentrations depending on formation geology and age of the well [49, 99]. Metal concentrations in produced water are usually higher than those found in sea water [83, 94]. The most commonly studied metals are Ba, Cd, Cr, Cu, Pb, Hg, Ni, Ag, and Zn (Table 3) [50]. Produced water contains other trace metals including Al, B, Fe, Li, Mn, Se, and Sr. Certain metals are of particular environmental concern as they may bioaccumulate and/or be toxic [50].

3.4. Dissolved and Dispersed Oil Components.
Dispersed and dissolved oil components are derived from the source rock and chemical additives in HF fluids, and their concentrations may be very high at some oilfields [86, 88, 100, 101]. BTEX, phenols, aliphatic hydrocarbons, carboxylic acid, and low molecular weight aromatics are classified as dissolved oil, while the more hydrophobic PAHs and heavy alkyl phenols are present in produced water as dispersed oil [100].

Produced water from the Marcellus (PA) and Barnett (TX) plays contains predominantly C_6–C_{16} hydrocarbons, while Eagle Ford (TX) produced water shows the highest concentration in the C_{17}–C_{30} range [102]. The structures of saturated hydrocarbons identified generally follow the trend of linear > branched > cyclic. Heterocyclic compounds, fatty alcohols, esters, and ethers have also been identified. The presence of various fatty acid phthalate esters in the Barnett and Marcellus produced water may be related to their use in HF fluids [102]. No polyaromatic hydrocarbons (PAHs) were observed in these shale plays [102].

3.5. Produced Solids.
Produced solids include clays, precipitated solids, waxes, microbial biomass, carbonates, sand and silt, corrosion and scale products, proppant, formation solids, and other suspended solids [49]. Their concentrations vary from one oilfield to another.

4. OGPW in the Terrestrial Environment: Releases, Effects, and Remediation

4.1. Releases.
The management of OGPW is largely monitored and controlled; however, accidental releases are inevitable. In addition, application of HF fluids to soil is considered an acceptable form of disposal in many states [103]. Inadvertent releases and intentional land application could potentially expose soil to hundreds of heterogeneous chemicals. The US EPA has studied potential scenarios that could lead to environmental contamination by HF fluids [104].

4.1.1. Pipe Overflows, Leaks, and Blowouts.
In a 2009 study it was revealed that 630 out of 4,000 legally permitted wells in Pennsylvania had drilling site leaks [105]. In 2011 a mechanical problem at a Pennsylvania natural gas well caused thousands of gallons of briny water and HF fluid to erupt from the well, overwhelm containment facilities, and flow into surrounding fields. Local families were ordered to evacuate their homes. After six days workers sealed the leak, replaced the wellhead, and got the well "under control" [3].

In 2014 a North Dakota oil well leaked HF fluid and oil, releasing between 2,100 and 2,940 gallons per day of OGPW and 8,400 gallons per day of oil [106]. In a 2015 North Dakota well blowout, 4,620 gallons of OGPW and 23,100 gallons of oil were released. Most of the spill was contained at the well site, but some escaped and contaminated nearby terrain [107]. From January 2006 to October 2014 more than 18 million gallons of OGPW and oil was spilled in North Dakota alone. Most individual spills were contained to the immediate drilling area, but many larger spills affected surrounding farms and waterways [36].

4.1.2. Deliberate Improper Disposal.
A petroleum subsidiary had permission to discharge drilling mud and boring waste to an oilfield sump near almond orchards in Shafter, California. State investigators, however, found that the fluid contained excess salinity, boron, benzene, and gasoline and diesel hydrocarbons believed to have been used in HF [19, 108].

4.1.3. Holding Ponds.
Gas and oil producers are increasingly reusing spent HF fluids. However, reuse involves storage in holding ponds and eventually diluting the OGPW with fresh water [109]. In 2009 a wastewater pit overflowed at a Pennsylvania gas well and an unknown quantity of OGPW entered a "high quality watershed." The company failed to report the spill and in 2010 a fine of $97,350 was levied against the company [3].

4.1.4. Natural Events. Natural disasters such as floods add to the potential for soil contamination by OGPW. During the late 2013 floods in Colorado, floodwaters in Weld County (where 20,000 oil and gas wells are located) surged into drilling centers and damaged pipes, overflowed wells, and shifted oil tanks from their foundations [110]. Approximately 35,000 gallons of oil and condensate were released. HF fluids from evaporation pits may have contaminated local soils and possibly been carried farther by the floodwaters [19].

4.2. Effects on the Soil Resource. Potential soil quality and plant impacts from OGPW include the following [100]:

(i) Excess sodicity can cause clays to deflocculate, thereby lowering the permeability of soil to air and water.

(ii) Excess soluble salts will cause plants to desiccate and die. Where levels of natural precipitation are low, salts may accumulate to excessive concentrations in soil.

(iii) Existing plant species may become displaced by new species as a result of chemical changes in soils resulting from contact with OGPW.

(iv) Salt-tolerant plants may increase in distribution.

In a greenhouse study Swiss chard (*Beta vulgaris* L.) and ryegrass (*Lolium perenne* L.) were grown in soils containing synthetic HF fluids [111]. The HF fluids increased soil pH, EC, and concentrations of total and extractable Zn, Cu, Cd, Pb, and As. Chard and ryegrass yields may have been reduced by high soil Zn and EC levels. The HF fluids may have resulted in lower levels of trace elements in plant tissue due to increased soil pH. In a greenhouse study Miller et al. [112] studied the effects of OGPW components on plant growth and found that diesel oil, KCl, NaOH, Cr, starches, and other compounds reduced yields of sweet corn (*Zea mays* L. var. *saccharata*) and/or green beans (*Phaseolus vulgaris* L.).

Six drilling fluids reduced yields of green beans and sweet corn when added to soils at differing ratios [113]. High levels of soluble salts or high percentage exchangeable Na^+ was considered to be the major cause of reduced plant growth. Adams [103] reported severe acute and chronic toxicity of mixed hardwood trees (*Quercus* spp., *Acer rubrum* L., and *Liriodendron tulipifera*), mixed shrub subcanopy (*Fagus grandifolia* (Ehrh.), *A. rubrum*, and *Sassafras albidum* (Nutt.)), and ground vegetation (*Vaccinium* L., *Smilax rotundifolia* L., and *Kalmia latifolia* L.) that resulted in 56% vegetation mortality after two years of land application of HF fluid. Soil Na^+ and Cl^- concentrations increased by approximately 50-fold as a result of land application of the fluids.

OGPW components in soil may substantially impact each other's fate and transport; for instance, the presence of biocides may decrease the potential for biodegradation, while viscosity-enhancing compounds may hinder the mobility of other compounds. Factors that affect the behavior of OGPW constituents in soil and therefore their potential impact on terrestrial life include the following [100]:

(i) Dilution of the OGPW in the receiving environment.

(ii) Immediate and long-term precipitation of metals and other contaminants.

(iii) Volatilization of low molecular weight hydrocarbons.

(iv) Physical-chemical reactions with other chemical species present in soil.

(v) Adsorption to particulate matter.

(vi) Biodegradation of organic constituents.

Some specific HF components and their possible fates are described below.

4.2.1. Gelling Agents. Gellants such as guar and cellulose occur naturally and are nontoxic and readily biodegradable. The same is true for the common acids and alcohols used as gelling agents [46]. It is likely that these compounds, when in contact with soil, will enhance microbiological growth.

Ethylene glycol is highly soluble in water; it adsorbs poorly to soil colloids and is thus highly mobile in the soil profile [114]. Volatilization of ethylene glycol from soil is not expected.

4.2.2. Friction Reducers. Polyacrylamide is readily biodegradable. McLaughlin [115] studied transformation kinetics of PAM and polyethylene glycol (PEG) in the presence of cocontaminants. Over time higher rates of disappearance occurred in raw, that is, biologically active, soil compared with sterilized soil, indicating that PEG disappearance is due to both sorption and biodegradation. In a study by Wen et al. [116], two PAM-degrading bacterial strains were isolated from soil in an oilfield contaminated by PAM; these were identified as *Bacillus cereus* and *Bacillus flexus*. No acrylamide, which is a known human carcinogen, mutagen, and teratogen, was produced during aerobic biodegradation of polyacrylamide [116, 117]; however, it has been suggested that acrylamide may be formed via heating or exposure of polyacrylamide to ultraviolet radiation [116, 118].

4.2.3. Cross-Linkers. Human exposure to boron and amines used in cross-linkers is of concern as they have known toxic effects and can be mobile in soil and groundwater; however, the amines are not known to persist in the environment [46].

4.2.4. Breakers. The use of enzymes as breakers for fracturing fluids is preferred over the use of oxidizers because enzymes are environmentally benign [68]. Their mobility, however, remains largely unknown [76].

4.2.5. Acids and Bases. Organic acids are potentially biodegradable depending on concentration. Strong acids or bases are known to cause adverse effects on soil. For example, strong acidity will result in leaching of bases such as Ca^{2+} and Mg^{2+} [119]; extremes of acidity will cause dissolution of soil solids [120]. The hydrated Na^+ ion in sodium hydroxide disperses soil aggregates and destroys soil structure. Extremes in pH may drastically alter microbial composition [121].

4.2.6. Biocides. Some common HF biocides are known to be volatile or sorb to soils and can persist in the environment, although their fates are largely unknown [76].

QACs have distinct physical/chemical properties which are conferred by their substituents, primarily alkyl chain length. The mechanism of QAC sorption to solids is complex, but both hydrophobic and ionic interactions probably occur [122]. The log K_{OC} values of several mono-, di-, and benzalkonium chlorides range between 0.28 and 2.97 [123, 124]. QACs are therefore expected to sorb to soil colloids and not leach to groundwater. Sorption of QACs on organic surfaces such as humic compounds and sediment increases as the alkyl chain length increases [122]. In the benzalkonium chlorides the benzyl group enhances adsorption.

QACs have been identified in sediments near wastewater discharge sites, suggesting that at least some are environmentally persistent [125, 126]. The degree of biodegradability is variable; biodegradation decreases with increasing length of the alkyl chain, and QACs that contain a benzyl group experience lower biodegradation rates [127]. QACs were found to be recalcitrant under methanogenic conditions [122]. Under nitrate reducing and fermentative conditions, benzalkonium chlorides (BACs) were transformed to alkyldimethyl amines via abiotic reactions [122]. Microorganisms have been isolated that are resistant to QACs and capable of QAC degradation [128–132]. In a study by Tezel [122] the bacterial community involved in aerobic degradation of BACs was mainly composed of species belonging to the genus *Pseudomonas*.

Generally, QAC sorption exceeds biodegradation in aerobic biological systems [46, 125, 129].

Glutaraldehyde (GA) and the phosphonium-based biocides are sometimes considered "green" alternatives to conventional biocides as they are less persistent in the environment. GA is readily biodegradable under both aerobic and anaerobic conditions [133, 134]. In aerobic batch experiments McLaughlin [115] studied decomposition of GA and didecyl dimethyl ammonium chloride (DDAC) to determine transformation kinetics in the presence of cocontaminants. DDAC underwent almost immediate sorption to soil. GA slowed the initial rate of PEG and PAM biodegradation. After one week, GA was completely eliminated from the aqueous phase due to sorption.

DDAC-degrading bacteria were isolated via enrichment culture with DDAC as the sole carbon source [130]. One isolate, *Pseudomonas fluorescens* TN4, degraded DDAC to produce decyldimethylamine and subsequently dimethylamine as intermediates. The TN4 strain also assimilated other QACs, alkyltrimethyl- and alkyl benzyl dimethyl ammonium salts, but not alkylpyridinium salts. TN4 was highly resistant to these QACs and degraded them by an N-dealkylation process [130].

Tetrakis hydroxymethyl phosphonium sulfate (THPS) has low K_{OW} (140) [135], K_{OC}, and K_H values, which suggests it will not sorb to soil but will leach to groundwater. Under abiotic conditions THPS is readily biodegradable [46]. THPS decomposes under natural conditions via hydrolysis, oxidation, and photodegradation [104]; it initially degrades to trihydroxymethyl phosphine (THP), releasing formaldehyde

and sulfuric acid [136]. Carbon dioxide, water, and inorganic phosphate are among the final products [137]. At pH 5–7 the half-lives of THPS exceed 30 days; at pH > 8 THPS degrades within 7 days. THPS is expected to volatilize from dry soil surfaces [138].

DBNPA has low log K_{OW} and K_{OC}, suggesting it will not sorb to soil and may leach to groundwater. Disappearance of DBNPA in soil may be due to hydrolysis, adsorption, chemical degradation, and/or microbial degradation. Sunlight also degrades DBNPA [139]. Hydrolysis reactions convert DBNPA into dibromoacetonitrile, followed by dibromoacetamide, dibromoacetic acid, glyoxylic acid, and oxalic acid. The most stable product of these products is dibromoacetic acid.

4.2.7. Corrosion Inhibitors. In general, corrosion inhibitors are highly soluble and biodegradable. Their low log K_{OW} and K_{OC} values indicate that these chemicals are not likely to sorb to soils, and there is potential for leaching to groundwater. This group contains compounds that are toxic and/or carcinogenic [46, 76].

Propargyl alcohol and thiourea are GHS Category 2 chemicals, making them among the most toxic chemicals used in HF fluids. Propargyl alcohol is considered readily biodegradable; it is also highly mobile in soil [140]. Volatilization of propargyl alcohol from moist soil surfaces is expected to be substantial, given an estimated Henry's Law constant of 1.1×10^{-6} atm-cu m/mole [140]. Volatilization from dry soils is also expected to occur. The half-life of propargyl alcohol in an alkaline silt loam soil (pH 7.8, 3.25% organic carbon) was 12.6 days and 13 days in an acidic sandy loam (pH 4.8, 0.94% organic carbon) [140, 141]. Thiourea is considered biodegradable and highly mobile in soil. Sorption of thiourea to organic matter of three different soil orders was characterized as low (spodosol) to moderate (entisol/alfisol) [142].

4.2.8. Iron Control Agents. Acetic acid, citric acid, sodium erythorbate, and mercaptoacetic acid (thioglycolic acid) are highly soluble in water. The low K_{OC} values of citric acid and thioglycolic acid indicate that they will not sorb markedly to soils but will be mobile in the profile. The pK_a of thioglycolic acid is 3.55, suggesting that it will exist almost entirely in the anionic form in soil and will therefore not sorb to clay and organic matter [46]. With the exception of acetic acid, these compounds are not expected to volatilize from OGPW based on their Henry's Law constants [46].

All iron control agents, in particular acetic acid, citric acid, and thioglycolic acid, tend to be readily degraded and are not persistent [143, 144]; however, some are known to be toxic [76]. Acetic acid, citric acid, and sodium erythorbate are of low toxicity to humans. Of the iron control agents, thioglycolic acid appears to be the greatest concern as a soil contaminant as it poses a toxicity risk based on an oral LD_{50} value of 114 mg kg^{-1} [145].

4.2.9. Surfactants. Most surfactants are highly soluble in water and readily biodegradable. Sodium lauryl sulfate has a moderately high K_{OC} value and is expected to have moderate to low mobility in soil [46, 76, 146, 147]. Sodium lauryl sulfate occurs in household products and is not anticipated to be a health risk due to its LD_{50} value.

4.2.10. Excess Salinity. Soil salinity imposes ion toxicity, nutrient (N, Ca, K, P, Fe, and Zn) deficiencies, nutritional imbalances, osmotic stress, and oxidative stress on plants [148, 149]. Soil salinity significantly reduces plant phosphorus (P) uptake because phosphate ions precipitate with Ca ions [150]. Some elements, such as Na, Cl, and B, impart specific toxic effects on plants. Excessive accumulation of Na in cell walls can lead to osmotic stress and cell death [151]. All these factors cause adverse effects on plant growth and development at physiological and biochemical levels [152] and at the molecular level [153]. Salinity hinders seed germination; seedling growth; enzyme activity; DNA, RNA, and protein synthesis; and mitosis [154, 155].

4.2.11. Hydrocarbons. The primary hydrocarbons which contribute to acute toxicity of OGPW are the aromatic and phenol fractions of dissolved hydrocarbons [156].

4.3. Soil Remediation. In situ remediation of HF-affected soil involves (1) removal of salts in the soil solution via leaching with irrigation or natural precipitation; (2) replacement of exchangeable Na^+ with Ca^{2+}; (3) removal or destruction of hydrocarbons; and (4) removal or immobilization of metals. Remediation practices on OGPW-contaminated soils often tend to be straightforward.

4.3.1. Treatment of Salinity and Sodicity. Simple soil dilution may relieve salinity problems following release of OGPW. In the study by Wolf et al. [157], OGPW occurred primarily at the soil surface. Mixing of the less-contaminated deeper soil with the surface soil resulted in dilution of contaminants. Ahmad et al. [158] and Lloyd [159] concluded that the salt concentration of drilling waste was the primary factor in determining the waste loading rate in soil systems.

Addition of inexpensive amendments is often successful in treating soil salinity and sodicity problems. Both inorganic amendments (e.g., $CaSO_4$ [160]) and organic materials (animal manures) have proven to be successful. The most commonly used dry amendments are gypsum ($CaSO_4 \cdot 2H_2O$) and calcium nitrate ($Ca(NO_3)_2$), although calcium chloride ($CaCl_2$) may be used if adequate drainage control is provided and leachate is managed [37]. Use of calcium amendments may require subsequent irrigation and leachate collection to move the calcium amendment into the affected soil layers for replacement of Na and to leach salts beyond the root zone. Sulfur may be applied, either as elemental S or as aluminum sulfate, to decrease pH.

Livestock manure can be successfully used as a soil amendment. Organic material creates macropores thus allowing for soil drainage; it furthermore greatly augments soil biological activity. Only well-decomposed or composted manure should be used in order to limit inputs of salts and to prevent proliferation of weeds. Addition of significant organic amendments such as chicken or some feedlot manures can increase soil salinity over several applications. Testing manure and compost for salinity is recommended [37, 161].

Organic amendments must be thoroughly mixed into soil upon application. Low-N organic matter such as cereal straw requires additional N for decomposition; therefore, a high-N fertilizer such as ammonium nitrate, ammonium sulfate, or calcium nitrate should be included [37].

Additional amendments may prove to be beneficial in treating OGPW-affected soil. For example, use of synthetic polymers (e.g., polyacrylamides) to stabilize aggregates has proved to be successful in improving the physical properties of Na-enriched soil [37, 162]. Given that soil biological activity may be drastically reduced in OGPW-contaminated soil, it is recommended that mycorrhizal fungi be applied [163].

Electrokinetic remediation has been suggested for treatment of saline soils [160]. This technology involves application of low density direct current between electrodes placed in the soil to mobilize contaminants which occur as charged species. This allows for separation and removal of Na^+, Cl^-, and other highly soluble ions. Electrodes can be installed horizontally or vertically in deep, directionally drilled tunnels or in trenches around sites contaminated by OGPW [164].

When the average EC of the uppermost soil is > 35,000 uS/cm, soil removal and replacement may be more economical than treatment [160]. It may be costly, however, to haul and dispose contaminated soil at a special waste landfill. There is also potential long-term liability of impacted soil placed in a landfill [165].

4.3.2. Treatment of Hydrocarbon Contamination. Hydrocarbon contamination of OGPW-affected soils is typically not expected to be significant, given the relatively low concentrations occurring in OGPW. However, in cases of a catastrophic release, microbial decomposition of oily wastes is encouraged. So-called bioremediation processes, if conducted properly, should result in few residuals and minimal alteration of the local environment [164].

In situ bioremediation systems introduce aerated, nutrient-enriched water into the contaminated zone through an array of injection wells, sprinklers, or trenches. Sufficient time is allowed for the reaction of indigenous microbial communities with the contaminants, and the treated water is eventually recovered downgradient. The recovered water may be further treated (e.g., passage over granular activated carbon) and reintroduced to the affected soil. Otherwise it may be discharged to a municipal wastewater treatment plant or to surface water [164, 166].

The affected soil should receive adequate nutrients (in particular, N and P) to promote microbial growth and activity. Also, it is essential that adequate oxygen be available, which may be provided by aeration of the flushing solution. Soil pH must be maintained near neutral in order to promote microbial proliferation [164].

In slurry biodegradation, contaminated soil is transferred from the affected area to a lined lagoon and mixed with water. The slurry is continuously stirred and aerated in the lagoon. Decomposition of organic contaminants takes place via aerobic microbial processes. Slurry biodegradation can treat a range of hydrocarbons including crude and refined petroleum products [167–169]. The presence of heavy metals and other potential toxins in OGPW (e.g., biocides) may inhibit microbial metabolism and require pretreatment.

A significant benefit in the use of slurry biodegradation is the enhanced rate of contaminant degradation, a direct result of improved contact between the microorganisms and hydrocarbons. The agitation of contaminants in the liquid phase provides for a high degree of solubilization of compounds and significant homogeneity [167].

Land treatment techniques for bioremediation, for example, landfarming, are commonly used for treatment of hydrocarbon-contaminated soil [170]. Contaminants treated include fuel, lubricating oil, and pesticides. Landfarming can be regarded as a combination of biodegradation and soil venting; microbial oxidation reactions occur in combination with volatilization.

A common field installation calls for the affected soil to be excavated and transferred to a prepared location (a land treatment unit or cell) which is designed for controlling the process. Treatment involves installation of layers ("lifts") of contaminated soil to the cell. The cell is usually graded at the base to provide for drainage and lined with clay and/or plastic to contain all runoff within the unit. It may also be provided with sprinklers or irrigation and drainage. Because of the high water application rates, LTUs are often bordered by berms [164].

A major benefit of the land treatment technique is that it allows for very close monitoring of process variables that control the decomposition of hydrocarbons [164, 170].

4.3.3. Treatment of Metals Contamination. Metals at OGPW-contaminated sites may occur in complex forms.

Metals may be extracted from soil via elutriation for eventual recovery, treatment, and disposal. Also known as soil flushing, contaminants are solubilized or similarly desorbed from solid forms and recovered [171]. Metal removal efficiencies during soil flushing depend not only on soil characteristics but also on metal concentration, chemistry of the metal(s), extractant chemistry, and overall processing conditions [171].

In situ extraction processes are applicable for either the vadose zone or the saturated zone. The flushing solution is applied to the affected site via sprinklers or irrigation, or by subsurface injection. A sufficient period is allowed for the applied reagents to percolate downward and react with contaminant metals. The contaminants are subsequently mobilized by solubilization. The elutriate is collected in strategically placed wells [164]. Metals which are minimally soluble in water often require acids, chelating agents, or other solvents for successful washing [172].

One drawback to soil flushing technology is the possible production of residuals. These include excess chelating agents, some of which may be toxic to biota. In addition, leaching of soil with dilute acids may destroy the biological portion of the soil, alter its chemical and physical properties, and create a relatively inert material [164].

Phytoremediation is a cost-effective, low-technology process defined as the engineered use of green plants to extract, accumulate, and/or detoxify environmental contaminants. Phytoremediation employs common plants including trees, vegetable crops, grasses, and even annual weeds to treat heavy metals in soil [173–175].

A simple and common application of phytoremediation is *phytoextraction*, which involves the use of hyperaccumulating plants to take up metals from the soil and concentrate them into roots and aboveground shoots. In certain cases contaminants can be concentrated thousands of times higher in the plant than in the soil. Following harvest of the extracting crop, the metal-rich plant biomass can be ashed to reduce its volume, and the residue can be processed as an "ore" to recover the contaminant metals. If recycling the metal is not economically feasible, the small amount of ash (compared to the original plant biomass or the large volume of contaminated soil) can be disposed [164, 173–176].

Phytoremediation is useful for soils contaminated with metals to shallow depths. This technology can work well in low permeability soils, where many technologies have a low success rate. It can also be used in combination with conventional cleanup technologies (e.g., "pump and treat" of groundwater). Phytoremediation can be an alternative to harsher remediation technologies such as soil flushing [164].

5. Pollution Prevention for OGPW

Once OGPW has been brought to the surface it is either disposed or reused. Given that OGPW is enriched in TDS, TSS, metals, dispersed oil, dissolved and volatile organic compounds, HF additives, and other contaminants to varying degrees (Table 3), significant management challenges face operators. OGPW must be managed in ways that both reduce the operational costs as well as are protective of the environment. OGPW management practices vary widely across the United States and in some instances across a single oil and gas field [177].

OGPW management falls under two broad categories: underground injection and surface management. Selection of a management option for OGPW at a site varies based on the following [178]:

(i) Chemical and physical properties of the OGPW.

(ii) Volumes, duration, and flow rate generated.

(iii) Desired end-use of the wastewater.

(iv) Treatment and disposal options allowed by state and federal regulations.

(v) Technical and economic feasibility of a particular option, including transportation.

(vi) Availability of suitable infrastructure for management.

(vii) Willingness of companies to employ a particular technology or management option, including concerns about potential liability.

Some common options available to oil and gas operators for managing OGPW are addressed below.

5.1. Limit Production of Water at the Surface. Technologies are available for managing water within the wellbore. These technologies do not reduce the volume of water entering the well but minimize the quantity of OGPW that rises to the surface.

5.1.1. Downhole Oil/Water Separation and Injection. Downhole oil/water separators (DHOWS) separate water from oil within the wellbore so that oil or gas with little water is brought to the surface. Significant quantities of water are disposed in nonproducing formations above or below the oil- or gas-producing formation using injection tools within the well [100]. The downhole separator assembly comprises several compact elements installed within the wellbore including (1) a separation tool, which separates OGPW from incoming hydrocarbons from the formation; (2) a pump that pressurizes water from the separator and injects the OGPW into the disposal zone; (3) a heavy-duty motor to perform the pumping; and (4) miscellaneous equipment such as downhole monitoring equipment and cables [177].

5.1.2. Downhole Gas/Water Separators. Devices similar to DHOWS are available for gas wells. A study by the Gas Research Institute identified 53 commercial field tests of downhole gas/water separators involving 34 operators in the USA and Canada [179]. Gas production rates increased in 57% of the tests; 47% of the field tests were considered successful [100].

5.1.3. Dual Completion Wells. Oil production may decline in a well as water generates a "cone" around the production perforations, limiting the volume of oil that can be recovered. This phenomenon may be reversed and managed by completing the well with two separate tubing strings and pumps. The primary completion is made at a depth corresponding to strong oil production, and a secondary completion is made lower in the strata at a depth experiencing significant water production. The two completions are separated by a packer. The oil collected above the packer is brought to the surface, and the water collected below the packer is injected into a lower formation [100, 180, 181]. Swisher [182] reports on the results of using a dual completion well compared to three wells with conventional completions in a Louisiana oilfield. The dual completion well costs about twice as much to install but took the same number of months to reach payout as the other wells. However, at payout, it was producing 55 bpd of oil compared to about 16 bpd from the other three wells. Wojtanowicz [183] provides additional examples of using dual completion wells from differing geological settings.

5.2. Injection of OGPW. Injection involves the emplacement of OGPW into porous geologic strata by pumping, via an injection well, into a formation capable of receiving and storing water. Injection wells are regulated by the Federal Underground Injection Control (UIC) program which was initiated under the Safe Drinking Water Act to prevent contamination of underground sources of drinking water (USDW).

Approximately 90% of OGPW from land-based oil and gas recovery operations in the United States is reinjected into underground formations. Hundreds of thousands of injection wells operate daily to manage produced water and flowback [178, 184]. OGPW may be injected back to its formation or into other suitable formations [185]. Injection is dependent upon several variables including the availability of receiving formation(s); the quality of OGPW being injected; the quality of water in the receiving formation; and the ultimate storage capacity of the receiving formation(s). These factors will influence what type of injection well can be used for managing OGPW.

Many papers have described the process of underground injection of wastewater. Only a brief review is provided here.

The US EPA classifies five different injection well categories (Table 4); three may be applicable for management of produced water. In general, OGPW is considered an exempt waste and therefore can be injected in Class II or Class V injection wells. Class II wells may be used to hold fluids associated with oil and natural gas production [52] and are classified either as disposal wells (IID) or as enhanced recovery wells (IIR). Wastewater resulting from OGPW treatment must be disposed in Class I injection wells. EPA defines Class I wells as technologically sophisticated wells that inject hazardous and nonhazardous wastes below the lowermost USDW. Injection occurs into deep, isolated rock formations that are separated from the lowermost USDW by layers of impermeable clay and rock [52]. Class V wells (i.e., shallow injection, subsurface drip irrigation) are injection wells not included in the other four classes. Their simple construction provides little protection against possible soil and groundwater contamination.

Underground injection of OGPW often requires transport, along with treatment to reduce fouling and bacterial growth. Storage over the long-term may be required.

5.3. Beneficial Use: Discharge OGPW to the Surface. Some important emerging opportunities for management of OGPW are (1) treatment and reuse as a water supply for public consumption, agriculture, and industry and (2) secondary industrial processes such as extraction of minerals [178]. The presence of certain constituents, however, may limit produced water use in selected areas [178]. Produced water may be discharged to the land as long as it meets both onshore and offshore discharge regulations [83].

OGPW may be discharged to the land surface to surface impoundments, for land application for crop use and industrial uses (i.e., oil and gas completion activities, truck wash station, dust suppression, and cooling tower water) [177]. A surface impoundment is defined as an excavation or diked area used for treatment, storage, or disposal of liquids [186]. Impoundments are usually constructed in low permeability soils. These vary in size from < 1 acres to several hundred acres. Based on an EPA national survey that characterized 180,000 impoundments, the oil and gas industry uses impoundments for storage (29%), disposal (67%), and treatment (4%) [186].

Impoundments are used for OGPW management including evaporation and/or infiltration; storage prior to injection or irrigation; or beneficial use such as livestock and wildlife watering ponds, constructed wetlands, fishponds, or a recreational pond.

TABLE 4: US EPA injection well classification system.

Well class	Injection well description	Approximate inventory
Class I	Inject hazardous wastes beneath the lowermost USDW	500
	Inject industrial nonhazardous liquid beneath the lowermost USDW	
	Inject municipal wastewater beneath the lowermost USDW	
Class II	Dispose of fluids associated with the production of oil and natural gas	147,000
	Inject fluids for enhanced oil recovery	
	Inject liquid hydrocarbons for storage	
Class III	Inject fluids for the extraction of minerals	17,000
Class IV	Inject hazardous or radioactive waste into or above USDW, 40 sites	
	This activity is banned. These wells can only inject as part of	
	an authorized cleanup	
Class V	Wells not included in the other classes. Inject nonhazardous liquid into or above a USDW	500,000 to >685,000

[52].

5.3.1. Evaporation Ponds. An evaporation pond is a large body of water that is designed to evaporate water by solar energy [187]. Such ponds are constructed to prevent sub-surface infiltration [188]. Ponds are a favorable technology in regions where annual rainfall is relatively low and evaporation rates are high. If the pond is constructed solely for evaporative loss, it is typically designed as a broad shallow pool that takes advantage of the large surface area. Areas with high winds and few natural windbreaks provide additional evaporative potential and may be considered in siting a pond. Ponds are usually covered with netting to prevent problems to migratory waterfowl caused by contaminants in OGPW [189].

As pure water evaporates from the pond, TDS increase in the remaining water. Over time the remaining water may become concentrated brine [177].

Over 80 million gallons of OGPW are managed daily by EPA under the Clean Water Act's National Pollutant Discharge Elimination System (NPDES) for beneficial reuses such as agricultural irrigation [100, 184]. The following is a summary of beneficial OGPW use practices in the USA.

5.3.2. Irrigation and Land Application. In the United States crop irrigation is the largest single use of freshwater, comprising 40% of all freshwater withdrawn, or 137 million gallons per day [190].

The determination of whether OGPW water can be used for agricultural purposes (i.e., irrigation, land application, and stock watering) depends both on the quality of the produced water and on characteristics of the recipient site. Relevant water and site variables include quantity of water required; length of time the water has been stored in impoundments prior to use; soil mineralogy, texture, and structure; and sensitivity of plant species [188].

The three most critical parameters regarding crop irrigation water quality requirements are salinity, sodicity, and elemental toxicity [100, 188]. As salinity rises above a species-specific salinity threshold, crop yields decrease. Irrigation water high in TDS diminishes the ability of roots to incorporate water, and reduces crop yield. The tolerance of various crops to salinity has been documented [191]. EC levels > 3,000 μS/cm are considered saline.

Excess sodium can damage soil physical properties. Irrigation water with SAR > 12 is considered sodic [100, 188]. Higher SAR values lead to soil dispersion and loss of structure and infiltration capability [119].

Under the Clean Water Act's Subpart E of 40 CFR Part 435, the NPDES permit system allows for the specialized reuse of wastewater from oil and gas facilities west of the 98th meridian. To qualify, the wastewater must contain < 35 mg/L of oil and grease and be used either for agriculture or for livestock watering [100]. The combination of the NPDES permitting allowance, the substantial requirement of water needed to perform HF, and the frequent geographical overlap between extraction sites and agricultural land has led to reuse of OGPW for irrigation in some states [100].

Raw water either is discharged directly to the land surface or is pretreated with amendments prior to application. Amendments are site-specific depending on soil properties, water chemistry, and plant species grown. Amendments could also be added to soil prior to, or following, application. Wolf et al. [157] found that addition of inorganic fertilizer, broiler litter, and Milorganite® to OGPW-treated soil markedly improved growth of Bermuda grass [*Cynodon dactylon* (L.) Pers.].

A program was created at Texas A&M University to develop a portable produced water treatment system that can be transported to oilfields to convert OGPW into irrigation or potable water. The goal was to produce water to levels of < 500 mg/L TDS and < 0.05 mg/L hydrocarbons [100, 192, 193]. Chevron Texaco developed a system to treat OGPW in Southern California [194]; 21 million gallons of oilfield water is recycled daily and sold to farmers who use it on about 45,000 acres of crops, about 10% of Kern County's farmland. The treated water is used for irrigation of fruit trees and other crops and for recharging shallow aquifers. An additional 360,000 bpd of water is further purified and used to make steam at a cogeneration facility. State and local officials praise the two-decade-old program as a national model for coping with the region's water shortages [195].

Thousands of acres in the Powder River Basin (WY) have been transformed to productive agricultural land using produced water [1, 196]. Livestock forage was irrigated using either OGPW on research plots or a blend of surface water and OGPW. Both treatments resulted in adequate crop production; however, the OGPW had to be applied at higher rates as plants did not utilize it as efficiently as the surface water blend [197]. Between watering intervals $CaSO_4$ and other supplements were applied to offset high SAR. After two years the rangeland was converted into highly productive grassland yielding livestock and wildlife benefits [100, 198].

The use of subsurface drip irrigation of OGPW is gaining popularity [199]. BeneTerra LLC has developed subsurface drip irrigation technology to provide produced water for crops [200]. Produced water is filtered, treated, and pumped through polyethylene tubing which spreads it uniformly through soil. The tubing is installed with a chisel plow to depths ranging from 18 to 48 inches. Haying operations can continue while the field is being irrigated. The drip irrigation systems are designed to utilize the native Ca and Mg present in the soil to offset the effects of Na. The salts percolate to a lower depth [1].

In some cases produced water may be treated for domestic supplies and drinking water. However, produced water from coal seams > 200-feet depth often has water that exceeds salinity levels appropriate for domestic use (i.e., about 3,000 mg/L). Also, water with high metal concentrations stains faucets and sinks. Water used by municipalities with treatment systems may be capable of removing certain harmful constituents or reducing their concentrations via existing processes [100, 188].

5.3.3. Aquaculture and Hydroponic Vegetable Culture. Greenhouse experiments were conducted to raise vegetables and fish using either produced water or potable water [201]. In a system using a combination of hydroponic plant cultivation and aquaculture, tomatoes grown with OGPW were smaller than those grown in potable water. The produced water tank grew a larger weight of tilapia (*Oreochromis niloticus/aureus*); however, some fish died, while none in the potable water tank died [100].

5.3.4. Constructed Wetlands. Constructed wetlands were developed approximately 50 years ago to exploit the ability of plants to treat contaminants in aqueous ecosystems [202]. Advantages of these systems include low construction and operation costs [188, 202] and public acceptance. Wetlands provide significant environmental benefits—they can be used by wetland birds and animals as well as by aquatic life. Wetlands can also be utilized for livestock and wildlife watering purposes [1, 203].

An artificial sedge wetland system was constructed to treat produced water [14]. After one year of operation the wetland effectively treated Fe and, to a lesser extent, Ba [188, 204]; however, SAR increased from 12.1 to 14.1. This was attributed to calcite precipitation without associated dissolution of Ca and Mg [204]. It was concluded that "clean water is needed to supplement sodicity and salinity treatment by vegetation and soil" [205].

5.3.5. Livestock Watering/Confined Animal Feeding Operations (CAFO). The water needs of CAFOs include animal consumption, irrigation of forage crops, and waste management. Livestock watering is one of the most common and proven beneficial uses of produced water [177].

The quality of OGPW presents the greatest constraint for use of livestock watering. Livestock are known to tolerate a range of contaminants in their drinking water. ALL [188] provides data showing acceptable TDS levels for livestock watering. In general, animals may tolerate higher TDS if they are gradually acclimated to the elevated levels. Water with TDS < 1,000 mg/L is considered excellent source water. Water with TDS from 1,000 to 7,000 mg/L can be used for livestock but may cause digestive problems [188]. Water with TDS of 10,000 is considered unsatisfactory for animal consumption.

5.3.6. Wildlife. Some Rocky Mountain area gas projects have created impoundments measuring at least several acres that collect and retain large volumes of produced water. These impoundments provide a source of drinking water for wildlife and offer habitat for fish and waterfowl and can provide additional recreational opportunities [100]. In some areas, watering ponds provide wintering areas for migrating waterfowl, neotropical birds, or other transient species. In severe drought conditions, watering ponds are used to provide water for large mammals and other wildlife. At Custer Lake, Wyoming, approximately 30,000 barrels of water per day are discharged to what would normally be a seasonal playa lake. Waterfowl and big game are reported to flourish there [177].

The quality of OGPW may limit the use of this management practice, as contaminants may adversely affect fish and wildlife. Research conducted by the USGS has demonstrated acute and chronic sodium bicarbonate toxicity to aquatic species [206]. Coal bed methane-produced water discharges containing Se in concentrations > $2 \mu g/L$ may cause bioaccumulation in sensitive species [1, 206]. Water with TDS > 10,000 mg/L is not of sufficient quality for wildlife consumption.

5.3.7. Reuse in Oil and Gas Operations. Water is required for a range of day-to-day operations in the oil and gas industries. Activities such as well completions and truck washing may not require water to be of high quality. Therefore, OGPW can be used with little concern for water quality. Minimally treated produced water may be reused by petroleum operators for drilling operations.

Lebas et al. [51] found that produced water with TDS levels as high as 285,000 mg/L (28.5% salinity) could generate proper cross-linked rheology for hydraulic fracturing consistent with wells that were fracturing with just 20,000 mg/L. A mixture of common drilling chemicals including carboxymethyl hydroxypropyl guar gum, a Zr-based cross-linker, sodium chlorite breakers, and nonemulsified surfactants was blended with 100% treated OGPW to generate HF fluid that performed as well as that expected from a fluid based on fresh water. The fluids were used to complete seven wells in New Mexico's Delaware basin. The study showed that OGPW possesses all the characteristics required for effective HF, that is, easy preparation, rapid hydration, low fluid loss, good proppant transport capacity, low pipe friction, and effective recovery from the reservoir. Lebas et al. [51] state that, in addition to preserving fresh water for agricultural and commercial use, employing produced water for HF can help reduce approximately 1400 truck-loads from the roads and all but eliminate the use of disposal wells.

Erskine et al. [207] report on the use of produced water for drilling projects in New Mexico's San Juan Basin. The authors state that a combination of advection and dilution reduced Cl concentrations of 10,000 mg/L to <1 mg/L in one year. In the Powder River Basin of Wyoming a water truck load-out facility utilizes produced water for oil and gas well operations, thus taking some pressure off the local water supply to meet this demand. In the Barnett Shale play (Texas) as much as 2 million gallons of make-up water is required for a fracture job. This water is subsequently produced back to the surface in the early stages of development. To reduce the cost of fracturing wells, the produced water is reclaimed and recycled by using it to fracture the next well. In the Battle Creek play (Montana), a zero-discharge system was developed to manage produced water through enhanced evaporation ponds coupled with recycling the produced water for well completion and dust prevention [177].

The main constraint to using produced water for oil and gas operations is the fact that the volume of water used may be modest when compared to the total volume of OGPW produced; therefore, it may be uneconomical to put practices in place solely for recycling produced water for operational uses. This can be overcome by formulating a set of water management practices, where the water is readily available for operations, but for additional purposes as needed [177].

5.3.8. Other Uses/Fire Control. In the Western USA, only limited surface and ground water resources may be available for firefighting. Application of large volumes of saline produced water may adversely impact soil quality to some degree; however, this impact is less devastating than that from a large fire. ALL [188] reports that firefighters near Durango, Colorado, used produced water impoundments as sources of water for firefighting.

5.4. Technologies for Treatment of OGPW. Where reuse of OGPW is practical, authorized by regulatory agencies, and cost-effective, it constitutes a beneficial use of what would otherwise be considered a waste [178]. Beneficial use of produced water may require significant treatment [83]. The primary objectives for treating produced water include desalinization; removal of dispersed oil and grease, suspended particles and sand, soluble organic compounds, dissolved gases, and naturally occurring radioactive material; disinfection; and softening (i.e., to remove excess water hardness) [83, 208].

The optimal wastewater treatment technologies available are not able to strip all toxic chemicals from OGPW and are often selectively implemented due to cost [100, 184]. A range of dedicated and combined physical, biological, and chemical treatment processes have been developed to treat OGPW. Some popular technologies are reviewed here.

5.4.1. Membrane Processes for Removal of TDS

(1) Membrane Filtration. Membrane separation processes available for treating OGPW include microfiltration (MF), ultrafiltration (UF), nanofiltration (NF), and reverse osmosis (RO) [209]. Membranes are microporous films with specific pore sizes which selectively separate components from a fluid. MF uses the largest pore size (0.1–3 mm) and is typically employed for removal of suspended solids and turbidity reduction. UF pore sizes are between 0.01 and 0.1 mm; this technology is employed for removal of color, odor, viruses, and colloidal organic matter [189, 210, 211]. UF is the most effective method for oil removal from produced water as compared with conventional separation methods [212]. UF is more efficient than MF for removal of hydrocarbons, suspended solids, and dissolved constituents from oilfield produced water [213].

NF is a successful technology for water softening and metals removal and is designed to remove contaminants as small as 0.001 mm [189]. NF is selective for multivalent ions such as Ca^{2+} and Mg^{2+} [214]. It is applicable for treating water containing TDS in the range of 500–25,000 mg/L. NF membranes have been employed for produced water treatment on both bench and pilot scales [209, 215]. Mondal and Wickramasinghe [216] studied the effectiveness of NF membranes for treatment of oilfield produced water. Effectiveness of treatment of brackish feed water was similar between NF and RO techniques [83].

(2) Ceramic Membranes. Ceramic UF/MF membranes have been used in a full-scale facility for treatment of produced water [189]. Treated product water was reported to be free of suspended solids and nearly all nondissolved organic carbon [217–223]. Ceramic UF/MF membranes have a lifespan of about 10 years. Chemicals are not required for this technology except during cleaning of membranes [83].

(3) Reverse Osmosis. RO is a pressure-driven membrane processes. Osmotic pressure of the feed solution is suppressed by applying hydraulic pressure whose forces permeate (i.e., clean water) to diffuse through a dense, nonporous membrane [224]. The major disadvantage of the technology is membrane fouling and scaling [189, 225]. Nicolaisen and Lien [226] reported successful RO treatment of oilfield OGPW in Bakersfield, California. The pilot system was operated for over 6 months and produced 20 gpm of clean water. A process for converting oilfield produced water into irrigation/drinking quality water consisted of air flotation, clarification, softening, filtration, RO, and water reconditioning [227]. A pilot plant handled water with approximately 7,000 mg/L TDS, 250 mg/L silica, and 170 mg/L soluble oil, ranging in pH from 7 to 11. The major source of fouling of RO membranes was from organics in the feed water, including organic sulfur compounds. A portion of these entered the RO system as TSS and some precipitated. The quality of treated water met the stringent California Title 22 Drinking Water Maximum Contaminant Levels [227].

RO membrane technology should be appropriate for treating oilfield produced water with appropriate pretreatment technology [200]. RO membrane systems generally have a life expectancy of 3–7 years [189].

(4) Electrodialysis/Electrodialysis Reversal. Electrodialysis (ED) and ED reversal (EDR) are well-established desalination technologies. These electrochemically driven processes separate dissolved ions from water through ion exchange membranes. A series of membranes containing electrically charged functional sites are arranged in an alternating mode between an anode and a cathode to remove charged substances from feed water [189]. EDR and ED technologies have been tested at the laboratory-scale for treatment of produced water. Sirivedhin et al. [87] reported that ED is an excellent produced water treatment technology; however, it works optimally for treating relatively low-saline produced water.

The ED/EDR membrane lifetime is between 4 and 5 years. Major limitations of this technology are regular membrane fouling and high treatment cost [189].

(5) Biological Aerated Filters. Biological aerated filtration (BAF) consists of permeable media under aerobic conditions to facilitate biochemical oxidation and removal of organic constituents in wastewater. Media do not exceed 10 cm in diameter to prevent clogging of pore spaces when sloughing occurs [228]. BAF can remove oil, ammonia, suspended solids, nitrogen, BOD, COD, heavy metals, Fe, soluble organics, trace organics, and H_2S from produced water [189, 229]. Removal efficiencies of up to 70% N, 80% oil, 60% COD, 95% BOD, and 85% TSS have been achieved with BAF treatment [229]. Water recovery is nearly 100% since waste generated is removed in solid form [230]. The method is most effective for produced water with Cl levels < 6600 mg/L [189].

BAF systems usually have a long lifespan; they do not require any chemicals or cleaning during normal operations. Accumulated sludges are captured in sedimentation basins. Solid waste disposal can account for up to 40% of the total cost of this technology [228].

(6) Vibrating Membrane Process. The vibrating membrane process VSEP® (*Vibratory Shear Enhanced Process*) limits membrane fouling, removing the main contaminants from wastewater without the addition of antiscalant chemical substances. The design greatly reduces the fouling common to all membrane processes [231]. The pressure vessel moves in a vigorous vibratory motion, tangential to membrane surface, thus creating shear waves which prevent membrane fouling [232, 233]. RO may be implemented as tertiary treatment.

5.5. Thermal Technologies for Removing Oil and Grease Content

5.5.1. Multistage Flash (MSF). MSF distillation involves evaporation of water by reducing atmospheric pressure instead of raising temperature. Feed water is preheated and flows into a chamber with reduced air pressure where it immediately flashes into steam [234]. Water recovery from MSF treatment is approx. 20%; it often requires posttreatment because it typically contains 2–10 mg/L TDS [189]. A setback

in operating MSF is scale formation on heat transfer surfaces which often requires the use of scale inhibitors and acids. MSF is a relatively cost-effective treatment method with plant life expectancy of more than 20 years [235].

5.5.2. Multieffect Distillation. The MED process involves application of energy that converts saline water into steam, which is condensed and recovered as pure water. Multiple effects are employed in order to improve the efficiency and minimize energy consumption. A major advantage of this system is the energy efficiency gained through the combination of several evaporator systems.

MED is suitable for treatment of high TDS produced water [189, 236]. Product water recovery from MED systems ranges from 20 to 67% depending on design [234]. Despite the high water recovery, MED has not been extensively used for water production like MSF because of scaling problems associated with early designs. Recently, falling film evaporators have been introduced to improve heat transfer rates and reduce the rate of scale formation [236]. Scale inhibitors and acids may be required to prevent scaling, and pH control is essential to prevent corrosion. MED has a lifespan of 20 years and can be applied to a wide range of feed water qualities, similar to MSF.

5.5.3. Vapor Compression Distillation. The VCD process is an established desalination technology for treating seawater and RO concentrate [189]. Vapor generated in the evaporation chamber is compressed thermally or mechanically, which raises the temperature and pressure of the vapor. The heat of condensation is returned to the evaporator and is used as a heat source. VCD can operate at temperatures below 70°C, which reduces scale formation problems [237]. Energy consumption of a VCD plant is significantly lower than that of MED and MSF. Although this technology is mainly associated with sea water desalination, various enhanced vapor compression technologies have been employed for produced water treatment [189].

5.5.4. Multieffect Distillation-Vapor Compression Hybrid. Multistage flash (MSF) distillation, vapor compression distillation (VCD) and multieffect distillation (MED) are extensively used thermal desalination technologies [234]; however, hybrid thermal desalination plants, that is, MED-VCD, have achieved higher efficiencies [236]. Increased production and enhanced energy efficiency are major advantages of this system [189]. GE has developed produced water evaporators which use mechanical vapor compression. These evaporators exhibit a number of advantages over conventional produced water treatment methods including reduction in chemical use, overall cost, fouling severity, and handling [238].

Membrane technologies are often preferred over thermal technologies; however, recent innovations in thermal process engineering have made the latter more competitive in treating highly contaminated water [83, 189, 235].

Gradient Corporation (Woburn, MA) is attempting to make HF a "water-neutral process" by reusing water for the HF process. The technology, carrier gas extraction (CGE), is a humidification and dehumidification technique;

it heats produced water into vapor and condenses it back to contaminant-free water. This process yields freshwater and saturated brine [239].

5.5.5. Freeze Thaw Evaporation. The FTEw process employs freezing, thawing, and conventional evaporation for produced water management. When produced water is cooled below 32°F but above its freezing point, relatively pure ice crystals and an unfrozen solution form. The solution contains high concentrations of dissolved constituents and is drained from the ice. The ice is collected and melted to produce clean water.

FTEw can remove >90% of TDS, TSS, volatile and semivolatile organics, total recoverable petroleum hydrocarbons, and heavy metals in produced water [220, 240]. FTE requires no chemical additives, infrastructure, or supplies that might restrict its use. It is easy to operate and monitor and has a life expectancy of approximately 20 years [189]. However, the technology can only work in a climate that has a substantial number of days with temperatures below freezing and requires significant land area. FTE technology generates a significant amount of concentrated brine and oil; therefore, waste management and disposal must be addressed.

5.5.6. Dewvaporation: AltelaRain^{SM} Process. The principle of operation of Dewvaporation is based on counter current heat exchange to produce distilled water [241]. Feed water is evaporated in one chamber and condenses in the opposite chamber of a heat transfer wall as distilled water. Approximately 100 bbl/day of produced water with salt concentrations > 60,000 mg/L TDS can be processed [165]. High removal rates of organics, heavy metals, and radionuclides from produced water have been reported for this technology. In one plant, Cl^- concentration was reduced from 25,300 to 59 mg/L, TDS was reduced from 41,700 to 106 mg/L, and benzene concentration was reduced from 450 mg/L to nondetectable after treatment with AlterRain^{SM} Dewvaporation technology [242].

5.6. Physical Separation Technologies

5.6.1. Hydrocyclones. Hydrocyclones physically separate solids from liquids; hydrocyclones can remove particles in the 5–15 mm range and have been widely used for treatment of OGPW [189, 243].

Hydrocyclones are used in combination with other technologies as a pretreatment process. They have a long lifespan and do not require chemical use or pretreatment of feed water. A major disadvantage of this technology is generation of substantial slurry of concentrated solid waste.

5.6.2. Gas Flotation. Flotation technology is extensively used for treatment of conventional oilfield produced water.

Flotation technology uses fine gas bubbles to separate suspended particles that are not easily separated by sedimentation. When gas is injected into produced water, suspended particulates and oil droplets become attached to air bubbles as they rise. This results in the formation of foam on the water surface which is skimmed off [244]. Flotation can be used to remove grease and oil, natural organic matter, volatile organics, and small particles from produced water [50, 189, 243, 245].

Two types of gas flotation technology are in use, that is, dissolved gas flotation and induced gas flotation, based on the method of gas bubble generation and resultant bubble sizes. Gas flotation can remove particles as small as 25 mm but cannot remove soluble oil constituents from water [189]. Flotation is most effective when gas bubble size is smaller than oil droplet size. It is expected to work best at low temperature since the process involves dissolving gas into a water stream. The technology does not require chemical use, except for coagulation chemicals that are added to enhance removal of target contaminants. Solids disposal is necessary for the sludge generated from this process.

5.6.3. Media Filtration. Filtration technology is used for removal of oil and grease and total organic carbon (TOC) from produced water [189]. Filtration is carried out using various media such as sand, gravel, anthracite, and walnut shells. This process is not affected by salinity levels and may be applied to any type of produced water. Media filtration technology is highly efficient for removal of oil and grease; efficiency of >90% has been reported [189]. Efficiency can be further enhanced if coagulants are added to the feed water prior to filtration. Media regeneration and solid waste disposal are drawbacks to this technology.

5.6.4. Adsorption. Adsorption is generally used as a polishing step in the OGPW treatment process rather than as a stand-alone technology, since adsorbents can be overloaded with organics. Adsorption is used to remove Mn, Fe, TOC, BTEX, oil, and more than 80% of heavy metals present in produced water [189]. A wide range of adsorbents is available including activated carbon, organoclays, activated alumina, and zeolites [246]. Adsorption processes are successfully applied to water treatment regardless of salinity. Replacement or regeneration of the sorption media may be required depending on feed water quality and media type [189, 246]. Chemicals are used to regenerate media when all active sites are blocked.

5.6.5. Ion Exchange Technology. Ion exchange technology is in demand for numerous industrial operations including treatment of OGPW. It is especially useful in the removal of monovalent and divalent ions and metals from produced water [247]. Ion exchange technology has a lifespan of about 8 years and requires pretreatment for solids removal. It also requires chemicals for resin regeneration and disinfection [181].

5.6.6. Macroporous Polymer Extraction Technology. Macroporous polymer extraction (MPPE) is a liquid-liquid extraction technology where the extraction liquid is immobilized within polymer particles impregnated with macropores. The particles have pore sizes of 0.1–10 mm and a porosity of 60–70%. The polymers were initially designed for absorbing oil from water but later applied to produced water treatment [248]. In the MPPE unit, produced water is passed through a column packed with MPPE particles containing a specific extraction liquid. The immobilized extraction liquid removes hydrocarbons from the produced water [71].

In a commercial unit, MPPE was used for removal of dissolved and dispersed hydrocarbons and achieved 99% removal of BTEX, PAHs, and aliphatic hydrocarbons at 300–800 mg/L influent concentration. It had a removal efficiency of 95–99% for aliphatics below C_{20} and it was reported that total aliphatic removal efficiency of 91–95% was feasible [249].

The hydrocarbons recovered from the MPPE process can be disposed or recycled. Stripped hydrocarbons can be condensed and separated from feed water by gravity, and product water is either discharged or reused. This technology can withstand produced water high in salinity and containing methanol, glycols, corrosion inhibitors, scale inhibitors, H_2S scavengers, demulsifiers, defoamers, and dissolved heavy metals [83].

6. Conclusions

This review paper has described the composition of oil and gas produced water including common HF additives and documented several key effects on soil properties. Beneficial use of OGPW must take into consideration its chemical make-up, the properties of the recipient soil, and long-term land use objectives. For example, it has been shown that untreated OGPW may be directly applied to the land surface with limited adverse effects. In others, where the water is to be used for domestic or industrial purposes, extensive treatment is required. There is a significant need for further field studies, in particular the study of complex mixtures and how interactions between individual OGPW chemicals influence their environmental fate.

Future OGPW management technologies are likely to focus on (1) capturing secondary value from repurposed water; (2) minimizing transportation; (3) minimizing energy inputs; and (4) reduced air emissions (including CO_2) all the while with a vision to reduce overall management costs.

References

[1] Aqwatec, *Produced Water Beneficial Use Case Studies*, Produced Water Treatment and Beneficial Use Information Center, 2015, http://aqwatec.mines.edu/produced_water/assessbu/case/.

[2] Unconventional Oil and Gas Report (UOGR), *Unconventional Resources*, 2015, http://www.ogj.com/unconventional-resources.html.

[3] Earthworks, "Hydraulic Fracturing 101. Hydraulic fracturing—What it is," 2015, https://www.earthworksaction.org/issues/detail/hydraulic_fracturing_101#.Vi4kGSv6G0I.

[4] L. Britt, "Fracture stimulation fundamentals," *Journal of Natural Gas Science and Engineering*, vol. 8, pp. 34–51, 2012.

[5] M. J. Economides and K. G. Nolte, *Reservoir Stimulation*, John Wiley & Sons, Chichester, UK, 3rd edition, 2000.

[6] R. A. Kerr, "Natural gas from shale bursts onto the scene," *Science*, vol. 328, no. 5986, pp. 1624–1626, 2010.

[7] H. A. Waxman, E. J. Markey, and D. DeGette, *United States House of Representatives. Committee on Energy and Commerce. Chemicals Used In Hydraulic Fracturing*, Government Printing Office, Washington, DC, USA, 2011.

[8] H. J. Wiseman, "Untested waters: the rise of hydraulic fracturing in oil and gas production and the need to revisit regulation," *Fordham Environmental Law Review*, vol. 20, pp. 115–169, 2008.

[9] M. Ratner and M. Tiemann, *An Overview of Unconventional Oil and Natural Gas: Resources and Federal Actions*, Congressional Research Service, Washington, DC, USA, 2015.

[10] E. Gruber, *Recycling Produced & Flowback Wastewater for Fracking*, 2013, http://blog.ecologixsystems.com/wp-content/uploads/2013/04/Recycling-Produced-and-Flowback-Water-for-Fracking.pdf.

[11] U.S. Environmental Protection Agency, "Assessment of the potential impacts of hydraulic fracturing for oil and gas on drinking water resources," EPA/600/R-15/047a, 2015, http://cfpub.epa.gov/ncea/hfstudy/recordisplay.cfm?deid=244651.

[12] C. D. Kassotis, D. E. Tillitt, C.-H. Lin, J. A. McElroy, and S. C. Nagel, "Endocrine-disrupting chemicals and oil and natural gas operations: potential environmental contamination and recommendations to assess complex environmental mixtures," *Environmental Health Perspectives*, 2015.

[13] J. Deutch, S. Holditch, F. Krupp et al., "The Secretary of the Energy Board Shale Gas Production Subcommittee Ninety-Day Report," 2001, https://www.edf.org/sites/default/files/11903_Embargoed_Final_90_day_Report%20.pdf.

[14] B. E. Fontenot, L. R. Hunt, Z. L. Hildenbrand et al., "An evaluation of water quality in private drinking water wells near natural gas extraction sites in the Barnett shale formation," *Environmental Science & Technology*, vol. 47, no. 17, pp. 10032–10040, 2013.

[15] J. S. Harkness, G. S. Dwyer, N. R. Warner, K. M. Parker, W. A. Mitch, and A. Vengosh, "Iodide, bromide, and ammonium in hydraulic fracturing and oil and gas wastewaters: environmental implications," *Environmental Science & Technology*, vol. 49, no. 3, pp. 1955–1963, 2015.

[16] T. G. Harvey, T. W. Matheson, and K. C. Pratt, "Chemical class separation of organics in shale oil by thin-layer chromatography," *Analytical Chemistry*, vol. 56, no. 8, pp. 1277–1281, 1984.

[17] A. L. Maule, C. M. Makey, E. B. Benson, I. J. Burrows, and M. K. Scammell, "Disclosure of hydraulic fracturing fluid chemical additives: analysis of regulations," *New Solutions*, vol. 23, no. 1, pp. 167–187, 2013.

[18] N. R. Warner, R. B. Jackson, T. H. Darrah et al., "Geochemical evidence for possible natural migration of Marcellus formation brine to shallow aquifers in Pennsylvania," *Proceedings of the National Academy of Sciences of the United States of America*, vol. 109, no. 30, pp. 11961–11966, 2012.

[19] B. Ong, "The potential impacts of hydraulic fracturing on agriculture," *European Journal of Sustainable Development*, vol. 3, no. 3, pp. 63–72, 2014.

[20] K. R. Gilmore, R. L. Hupp, and J. Glathar, "Transport of hydraulic fracturing water and wastes in the susquehanna river basin, Pennsylvania," *Journal of Environmental Engineering*, vol. 140, no. 5, pp. 10–21, 2014.

[21] D. S. Lee, J. D. Herman, D. Elsworth, H. T. Kim, and H. S. Lee, "A critical evaluation of unconventional gas recovery from the marcellus shale, northeastern United States," *KSCE Journal of Civil Engineering*, vol. 15, no. 4, pp. 679–687, 2011.

[22] A. Bergman, J. J. Heindel, S. Jobling, K. A. Kidd, and R. T. Zoeller, "State of the science of endocrine disrupting chemicals 2012," World Health Organization, 2013, http://www.who.int/ceh/publications/endocrine/en/.

[23] E. Diamanti-Kandarakis, J.-P. Bourguignon, L. C. Giudice et al., "Endocrine-disrupting chemicals: an endocrine society scientific statement," *Endocrine Reviews*, vol. 30, no. 4, pp. 293–342, 2009.

[24] R. Zoeller, Å. Bergman, G. Becher et al., "A path forward in the debate over health impacts of endocrine disrupting chemicals," *Environmental Health*, vol. 13, article 118, 2014.

[25] S. B. C. Shonkoff, J. Hays, and M. L. Finkel, "Environmental public health dimensions of shale and tight gas development," *Environmental Health Perspectives*, vol. 122, no. 8, pp. 787–795, 2014.

[26] S. Souther, M. W. Tingley, V. D. Popescu et al., "Biotic impacts of energy development from shale: research priorities and knowledge gaps," *Frontiers in Ecology and the Environment*, vol. 12, no. 6, pp. 330–338, 2014.

[27] O. Braga, G. A. Smythe, A. I. Schafer, and A. J. Feitz, "Steroid estrogens in primary and tertiary wastewater treatment plants," *Water Science and Technology*, vol. 52, no. 8, pp. 273–278, 2005.

[28] C. G. Campbell, S. E. Borglin, F. B. Green, A. Grayson, E. Wozei, and W. T. Stringfellow, "Biologically directed environmental monitoring, fate, and transport of estrogenic endocrine disrupting compounds in water: a review," *Chemosphere*, vol. 65, no. 8, pp. 1265–1280, 2006.

[29] P. Westerhoff, Y. Yoon, S. Snyder, and E. Wert, "Fate of endocrine-disruptor, pharmaceutical, and personal care product chemicals during simulated drinking water treatment processes," *Environmental Science & Technology*, vol. 39, no. 17, pp. 6649–6663, 2005.

[30] K. J. Ferrar, D. R. Michanowicz, C. L. Christen, N. Mulcahy, S. L. Malone, and R. K. Sharma, "Assessment of effluent contaminants from three facilities discharging marcellus shale wastewater to surface waters in Pennsylvania," *Environmental Science & Technology*, vol. 47, no. 7, pp. 3472–3481, 2013.

[31] N. R. Warner, C. A. Christie, R. B. Jackson, and A. Vengosh, "Impacts of shale gas wastewater disposal on water quality in Western Pennsylvania," *Environmental Science and Technology*, vol. 47, no. 20, pp. 11849–11857, 2013.

[32] Natural Resources Defense Council (NRDC), *Fracking's Most Wanted: Lifting the Veil on Oil and Gas Company Spills and Violations*, IP:15-01-A, 2015, http://www.nrdc.org/land/drilling/files/fracking-company-violations-IP.pdf.

[33] M. Soraghan, *Oil and Gas: Spills up 17 Percent in U.S. in 2013*, Environment & Energy Publishing, 2014, http://www.eenews.net/stories/1059999364.

[34] S. A. Gross, H. J. Avens, A. M. Banducci, J. Sahmel, J. M. Panko, and B. E. Tvermoes, "Analysis of BTEX groundwater concentrations from surface spills associated with hydraulic fracturing operations," *Journal of the Air & Waste Management Association*, vol. 63, no. 4, pp. 424–432, 2013.

[35] P. F. Ziemkiewicz, J. D. Quaranta, A. Darnell, and R. Wise, "Exposure pathways related to shale gas development and procedures for reducing environmental and public risk," *Journal of Natural Gas Science and Engineering*, vol. 16, pp. 77–84, 2014.

[36] D. Sontag and R. Gebeloff, "The downside of the boom," *The New York Times*, 2014, http://www.nytimes.com/images/2014/11/23/nytfrontpage/scan.pdf.

[37] Alberta Environment, "Salt Contamination Assessment & Remediation Guidelines," Environmental Sciences Division, Environmental Service, 2001, http://environment.gov.ab.ca/info/library/6144.pdf.

[38] R. B. Jackson, A. Vengosh, T. H. Darrah et al., "Increased stray gas abundance in a subset of drinking water wells near Marcellus shale gas extraction," *Proceedings of the National Academy of Sciences of the United States of America*, vol. 110, no. 28, pp. 11250–11255, 2013.

[39] S. G. Osborn, A. Vengosh, N. R. Warner, and R. B. Jackson, "Methane contamination of drinking water accompanying gas-well drilling and hydraulic fracturing," *Proceedings of the National Academy of Sciences of the United States of America*, vol. 108, no. 20, pp. 8172–8176, 2011.

[40] A. Vengosh, R. B. Jackson, N. Warner, T. H. Darrah, and A. Kondash, "A critical review of the risks to water resources from unconventional shale gas development and hydraulic fracturing in the United States," *Environmental Science & Technology*, vol. 48, no. 15, pp. 8334–8348, 2014.

[41] S. L. Brantley, D. Yoxtheimer, S. Arjmand et al., "Water resource impacts during unconventional shale gas development: the Pennsylvania experience," *The International Journal of Coal Geology*, vol. 126, pp. 140–156, 2014.

[42] G. A. Burton Jr., N. Basu, B. R. Ellis, K. E. Kapo, S. Entrekin, and K. Nadelhoffer, "Hydraulic 'fracking': are surface water impacts an ecological concern?" *Environmental Toxicology and Chemistry*, vol. 33, no. 8, pp. 1679–1689, 2014.

[43] Environmental and Energy Study Institute (EESI), *Shale Gas and Oil Terminology Explained: Technology, Inputs & Operations*, EESI, Washington, DC, USA, 2011.

[44] Ostroff Law, "Dangerous fracking chemicals," 2015, http://frackinginjurylaw.com/dangerous-fracking-chemicals/.

[45] FracFocus.org, "Chemical Use in Hydraulic Fracturing," 2015, http://fracfocus.org/, http://www.fiercewater.com/story/business-booming-fracking-wastewater/2015-07-28.

[46] W. T. Stringfellow, J. K. Domen, M. K. Camarillo, W. L. Sandelin, and S. Borglin, "Physical, chemical, and biological characteristics of compounds used in hydraulic fracturing," *Journal of Hazardous Materials*, vol. 275, pp. 37–54, 2014.

[47] US Environmental Protection Agency, "Hydraulic fracturing fluids," in *Evaluation of Impacts to Underground Sources of Drinking Water by Hydraulic Fracturing of Coalbed Methane Reservoirs*, EPA 816-R- 04-003, US Environmental Protection Agency, Washington, DC, USA, 2004.

[48] Halliburton Energy Services, *GBW-30 Breaker MSDS*, Halliburton Energy Services, Duncan, Okla, USA, 2009.

[49] A. Fakhru'l-Razi, A. Pendashteh, L. C. Abdullah, D. R. A. Biak, S. S. Madaeni, and Z. Z. Abidin, "Review of technologies for oil and gas produced water treatment," *Journal of Hazardous Materials*, vol. 170, no. 2-3, pp. 530–551, 2009.

[50] J. P. Ray and F. Rainer Engelhardt, *Produced Water: Technological/Environmental Issues and Solutions*, vol. 46 of *Environmental Science Research*, Springer, New York, NY, USA, 1992.

[51] R. A. Lebas, T. W. Shahan, P. Lord, and D. Luna, "Development and use of high-TDS recycled produced water for crosslinked-gel-based hydraulic fracturing," in *Proceedings of the SPE Hydraulic Fracturing Technology Conference*, SPE-163824-MS, Society of Petroleum Engineers, The Woodlands, Tex, USA, February 2013, http://www.ftwatersolutions.com/pdfs/ProducedWaterPaper.pdf.

[52] U.S. Environmental Protection Agency, "Underground Injection Control Well Classes," 2002, http://water.epa.gov/type/groundwater/uic/wells.cfm.

[53] B. R. Hansen and S. R. H. Davies, "Review of potential technologies for the removal of dissolved components from produced water," *Chemical Engineering Research and Design*, vol. 72, pp. 176–188, 1994.

[54] D. Rahm, "Regulating hydraulic fracturing in shale gas plays: the case of Texas," *Energy Policy*, vol. 39, no. 5, pp. 2974–2981, 2011.

[55] B. G. Rahm, J. T. Bates, L. R. Bertoia, A. E. Galford, D. A. Yoxtheimer, and S. J. Riha, "Wastewater management and Marcellus Shale gas development: trends, drivers, and planning implications," *Journal of Environmental Management*, vol. 120, pp. 105–113, 2013.

[56] FracFocus, "Hydraulic fracturing—the process," 2015, https://fracfocus.org/hydraulic-fracturing-how-it-works/hydraulic-fracturing-process.

[57] PetroWiki, "Fracturing fluids and additives," 2015, http://petrowiki.org/Fracturing_fluids_and_additives.

[58] R. D. Vidic, S. L. Brantley, J. M. Vandenbossche, D. Yoxtheimer, and J. D. Abad, "Impact of shale gas development on regional water quality," *Science*, vol. 340, no. 6134, 2013.

[59] S. A. Holditch, *Criteria for Propping Agent Selection*, Norton Company, Dallas, Tex, USA, 1979.

[60] Saint-Gobain, Products, 2015, http://www.proppants.saint-gobain.com/products.

[61] PetroWiki, "Propping agents and fracture conductivity," 2015, http://petrowiki.org/Propping_agents_and_fracture_conductivity.

[62] C. Montgomery, "Fracturing fluid components," in *Effective and Sustainable Hydraulic Fracturing*, A. P. Bunger, J. McLennan, and R. Jeffrey, Eds., chapter 2, InTech, Vienna, Austria, 2013.

[63] R. Hodge, "Crosslinked and Linear Gel Composition—Chemical and Analytical Methods," EPA's Study of Hydraulic Fracturing and Its Potential Impact on Drinking Water Resources, 2015, http://www2.epa.gov/hfstudy/crosslinked-and-linear-gel-composition-chemical-and-analytical-methods.

[64] New York State Department of Environmental Conservation (NYS DEC), *Revised Draft Supplemental Generic Environmental Impact Statement on the Oil, Gas and Solution Mining Regulatory Program, Well Permit Issuance for Horizontal Drilling and High-Volume Hydraulic Fracturing in the Marcellus Shale and Other Low-Permeability Gas Reservoirs*, New York State Department of Environmental Conservation, Albany, NY, USA, 2011, http://www.dec.ny.gov/energy/75370.html.

[65] ALL Consulting, *Ground Water Protection Council, Modern Shale Gas Development in the United States: A Primer*, U.S. Department of Energy, Office of Fossil Energy, Washington, DC, USA, 2009.

[66] G. E. King, "Hydraulic fracturing 101: what every representative, environmentalist, regulator, reporter, investor, university researcher, neighbor and engineer should know about estimating frac risk and improving frac performance in unconventional gas and oil wells," in *Proceedings of the SPE Hydraulic Fracturing Technology Conference*, Society of Petroleum Engineers, The Woodlands, Tex, USA, February 2012.

[67] URS Corporation, "Water-related issues associated with gas production in the Marcellus Shale: additives use, flowback quality and quantities, regulations, on-site treatment, green technologies, alternate water sources, water well-testing," Contract PO 10666, URS Corporation, Fort Washington, Pa, USA, 2011.

[68] R. Barati and J.-T. Liang, "A review of fracturing fluid systems used for hydraulic fracturing of oil and gas wells," *Journal of Applied Polymer Science*, vol. 131, no. 16, Article ID 40735, 2014.

[69] G. Zimmermann, G. Blöcher, A. Reinicke, and W. Brandt, "Rock specific hydraulic fracturing and matrix acidizing to enhance a geothermal system—concepts and field results," *Tectonophysics*, vol. 503, no. 1-2, pp. 146–154, 2011.

[70] R. McCurdy, "High rate hydraulic fracturing additives in non-marcellus unconventional shale," in *Proceedings of the Technical Workshops for the Hydraulic Fracturing Study: Chemical & Analytical Methods*, U.S. Environmental Protection Agency, Arlington, Va, USA, February 2011.

[71] Akzo Nobel MPP Systems, *Macro-Porous Polymer Extraction for Offshore Produced Water Removes Dissolved and Dispersed Hydrocarbons*, Business Briefing: Exploration & Production: The Oil & Gas Review, Pasadena, Calif, USA, 2004.

[72] A. Aminto and M. S. Olson, "Four-compartment partition model of hazardous components in hydraulic fracturing fluid additives," *Journal of Natural Gas Science and Engineering*, vol. 7, pp. 16–21, 2012.

[73] A. A. Al-Zahrani, "Innovative method to mix corrosion inhibitor in emulsified acids," in *Proceedings of the 6th International Petroleum Technology Conference*, Beijing, China, March 2013.

[74] A. Rostami and H. A. Nasr-El-Din, "Review and evaluation of corrosion inhibitors used in well stimulation," in *Proceedings of the SPE International Symposium on Oilfield Chemistry*, Society of Petroleum Engineers, April 2009.

[75] J. Yang, V. Jovancicevic, S. Mancuso, and J. Mitchell, "High performance batch treating corrosion inhibitor," in *Proceedings of the Corrosion Conference & Expo*, NACE International, Nashville, Tenn, USA, March 2007.

[76] M. McHugh, *The natural degradation of hydraulic fracturing fluids in the shallow subsurface [Ph.D. thesis]*, The Ohio State University, Columbus, Ohio, USA, 2015.

[77] M. M. Brezinski, T. R. Gardner, W. M. Harms, J. L. Lane, and K. L. King, "Controlling iron in aqueous well fracturing fluids," US Patent, vol. 5, pp. 674, 817, Halliburton Energy Services, 1997.

[78] K. C. Taylor, H. A. Nasr-El-Din, M. J. Al-Alawi, and S. Aramco, "Systematic study of iron control chemicals used during well stimulation," *Society of Petroleum Engineers Journal*, vol. 4, pp. 19–24, 1999.

[79] International Energy Agency (IEA), *Water Management Associated with Hydraulic Fracturing*, AP Guidance Document HF2, International Energy Agency (IEA), Washington, DC, USA, 2010.

[80] W. R. Dill and G. Fredette, "Iron control in the Appalachian basin," in *Proceedings of the SPE Eastern Regional Meeting*, Society of Petroleum Engineers, Pittsburgh, Pa, USA, November 1983.

[81] Z. Zhou and D. H. S. Law, *Swelling Clays in Hydrocarbon Reservoirs: The Bad, the Less Bad, and the Useful*, Alberta Research Council, Alberta, Canada, 1998.

[82] T. Strømgren, S. E. Sørstrøm, L. Schou et al., "Acute toxic effects of produced water in relation to chemical composition and dispersion," *Marine Environmental Research*, vol. 40, no. 2, pp. 147–169, 1995.

[83] E. T. Igunnu and G. Z. Chen, "Produced water treatment technologies," *International Journal of Low-Carbon Technologies*, vol. 9, no. 3, pp. 157–177, 2014.

[84] R. R. Reynolds, "Produced water and associated issues: a manual for the independent operator," *Oklahoma Geological Survey Open-File Report*, vol. 6, pp. 1–56, 2003.

[85] L.-H. Chan, A. Starinsky, and A. Katz, "The behavior of lithium and its isotopes in oilfield brines: evidence from the Heletz-Kokhav field, Israel," *Geochimica et Cosmochimica Acta*, vol. 66, no. 4, pp. 615–623, 2002.

[86] T. Hayes and D. Arthur, "Overview of emerging produced water treatment technologies," in *Proceedings of the 11th Annual International Petroleum Environmental Conference*, Albuquerque, NM, USA, 2004.

[87] T. Sirivedhin, J. McCue, and L. Dallbauman, "Reclaiming produced water for beneficial use: salt removal by electrodialysis," *Journal of Membrane Science*, vol. 243, no. 1-2, pp. 335–343, 2004.

[88] C. M. Hudgins Jr., "Chemical use in North Sea oil and gas E&P," *Journal of Petroleum Technology*, vol. 46, no. 1, pp. 67–74, 1994.

[89] M. E. Blauch, R. R. Myers, T. R. Moore, and B. A. Lipinski, "Marcellus Shale postfrac flowback waters—where is all the salt coming from and what are the implications?" in *Proceedings of the Society of Petroleum Engineers Eastern Regional Meeting*, SPE 125740, pp. 1–20, Charleston, SC, USA, 2009.

[90] L. O. Haluszczak, A. W. Rose, and L. R. Kump, "Geochemical evaluation of flowback brine from Marcellus gas wells in Pennsylvania, USA," *Applied Geochemistry*, vol. 28, pp. 55–61, 2013.

[91] C. W. Poth, *Occurrence of Brine in Western Pennsylvania*, Pennsylvania Geological Survey, Report 47, 4th series, Mineral Resource, 1962.

[92] P. E. Dresel, *The geochemistry of oilfield brines from western pennsylvania [M.S. thesis]*, Pennsylvania State University, 1985.

[93] P. E. Dresel and A. W. Rose, "Chemistry and origin of oil and gas well brines in Western Pennsylvania," Pennsylvania Geological Survey, 4th Series, Open-File Report OFOG 10-01.0, 2010.

[94] R. W. Roach, R. S. Carr, and C. L. Howard, *An Assessment of Produced Water Impacts at Two Sites in the Galveston Bay System*, United States Fish and Wildlife Service, Clear Lake Field Office, Houston, Tex, USA, 1993.

[95] K. S. E. Al-Malahy and T. Hodgkiess, "Comparative studies of the seawater corrosion behaviour of a range of materials," *Desalination*, vol. 158, no. 1–3, pp. 35–42, 2003.

[96] E. C. Chapman, R. C. Capo, B. W. Stewart et al., "Geochemical and strontium isotope characterization of produced waters from Marcellus shale natural gas extraction," *Environmental Science & Technology*, vol. 46, no. 6, pp. 3545–3553, 2012.

[97] M. A. Cluff, *Microbial aspects of shale flowback fluids and response to hydraulic fracturing fluids [M.S. thesis]*, The Ohio State University, Columbus, Ohio, USA, 2013.

[98] T. Hayes, "Sampling and Analysis of Water Streams Associated with the Development of Marcellus Shale Gas," 2009, http://energyindepth.org/wp-content/uploads/marcellus/2012/11/MSCommission-Report.pdf.

[99] T. I. R. Utvik, "Composition, characteristics of produced water in the North Sea," in *Proceedings of the Produced Water Workshop*, Aberdeen, UK, March 2003.

[100] J. A. Veil, M. G. Puder, D. Elcock et al., *A White Paper Describing Produced Water from Production of Crude oil, Natural Gas, and Coal Bed Methane*, US. D.O.E, Argonne National Laboratory, 2004.

[101] Nature Technology Group (NTG), *Introduction to Produced Water Treatment*, Nature Technology Solutions, 2005.

[102] S. J. Maguire-Boyle and A. R. Barron, "Organic compounds in produced waters from shale gas wells," *Environmental Sciences: Processes and Impacts*, vol. 16, no. 10, pp. 2237–2248, 2014.

[103] M. B. Adams, "Land application of hydrofracturing fluids damages a deciduous forest stand in West Virginia," *Journal of Environmental Quality*, vol. 40, no. 4, pp. 1340–1344, 2011.

[104] U.S. Environmental Protection Agency, "Presidential green chemistry challenge award recipients 1996–2012," Tech. Rep. 744F12001, EPA, Washington, DC, USA, 2012.

[105] D. J. Rozell and S. J. Reaven, "Water pollution risk associated with natural gas extraction from the Marcellus Shale," *Risk Analysis*, vol. 32, no. 8, pp. 1382–1393, 2012.

[106] Reuters, "Update 2-Oil well in North Dakota out of control, leaking," 2014, http://www.reuters.com/article/2014/02/14/energy-crude-blowout-idUSL2N0LJ15820140214.

[107] A. Dalrymple, "ND well blowout leaks 23,000 gallons of oil, but company says minimal damage," InForum, 2015, http://www.eeia.myindustrytracker.com/en/article/73089/nd-well-blowout-leaks-23-000-gallons-of-oil-but-company-says-minimal-damage.

[108] M. Grossi, "Fracking probe extends in Central Valley," Fresno Bee, November 2013.

[109] K. Bullis, "One Way to Solve Fracking's Dirty Problem," MIT Technology Review, 2013.

[110] J. Healy, "After the floods in Colorado, a deluge of worry about leaking oil," *The New York Times*, 2013.

[111] D. W. Nelson, S. L. Liu, and L. E. Somers, "Extractability and plant uptake of trace elements from drilling fluids," *Journal of Environmental Quality*, vol. 13, no. 4, pp. 562–566, 1983.

[112] R. W. Miller, S. Honarvar, and B. Hunsaker, "Effects of drilling fluids on soils and plants. I. Individual fluid components," *Journal of Environmental Quality*, vol. 9, no. 4, pp. 547–552, 1980.

[113] R. W. Miller and P. Pesaran, "Effects of drilling fluids on soils and plants: II. Complete drilling fluid mixtures," *Journal of Environmental Quality*, vol. 9, no. 4, pp. 552–556, 1980.

[114] Toxnet, "Ethylene glycol," National Library of Medicine, Toxicology Data Network, 2015, http://toxnet.nlm.nih.gov/cgi-bin/sis/search/a?dbs+hsdb:@term+@DOCNO+5012.

[115] M. McLaughlin, "Fate of hydraulic fracturing chemicals in agricultural topsoil," in *Proceedings of the American Chemical Society National Meeting*, Denver, Colo, USA, March 2015, http://presentations.acs.org/common/presentation-detail.aspx/Spring2015/ENVR/ENVR021a/2139634.

[116] Q. Wen, Z. Chen, Y. Zhao, H. Zhang, and Y. Feng, "Biodegradation of polyacrylamide by bacteria isolated from activated sludge and oil-contaminated soil," *Journal of Hazardous Materials*, vol. 175, no. 1–3, pp. 955–959, 2010.

[117] L. L. Liu, Z. P. Wang, K. F. Lin, and W. M. Cai, "Microbial degradation of polyacrylamide by aerobic granules," *Environmental Technology*, vol. 33, no. 9, pp. 1049–1054, 2012.

[118] M. T. Bao, Q. G. Chen, Y. M. Li, and G. C. Jiang, "Biodegradation of partially hydrolyzed polyacrylamide by bacteria isolated from production water after polymer flooding in an oil field," *Journal of Hazardous Materials*, vol. 184, no. 1–3, pp. 105–110, 2010.

[119] N. C. Brady and R. R. Weil, *The Nature and Properties of Soils*, Prentice-Hall, New York, NY, USA, 14th edition, 2007.

[120] M. C. Steele and J. Pichtel, "Ex-situ remediation of a metal-contaminated Superfund soil using selective extractants," *Journal of Environmental Engineering*, vol. 124, no. 7, pp. 639–645, 1998.

[121] A. J. Francis, "Effects of acidic precipitation and acidity on soil microbial processes," *Water, Air, and Soil Pollution*, vol. 18, no. 1–3, pp. 375–394, 1982.

[122] U. Tezel, *Fate and effect of quaternary ammonium compounds in biological systems [Ph.D. thesis]*, Georgia Institute of Technology, Atlanta, Ga, USA, 2009.

[123] G.-G. Ying, "Fate, behavior and effects of surfactants and their degradation products in the environment," *Environment International*, vol. 32, no. 3, pp. 417–431, 2006.

[124] C. Hansch, A. Leo, and D. H. Hoekman, *Exploring QSAR*, American Chemical Society (ACS), Washington, DC, USA, 1995.

[125] N. Kreuzinger, M. Fuerhacker, S. Scharf, M. Uhl, O. Gans, and B. Grillitsch, "Methodological approach towards the environmental significance of uncharacterized substances—quaternary ammonium compounds as an example," *Desalination*, vol. 215, no. 1–3, pp. 209–222, 2007.

[126] C. A. M. Bondi, "Applying the precautionary principle to consumer household cleaning product development," *Journal of Cleaner Production*, vol. 19, no. 5, pp. 429–437, 2011.

[127] M. T. García, I. Ribosa, T. Guindulain, J. Sánchez-Leal, and J. Vives-Rego, "Fate and effect of monoalkyl quaternary ammonium surfactants in the aquatic environment," *Environmental Pollution*, vol. 111, no. 1, pp. 169–175, 2001.

[128] C. G. van Ginkel, J. B. van Dijk, and A. G. M. Kroon, "Metabolism of hexadecyltrimethylammonium chloride in Pseudomonas Strain-B1," *Applied and Environmental Microbiology*, vol. 58, no. 9, pp. 3083–3087, 1992.

[129] R. S. Boethling, "Environmental aspects of cationic surfactants," in *Cationic Surfactants: Analytical and Biological Evaluation*, J. Cross and E. J. Singer, Eds., vol. 53, Marcel Dekker, New York, NY, USA, 1994.

[130] T. Nishihara, T. Okamoto, and N. Nishiyama, "Biodegradation of didecyldimethylammonium chloride by *Pseudomonas fluorescens* TN4 isolated from activated sludge," *Journal of Applied Microbiology*, vol. 88, no. 4, pp. 641–647, 2000.

[131] M. A. Patrauchan and P. J. Oriel, "Degradation of benzyldimethylalkylammonium chloride by *Aeromonas hydrophila* sp," *Applied Microbiology*, vol. 94, pp. 266–272, 2003.

[132] S. Takenaka, T. Tonoki, K. Taira, S. Murakami, and K. Aoki, "Adaptation of *Pseudomonas* sp. strain 7-6 to quaternary ammonium compounds and their degradation via dual pathways," *Applied and Environmental Microbiology*, vol. 73, no. 6, pp. 1799–1802, 2007.

[133] D. McIlwaine, "Challenging traditional biodegradation tests: the biodegradation of glutaraldehyde," 2002, http://ipec.utulsa.edu/Conf2002/mcilwaine_136.pdf.

[134] L. Laopaiboon, N. Phukoetphim, K. Vichitphan, and P. Laopaiboon, "Biodegradation of an aldehyde biocide in rotating biological contactors," *World Journal of Microbiology and Biotechnology*, vol. 24, no. 9, pp. 1633–1641, 2008.

[135] Toxnet, "Tetrakis(hydroxymethyl) phosphonium sulfate," National Library of Medicine, Toxicology Data Network, 2015, http://toxnet.nlm.nih.gov/cgi-bin/sis/search/a?dbs+hsdb:@term+@DOCNO+4215.

[136] A. N. Shamim, *Scoping Document: Product Chemistry/Environmental Fate/Ecotoxicity of: Tetrakis (Hydroxymethyl) Phosphonium Sulfate*, U.S. Environmental Protection Agency (U.S. EPA), Washington, DC, USA, 2011.

[137] S. Groome, "Tetrakis (hydroxymethyl) phosphonium sulfate THPS," Chemistry Document EPA-HQ-OPP-2011-0067, U.S. Environmental Protection Agency, Washington, DC, USA, 2011.

[138] W. J. Lyman, *Environmental Exposure from Chemicals*, vol. 1, Edited by W. B. Neely and G. E. Blau, CRC Press, Boca Raton, Fla, USA, 1985.

[139] J. H. Exner, G. A. Burk, and D. Kyriacou, "Rates and products of decomposition of 2,2-dibromo-3-nitrilopropionamide," *Journal of Agricultural and Food Chemistry*, vol. 21, no. 5, pp. 838–846, 1973.

[140] Wiser, "Propargyl Alcohol," 2015, http://webwiser.nlm.nih.gov/WebWISER/pda/getSubstanceData.do;jsessionid=E0B34CF5-6185EB60A4EA84DB4E4A255A?substanceId=146&menuItemID=76&identifier=Propargyl+Alcohol&identifierType=Name.

[141] United States Environmental Protection Agency, "Environmental Impact and Benefit Assessment for the Final Effluent Limitation Guidelines and Standards for the Airport Deicing Category," 2006, http://water.epa.gov/scitech/wastetech/guide/airport/upload/Environmental-Impact-and-Benefit-Assessment-for-the-Final-Effluent-Limitation-Guidelines-and-Standards-for-the-Airport-Deicing-Category.pdf.

[142] World Health Organization (WHO), "Thiourea," Concise International Chemical Assessment Document 49, 2003, http://www.who.int/ipcs/publications/cicad/en/cicad49.pdf.

[143] Chemicals Inspection Testing Institute (CITI), *Biodegradation and Bioaccumulation Data of Existing Chemicals based on the CSCL Japan*, Japan Chemical Industry Ecology—Toxicology and Information Center, Saitama, Japan, 1992.

[144] Toxnet, "Mercaptoacetic acid," National Library of Medicine, Toxicology Data Network, 2015, http://toxnet.nlm.nih.gov/cgi-bin/sis/search/a?dbs+hsdb:@term+@DOCNO+2702.

[145] R. J. Lewis Sr., Ed., *Sax's Dangerous Properties of Industrial Materials*, Wiley-Interscience, John Wiley & Sons, Hoboken, NJ, USA, 11th edition, 2004.

[146] Toxnet, "Hazardous Substance Data Bank (HSDB)," National Library of Medicine, Toxicology Data Network, 2015, http://toxnet.nlm.nih.gov/cgi-bin/sis/htmlgen?HSDB.

[147] European Chemicals Agency (ECHA) and International Uniform Chemical Information Database (IUCLID), *CD-ROM Year 2000 Edition*, European Chemicals Agency (ECHA), Helsinki, Finland, 2000.

[148] P. Shrivastava and R. Kumar, "Soil salinity: a serious environmental issue and plant growth promoting bacteria as one of the tools for its alleviation," *Saudi Journal of Biological Sciences*, vol. 22, no. 2, pp. 123–131, 2015.

[149] M. Ashraf, "Some important physiological selection criteria for salt tolerance in plants," *Flora*, vol. 199, no. 5, pp. 361–376, 2004.

[150] A. Bano and M. Fatima, "Salt tolerance in *Zea mays* (L). following inoculation with *Rhizobium* and *Pseudomonas*," *Biology and Fertility of Soils*, vol. 45, no. 4, pp. 405–413, 2009.

[151] R. Munns, "Comparative physiology of salt and water stress," *Plant, Cell and Environment*, vol. 25, no. 2, pp. 239–250, 2002.

[152] R. Munns and R. A. James, "Screening methods for salinity tolerance: a case study with tetraploid wheat," *Plant and Soil*, vol. 253, no. 1, pp. 201–218, 2003.

[153] M. Tester and R. Davenport, "Na$^+$ tolerance and Na$^+$ transport in higher plants," *Annals of Botany*, vol. 91, no. 5, pp. 503–527, 2003.

[154] S. Tabur and K. Demir, "Role of some growth regulators on cytogenetic activity of barley under salt stress," *Plant Growth Regulation*, vol. 60, no. 2, pp. 99–104, 2010.

[155] M. G. Javid, A. Sorooshzadeh, F. Moradi, S. A. M. M. Sanavy, and I. Allahdadi, "The role of phytohormones in alleviating salt stress in crop plants," *Australian Journal of Crop Science*, vol. 5, no. 6, pp. 726–734, 2011.

[156] T. K. Frost, S. Johnsen, and T. I. Utvik, *Environmental Effects of Produced Water Discharges to the Marine Environment*, OLF, Norway, 1998, http://www.olf.no/static/en/rapporter/producedwater/summary.html.

[157] D. C. Wolf, K. R. Brye, and E. E. Gbur, "Using soil amendments to increase Bermuda grass growth in soil contaminated with hydraulic fracturing drilling fluid," *Soil and Sediment Contamination*, vol. 24, no. 8, pp. 846–864, 2015.

[158] M. Ahmad, S. Soo Lee, J. E. Yang, H.-M. Ro, Y. Han Lee, and Y. Sik Ok, "Effects of soil dilution and amendments (mussel shell, cow bone, and biochar) on Pb availability and phytotoxicity in military shooting range soil," *Ecotoxicology and Environmental Safety*, vol. 79, pp. 225–231, 2012.

[159] D. Lloyd, "Drilling waste disposal in Alberta," in *Proceedings of the National Conference on Disposal of Drilling Wastes*, Environmental and Ground Water Institute, Norman, Okla, USA, May 1985.

[160] D. Anderson, "Effective saltwater remediation," NDPC Brine Task Force, Oasis Petroleum, 2015, https://www.ndoil.org/image/cache/AndersonDustin.pdf.

[161] E. de Jong, *The Final Report of a Study Conducted on the Reclamation of Brine and Emulsion Spills in the Cultivated Area of Saskatchewan*, Saskatchewan Department of Mineral Resources and the Canadian Petroleum Association, Saskatchewan Division, Saskatchewan, Canada, 1979.

[162] M. E. Sumner, "Sodic soils: new perspectives," *Australian Journal of Soil Research*, vol. 31, no. 6, pp. 683–750, 1993.

[163] J. Sharma, A. V. Ogram, and A. Al-Agely, *Mycorrhizae: Implications for Environmental Remediation and Resource Conservation*, IFAS Extension ENH-1086, University of Florida, Gainesville, Fla, USA, 2015, http://ufdcimages.uflib.ufl.edu/IR/00/00/17/79/00001/EP35100.pdf.

[164] J. Pichtel, *Fundamentals of Site Remediation for Metal- and Hydrocarbon-Contaminated Soils*, Government Institutes, Rockville, Md, USA, 2nd edition, 2007.

[165] Energy & Environmental Research Center (EERC), "Spills Cleanup Primer," BakkenSmart, 2015, http://www.northdakotaoilcan.com/home-menu/news-info/resources/spills-cleanup-primer/.

[166] M. Farhadian, C. Vachelard, D. Duchez, and C. Larroche, "In situ bioremediation of monoaromatic pollutants in groundwater: a review," *Bioresource Technology*, vol. 99, no. 13, pp. 5296–5308, 2008.

[167] I. V. Robles-González, F. Fava, and H. M. Poggi-Varaldo, "A review on slurry bioreactors for bioremediation of soils and sediments," *Microbial Cell Factories*, vol. 7, article 5, 2008.

[168] M. E. Fuller, J. Kruczek, R. L. Schuster, P. L. Sheehan, and P. M. Arienti, "Bioslurry treatment for soils contaminated with very high concentrations of 2,4,6-trinitrophenylmethylnitramine (tetryl)," *Journal of Hazardous Materials*, vol. 100, no. 1–3, pp. 245–257, 2003.

[169] D. Dean-Ross, "Biodegradation of selected PAH from sediment in bioslurry reactors," *Bulletin of Environmental Contamination and Toxicology*, vol. 74, no. 1, pp. 32–39, 2005.

[170] O. Ward, A. Singh, and J. Van Hamme, "Accelerated biodegradation of petroleum hydrocarbon waste," *Journal of Industrial Microbiology and Biotechnology*, vol. 30, no. 5, pp. 260–270, 2003.

[171] C. N. Mulligan, R. N. Yong, and B. F. Gibbs, "Remediation technologies for metal-contaminated soils and groundwater: an evaluation," *Engineering Geology*, vol. 60, no. 1–4, pp. 193–207, 2001.

[172] S. A. Wasay, S. Barrington, and S. Tokunaga, "Organic acids for the in situ remediation of soils polluted by heavy metals: soil flushing in columns," *Water, Air, and Soil Pollution*, vol. 127, no. 1–4, pp. 301–314, 2001.

[173] M. N. V. Prasad, "Phytoremediation of metals in the environment for sustainable development," *Proceedings of the National Academy of Sciences, India Section B: Biological Sciences*, vol. 70, no. 1, pp. 71–98, 2004.

[174] I. Alkorta, J. Hernández-Allica, J. M. Becerril, I. Amezaga, I. Albizu, and C. Garbisu, "Recent findings on the phytoremediation of soils contaminated with environmentally toxic heavy metals and metalloids such as zinc, cadmium, lead, and arsenic," *Reviews in Environmental Science and Biotechnology*, vol. 3, no. 1, pp. 71–90, 2004.

[175] C. Garbisu, J. Hernández-Allica, O. Barrutia, I. Alkorta, and J. M. Becerril, "Phytoremediation: a technology using green plants to remove contaminants from polluted areas," *Reviews on Environmental Health*, vol. 17, no. 3, pp. 173–188, 2002.

[176] M. N. V. Prasad, "Phytoremediation of metals and radionuclides in the environment: the case for natural hyperaccumulators, metal transporters, soil-amending chelators and transgenic plants," in *Heavy Metal Stress in Plants*, M. N. V. Prasad, Ed., chapter 14, pp. 345–391, Springer, Berlin, Germany, 1999.

[177] U.S. Department of Energy, *A Guide to Practical Management of Produced Water from Onshore Oil and Gas Operations in the United States*, Interstate Oil and Gas Compact Commission and ALL Consulting, 2006, http://www.all-llc.com/publicdownloads/ALL-PWGuide.pdf.

[178] National Petroleum Council (NPC), "Management of Produced Water from Oil and Gas Wells, Paper #2-17," NPC North American Resource Development Study, 2011, https://www.npc.org/Prudent_Development-Topic_Papers/2-17_Management_of_Produced_Water_Paper.pdf.

[179] GRI, "Technology assessment and economic evaluation of downhole gas/water separation and disposal tools," Tech. Rep. GRI-99/0218, Radian Corporation for the Gas Research Institute, Des Plaines, Ill, USA, 1999.

[180] E. I. Shirman and A. K. Wojtanowicz, "More oil using downhole water-sink technology: a feasibility study," SPE 66532, SPE Production and Facilities, 2002.

[181] A. K. Wojtanowicz, E. I. Shirman, and H. Kurban, "Downhole Water Sink (DWS) completions enhance oil recovery in reservoirs with water coning problem," in *Proceedings of the SPE Annual Technical Conference and Exhibition*, SPE 56721, Houston, Tex, USA, October 1999.

[182] M. Swisher, "Summary of DWS application in Northern Louisiana," in *Proceedings of the Downhole Water Separation Technology Workshop*, Baton Rouge, La, USA, March 2000.

[183] A. K. Wojtanowicz, "Smart dual completions for downhole water control in oil and gas wells," in *Proceedings of the Produced Water Management Workshop*, Houston, Tex, USA, April 2003.

[184] L. Shariq, "Uncertainties associated with the reuse of treated hydraulic fracturing wastewater for crop irrigation," *Environmental Science and Technology*, vol. 47, no. 6, pp. 2435–2436, 2013.

[185] J. D. Arthur, B. G. Langhus, and C. Patel, *Technical Summary of Oil & Gas Produced Water Treatment Technologies*, NETL, Tulsa, Okla, USA, 2005.

[186] U.S. Environmental Protection Agency, "Design, construction, and operation of hazardous and nonhazardous waste surface impoundments," Tech. Rep. EPA/530/SW-91/054a, EPA, Cincinnati, Ohio, USA, 1991.

[187] V. Velmurugan and K. Srithar, "Prospects and scopes of solar pond: a detailed review," *Renewable & Sustainable Energy Reviews*, vol. 12, no. 8, pp. 2253–2263, 2008.

[188] ALL Consulting, *Handbook on Coal Bed Methane Produced Water: Management and Beneficial Use Alternatives*, ALL Consulting, Tulsa, Okla, USA, 2003.

[189] Colorado School of Mines, "Technical assessment of produced water treatment technologies. An integrated framework for treatment and management of produced water," RPSEA Project 07122-12, Golden, Colo, USA, 2009.

[190] Agricultural Water Clearing House (AWCH), *FAQ—Water Supply, Sources, & Agricultural Use*, 2015, http://www.agwaterconservation.colostate.edu/FAQs_WaterSupplySourcesAgriculturalUse.aspx#A1.

[191] A. Horpestad, D. Skaar, and H. Dawson, "Water quality impacts from CBM development in the powder river basin, Wyoming and Montana," Water Quality Technical Report, 2001.

[192] D. B. Burnett and J. A. Veil, "Decision and risk analysis study of the injection of desalination by-products into oil-and gas-producing zones," in *Proceedings of the SPE International Symposium and Exhibition on Formation Damage Control*, SPE-86526-MS, Lafayette, La, USA, February 2004.

[193] The Ground Water Protection Council, "Making water produced during oil and gas operations a managed resource for beneficial uses," in *Proceedings of the Ground Water Protection Council Produced Water Conference*, Colorado Springs, Colo, USA, October 2002, http://www.gwpc.org/sites/default/files/events/Agenda.pdf.

[194] International Association of Oil and Gas Producers, "Guidelines for produced water injection," Report 2.80/302, 2002, http://www.ogp.org.uk/pubs/302.pdf.

[195] J. Cart, "Central Valley's growing concern: crops raised with oil field water," *Los Angeles Times*, 2014, http://www.latimes.com/local/california/la-me-drought-oil-water-20150503-story.html#page=1.

[196] Wyoming Department of Environmental Quality, UIC Program, 2015, http://deq.wyoming.gov/wqd/underground-injection-control/.

[197] A. J. DeJoia, "Developing sustainable practices for CBM-produced water irrigation," in *Proceedings of the Ground Water Protection Council Produced Water Conference*, Colorado Springs, Colo, USA, October 2002, http://www.gwpc.org/sites/default/files/events/Agenda.pdf.

[198] R. J. Paetz and S. Maloney, "Demonstrated economics of managed irrigation for CBM produced water," in *Proceedings of the Ground Water Protection Council Produced Water Conference*, Colorado Springs, Colo, USA, October 2002, http://www.gwpc.org/sites/default/files/events/Agenda.pdf.

[199] C. R. Bern, G. N. Breit, R. W. Healy, J. W. Zupancic, and R. Hammack, "Deep subsurface drip irrigation using coal-bed sodic water: part I. Water and solute movement," *Agricultural Water Management*, vol. 118, pp. 122–134, 2013.

[200] R. Chhabra, *Soil Salinity and Water Quality*, Balkema, Rotterdam, Netherlands, 1996.

[201] L. Jackson and J. Myers, "Alternative use of produced water in aquaculture and hydroponic systems at naval petroleum reserve no. 3," in *Proceedings of the Ground Water Protection Council Produced Water Conference*, Colorado Springs, Colo, USA, October 2002, http://www.gwpc.org/sites/default/files/events/Agenda.pdf.

[202] P. F. Cooper, G. F. Job, M. B. Green, and R. B. E. Shutes, *Reed Beds and Constructed Wetlands for Wastewater Treatment*, Water Research Centre Publications, Swindon, UK, 1996.

[203] J. R. Kuipers, "Technology-based effluent limitations for coalbed methane produced wastewater discharges in the Powder River basin of Montana and Wyoming," Draft Report, Northern Plains Resource Council, Billings, Mont, USA, 2004.

[204] F. Sanders, S. Gustin, and P. Pucel, *Natural Treatment of CBM Produced Water: Field Observations*, CBM Associates, 2001.

[205] J. Bauder, *The Role and Potential of Use Selected Plants, Plant Communities, Artificial, Constructed, and Natural Wetlands in Mitigation of Impaired Water for Riparian Zone Remediation*, Montana State University, Water Quality & Irrigation Management, Bozeman, Mont, USA, 2002.

[206] K. E. Lynch, "Agency collection activities: coalbed methane extraction sector survey," Survey Docket ID No. EPA-HQ-OW-2006-0771, Trout Unlimited to EPA Docket Center, 2008.

[207] D. W. Erskine, P. W. Bergman, and D. L. Wacker, "Use of produced water for oil and gas drilling in the San Juan Basin," in *Proceedings of the 9th Annual Petroleum Environmental Conference*, Albuquerque, NM, USA, October 2002.

[208] A. J. Daniel, B. G. Langhus, and C. Patel, *Technical Summary of Oil & Gas Produced Water Treatment Technologies*, NETL, 2005.

[209] P. Xu and J. E. Drewes, "Viability of nanofiltration and ultra-low pressure reverse osmosis membranes for multi-beneficial use of methane produced water," *Separation and Purification Technology*, vol. 52, no. 1, pp. 67–76, 2006.

[210] S. S. Madaeni, "The application of membrane technology for water disinfection," *Water Research*, vol. 33, no. 2, pp. 301–308, 1999.

[211] R. Han, S. Zhang, D. Xing, and X. Jian, "Desalination of dye utilizing copoly(phthalazinone biphenyl ether sulfone) ultrafiltration membrane with low molecular weight cut-off," *Journal of Membrane Science*, vol. 358, no. 1-2, pp. 1–6, 2010.

[212] Y. He and Z. W. Jiang, "Technology review: treating oilfield wastewater," *Filtration & Separation*, vol. 45, no. 5, pp. 14–16, 2008.

[213] T. Bilstad and E. Espedal, "Membrane separation of produced water," *Water Science and Technology*, vol. 34, no. 9, pp. 239–246, 1996.

[214] S. Judd and B. Jefferson, *Membranes for Industrial Wastewater Recovery and Re-Use Oxford*, Elsevier, Philadelphia, Pa, USA, 2013.

[215] C. Bellona and J. E. Drewes, *Reuse of Produced Water Using Nanofiltration and Ultra-Low Pressure Reverse Osmosis to Meet Future Water Demands*, Ground Water Protection Council, Oklahoma City, Okla, USA, 2015, http://www.gwpc.org/sites/default/files/event-sessions/Jorge_Drewes_PWC2002_0.pdf.

[216] S. Mondal and S. R. Wickramasinghe, "Produced water treatment by nanofiltration and reverse osmosis membranes," *Journal of Membrane Science*, vol. 322, no. 1, pp. 162–170, 2008.

[217] R. S. Faibish and Y. Cohen, "Fouling and rejection behavior of ceramic and polymer-modified ceramic membranes for ultrafiltration of oil-in-water emulsions and microemulsions," *Colloids and Surfaces A: Physicochemical and Engineering Aspects*, vol. 191, no. 1-2, pp. 27–40, 2001.

[218] R. S. Faibish and Y. Cohen, "Fouling-resistant ceramic-supported polymer membranes for ultrafiltration of oil-in-water microemulsions," *Journal of Membrane Science*, vol. 185, no. 2, pp. 129–143, 2001.

[219] K. Konieczny, M. Bodzek, and M. Rajca, "A coagulation-MF system for water treatment using ceramic membranes," *Desalination*, vol. 198, no. 1–3, pp. 92–101, 2006.

[220] J. Boysen, "The freeze-thaw/evaporation (FTE) process for produced water treatment, disposal and beneficial uses," in *Proceedings of the 14th Annual International Petroleum Environmental Conference*, Houston, Tex, USA, 2007.

[221] M. Ebrahimi, D. Willershausen, K. S. Ashaghi et al., "Investigations on the use of different ceramic membranes for efficient oil-field produced water treatment," K-State Research Exchange, 2015, http://krex.k-state.edu/dspace/bitstream/handle/2097/3520/EbrahimiDesalination2010.pdf?sequence=1.

[222] A. Lobo, Á. Cambiella, J. M. Benito, C. Pazos, and J. Coca, "Ultrafiltration of oil-in-water emulsions with ceramic membranes: influence of pH and crossflow velocity," *Journal of Membrane Science*, vol. 278, no. 1-2, pp. 328–334, 2006.

[223] G. Gutiérrez, A. Lobo, D. Allende et al., "Influence of coagulant salt addition on the treatment of oil-in-water emulsions by centrifugation, ultrafiltration, and vacuum evaporation," *Separation Science and Technology*, vol. 43, no. 7, pp. 1884–1895, 2008.

[224] K. S. Spiegler and O. Kedem, "Thermodynamics of hyperfiltration (reverse osmosis): criteria for efficient membranes," *Desalination*, vol. 1, no. 4, pp. 311–326, 1966.

[225] W. Mark, *The Guidebook to Membrane Desalination Technology: Reverse Osmosis, Nanofiltration and Hybrid Systems Process, Design, Applications and Economic*, L'Aquila Desalination Publications, 1st edition, 2007.

[226] B. Nicolaisen and L. Lien, "Treating oil and gas produced water using membrane filtration technology," in *Proceedings of the Produced Water Workshop*, Aberdeen, UK, 2003.

[227] F. T. Tao, S. Curtice, R. D. Hobbs et al., "Conversion of oilfield produced water into an irrigation/drinking quality water," in *Proceedings of the SPE/EPA Exploration and Production Environmental Conference*, SPE-26003-MS, San Antonio, Tex, USA, March 1993.

[228] US Environmental Protection Agency, *Onsite Wastewater Treatment and Disposal Systems Design Manual*, US Environmental Protection Agency, Washington, DC, USA, 1980.

[229] D. Su, J. Wang, K. Liu, and D. Zhou, "Kinetic performance of oil-field produced water treatment by biological aerated filter," *Chinese Journal of Chemical Engineering*, vol. 15, no. 4, pp. 591–594, 2007.

[230] H. L. Ball, "Nitrogen reduction in an on-site trickling filter/upflow filters wastewater treatment system," in *Proceedings of the 7th International Symposium on Individual and Small Community Sewage Systems*, American Society of Agricultural Engineers, Atlanta, Ga, USA, December 1994.

[231] W. Shi and M. M. Benjamin, "Membrane interactions with NOM and an adsorbent in a vibratory shear enhanced filtration process (VSEP) system," *Journal of Membrane Science*, vol. 312, no. 1-2, pp. 23–33, 2008.

[232] A. I. Zouboulis and M. D. Petala, "Performance of VSEP vibratory membrane filtration system during the treatment of landfill leachates," *Desalination*, vol. 222, no. 1–3, pp. 165–175, 2008.

[233] V. Piemonte, M. Prisciandaro, L. di Paola, and D. Barba, "Membrane processes for the treatment of produced waters," *Chemical Engineering Transactions*, vol. 43, 2015.

[234] O. A. Hamed, "Evolutionary developments of thermal desalination plants in the Arab gulf region," in *Proceedings of the Beirut Conference*, Beirut, Lebanon, June 2004, http://www.researchgate.net/publication/228945604_Evolutionary_developments_of_thermal_desalination_plants_in_the_Arab_Gulf_region.

[235] United States Bureau of Reclamation, *Desalting Handbook for Planners*, Desalination and Water Purification Research and Development Program Report no. 72, United States Department of the Interior, Bureau of Reclamation, Washington, DC, USA, 3rd edition, 2003.

[236] GWI, IDA Worldwide Desalting Plants Inventory Report no. 19 (Global water intelligence) Gnarrenburg, Germany, 2006.

[237] A. D. Khawaji, I. K. Kutubkhanah, and J.-M. Wie, "Advances in seawater desalination technologies," *Desalination*, vol. 221, no. 1–3, pp. 47–69, 2008.

[238] B. Heins, *World's First SAGB Facility Using Evaporators, Drum Boilers, and Zero Discharge Crystallizers to Treat Produced Water*, Efficiency and Innovation Forum for Oil Patch, Calgary, Canada, 2005.

[239] J. Brandt, *Business is Booming for Fracking Wastewater*, Fierce Water, 2015.

[240] J. E. Boysen, J. A. Harju, B. Shaw et al., "The current status of commercial deployment of the freeze thaw evaporation treatment of produced water," in *Proceedings of the SPE/EPA Exploration and Production Environmental Conference*, SPE-52700-MS, pp. 1–3, Austin, Tex, USA, March 1999.

[241] AltelaRain TM System ARS-4000, *New Patented Technology for Cleaning Produced Water On-Site*, Altela Information, 2007.

[242] N. A. Godshall, "AltelaRain[SM] produced water treatment technology: making water from waste," in *Proceedings of the International Petroleum Environmental Conference (IPEC '06)*, pp. 1–9, ALTELATM, San Antonio, Tex, USA, October 2006.

[243] Jain Irrigation Systems, *Sand Separator—Jain Hydro Cyclone Filter*, Jain Irrigation Systems, Jalgaon, India, 2010.

[244] A. L. Cassidy, "Advances in flotation unit design for produced water treatment," in *Proceedings of the Production Operations Symposium*, SPE-25472-MS, pp. 21–23, Oklahoma City, Okla, USA, March 1993.

[245] M. Çakmakce, N. Kayaalp, and I. Koyuncu, "Desalination of produced water from oil production fields by membrane processes," *Desalination*, vol. 222, no. 1–3, pp. 176–186, 2008.

[246] F. R. Spellman, *Handbook of Water and Wastewater Treatment Plant Operations*, CRC Press, Boca Raton, Fla, USA, 2003.

[247] D. A. Clifford, "Ion exchange and inorganic adsorption," in *Water Quality and Treatment*, R. D. Letterman, Ed., McGraw-Hill, New York, NY, USA, 1999.

[248] D. T. Meijer and C. Madin, "Removal of dissolved and dispersed hydrocarbons from oil and gas produced water with mppe technology to reduce toxicity and allow water reuse," *APPEA Journal*, pp. 1–11, 2010.

[249] H. M. Pars and D. T. Meijer, "Removal of dissolved hydrocarbons from production water by macro porous polymer extraction (MPPE)," in *Proceedings of the SPE International Conference on Health, Safety, and Environment in Oil and Gas Exploration and Production*, SPE-46577-MS, Society of Petroleum Engineers, Caracas, Venezuela, June 1998.

5

Vertical Phosphorus Migration in a Biosolids-Amended Sandy Loam Soil in Laboratory Settings: Concentrations in Soils and Leachates

Yulia Markunas,[1] **Vadim Bostan,**[1] **Andrew Laursen,**[1] **Michael Payne,**[2] **and Lynda McCarthy**[1]

[1]*Department of Chemistry and Biology, Ryerson University, 350 Victoria Street, Toronto, ON, Canada M5B 2K3*
[2]*Black Lake Environmental, 246 Black Lake Route, Perth, ON, Canada K7H 3C5*

Correspondence should be addressed to Vadim Bostan; vbostan@ryerson.ca

Academic Editor: Rafael Clemente

The impacts of biosolids land application on soil phosphorus and subsequent vertical migration to tile drainage were assessed in a laboratory setup. Soil, representing typical "nonresponse" Ontario soil as specified by Ontario Ministry of Agriculture, Food, and Rural Affairs (OMAFRA), was amended with anaerobically digested biosolids at a rate of 8 Mg ha^{-1} (dry weight). Over five months, these amended soil samples from two different depths were sequentially fractionated to determine various inorganic and organic phosphorus pools in order to evaluate phosphorus vertical migration within a soil profile. Soil leachate was analyzed for soluble reactive phosphorus. The results indicated that biosolids application did not significantly affect phosphorus concentrations in soil and did not cause phosphorus vertical migration. The concentrations of soluble reactive phosphorus also were not significantly affected by biosolids.

1. Introduction

Even though the term *biosolids*, as it is understood now, only appeared at the end of 20th century, the concept of human waste application to agricultural lands has been known for thousands of years [1]. Agricultural land application is considered a viable way to recycle municipal biosolids. Biosolids are valued as a soil conditioner [1] and as a source of macro- and micronutrients and organic matter necessary for healthy crops [2, 3].

Even though municipalities, farmers, and the general public may benefit from the use of biosolids on agricultural lands, there are some environmental concerns related to this practice. One of these concerns is the relatively high level of total phosphorus in biosolids and its potential migration to surface- and groundwater from biosolids-amended lands [4] as its overabundance in surface water can lead to eutrophication [5]. Many wastewater treatment plants (WWTPs) use alum or $FeCl_3$ in tertiary treatment to precipitate phosphorus,

which becomes part of the biosolids material. Further, stabilization of raw sewage or biosolids by alkaline stabilization may result in some phosphorus precipitation as $Ca_3(PO_4)_2$. As a result the phosphorus may be in forms that are not biologically available, or in forms that retard migration from point of application [6]. Therefore, total phosphorus in biosolids may not be relevant in considering the risk posed by phosphorus loss from fields to aquatic systems. Rather, to properly assess potential risk of biosolids-derived phosphorus to receiving waters, it may be more important to consider the various fractions of phosphorus in biosolids and their behaviour in soil (e.g., migration and transformation).

Generally, phosphorus has the potential of being transported to aquatic systems through three major pathways: surface runoff and erosion, subsurface flow, and tile drainage [4, 7]. To a smaller extent, it can also be transported by wind erosion and deposition [8]. Runoff and erosion are important pathways of soil phosphorus loss [9, 10], and the majority of studies on phosphorus loss from agricultural land are devoted to losses through runoff and erosion [11–16]. Application of municipal biosolids, specifically, can increase total phosphorus (TP) in runoff from soil in laboratory mesocosms [17], although loss of P in runoff in agricultural soils may be dependent on the stabilization method used to produce the biosolids [18–20]. These losses represent only horizontal migration of phosphorus through surface runoff, while vertical migration of phosphorus through the soil layers can also contribute to the phosphorus escape from agricultural lands [21]. It is especially significant in areas with little slope, areas with shallow groundwater, dry areas, or less weathered soils and organic soils that have a low phosphorus-sorbing capacity [22, 23]. Vertical migration represents a potential route for phosphorus loss in agricultural areas, yet few studies have considered vertical phosphorus migration, and none were found to have considered vertical migration and evolution of the various phosphorus fractions relevant to biosolids-amended soils.

The purpose of this study was to determine, in a laboratory setting, how land application of biosolids might affect various phosphorus fractions in soil, time-related transformations among fractions, and vertical loss of phosphorus to leachate.

2. Material and Methods

2.1. Experimental Setup.
In order to study vertical P migration, a series of soil columns were constructed in the laboratory [24]. Laboratory soil columns consisted of a series of eight plastic columns (7 cm diameter, 60 cm height). These columns had rubber end-caps, and a plastic funnel (7 cm outer diameter) was inserted into each column with the stem of the funnel protruding through a hole in the end-cap. The bottom 10 cm of the column (immediately above the funnel) was filled with river gravel to improve the drainage of the leachate percolating through the soil and to prevent soil from escaping through the funnel. Above the gravel, each column was filled with 40 cm of sandy loam soil. The soil had a bulk density of 1.4 with 3% organic matter and a pH of 6.5 and has a good water drainage. It is commercially available from "Circle Farms" in Brantford Ontario and it was selected as it is a good representative for the farm soils in southern Ontario. The four randomly selected columns were amended with 3.65 g of anaerobically digested biosolids (dry weight), equivalent to a rate of 8 Mg (dry weight) ha^{-1}. This was achieved by adding 260 mL of biosolids at 1.4% dry matter on top of the soil in each column selected for biosolids application and incorporated into the top 5 cm. Reference columns were watered with the same volume of distilled water (260 mL). The biosolids used in the current research were produced at a Southern Ontario wastewater treatment plant. The sewage treatment process at this plant utilizes a conventional secondary activated sludge process with chemical phosphorus removal and anaerobic sludge digestion. Secondary treatment involves phosphorus removal via precipitation with iron (in form of ferric chloride) followed by the addition of sodium hypochlorite as a disinfectant to the treated water. Consequently, precipitated $FePO_3$ becomes a constituent of the activated sludge and ultimately the biosolids. Anaerobic

digestion of the sludge occurs in airtight reactors over a two-week period. Biosolids produced at this wastewater treatment plant are either used on agricultural land or dewatered and landfilled. In Ontario, 8 Mg of biosolids (dry weight) per ha of land per 5 years is the common application rate [25] and was used in other similar studies [26].

2.2. Sampling.
An initial set of soil samples (before biosolids application) was collected for analysis of phosphorus fractions, including the Olsen sodium bicarbonate extractable phosphorus. Subsequent sample collections were performed with decreasing periodicity: two weeks, one month, two months, three months, and five months after biosolids application. The duration of the experiment approximated a growing season for southern Ontario. At each sampling period, approximately 2 g of soil samples was collected from two different depths within the soil columns (3 cm from the top and 35 cm from the top). Due to the 1.4 bulk density of the soil, each soil sample had a volume of approximately 3 cm^3. The samples were collected by drilling holes in the plastic columns, extracting one soil aliquot per each column, at each of the two depths, and subsequently sealing the holes with silicone glue. Water (approximately 200 mL per column) was added to columns on a weekly basis. Leachate samples were collected with decreasing periodicity: 1 day, 2 weeks, 1.5 months, 2 months, 3 months, and 5 months after biosolids application. All sampling events happened next day after water was added to the columns [24].

2.3. Chemical Analysis.
Biosolids, nonamended soils, and biosolids-amended soils were analyzed (reagents and concentrations are indicated after each of the forms) for various inorganic (water-soluble, distilled water; loosely bound, 1 M NH_4Cl; metal-bound, 1 M NaOH; and calcium-bound, 0.5 M HCl) and organic (labile, 0.5 M $NaHCO_3$; moderately labile, 1 M HCl; and nonlabile, ash at 555°C) phosphorus forms using the sequential fractionation procedure. For the water-soluble phosphorus, 0.5 g of soil sample (dry weight) was placed into a 50 mL Nalgene centrifuge tube and 25 mL of distilled water was added. The supernatant was separated from the solid residue by centrifugation, filtered through a 0.22 μm filter (paper filter), and analyzed colorimetrically using the ascorbic acid, molybdate method. The residual soil was kept for the next fractionations step, where persulfate digestion for nonlabile phosphorus was replaced by strong acid digestion [27]. The leachate from the columns was analyzed for soluble reactive phosphorus (SRP) according to the procedure described by Kovar and Pierzynski [27] which is an adaptation of the classical ascorbic acid method [28]. In addition, the Olsen phosphorus soil test was determined using 0.5 M $NaHCO_3$ according to the standard procedure [27].

2.4. Modeling P Concentrations.
Based on the results of initial soil and biosolids phosphorus analysis, a simple model was created to predict soil phosphorus increases after biosolids application. The expected phosphorus concentrations (c_f) were calculated for each form of phosphorus based upon the initial concentration of each form in the soil (c_1), the

concentration of each form in the biosolids (c_b), the total mass of soil (M_s), and the quantity of biosolids added (M_b) as follows.

$$c_f = c_1 + c_b * \frac{M_b}{M_s}. \tag{1}$$

2.5. Statistical Analysis. A generalized linear mixed model (GLMM) was used to compare the quantities of various forms of phosphorus in biosolids-amended soil versus reference soil, in surface soil versus bottom soil, and over time. Moreover, a GLMM can model response data with non-Gaussian distributions (e.g., exponential, lognormal) and can model a diversity of residuals correlation structures related to the longitudinal variable (e.g., Gaussian, compound symmetry, linear, power, and exponential). General linear mixed model ANOVAs (PROC GLIMMIX, SAS 9.4, SAS Institute, Cary, NC, USA) were used to detect the primary effects of biosolids application and location, as well as the interaction between these factors and time, to determine if concentrations of various forms of P followed different trajectories in response to treatment (biosolids versus reference) or location (top versus bottom). Such models can be used to fit longitudinal models for data, regardless of whether the response is linear with respect to the longitudinal variable (here, time). GLIMMIX models were run iteratively, with different combinations of specified covariance structure and response variable distributions. The final models were selected on the basis of fit (based upon Akaike Information Criterion (AIC)) and residuals distribution.

3. Results and Discussion

The results of the initial soil phosphorus analysis (Figure 1(a)) revealed that the total phosphorus concentration was 0.226 mg/g of soil. This value is in range with other agriculture soils (0.050 mg/g and up to 1.000 mg/g) [29, 30]. Potentially readily available phosphorus fractions represented by inorganic water-soluble and organic labile fractions were below the detection limit (BDL) of the method which is 0.005 mg/g. This may indicate a significant sorption capacity of the soil, although this was not explicitly measured. Inorganic loosely bound (0.055 mg/g of soil) and organic moderately labile fractions (0.022 mg/g of soil), however, represent the major part of the bioavailable fractions. They constituted more than 31% of the total phosphorus concentration in initial soil. This pool of moderately available phosphorus is almost double compared to other studies [31], reinforcing the idea that phosphorus sorption capacity is high for this soil. Relatively unavailable phosphorus fractions, such as inorganic calcium-bound (0.019 mg/g of soil) and organic nonlabile (0.049 mg/g of soil) fractions, were also minor. Their cumulative contribution to the total phosphorus concentration in initial soil was approximately 13%. The other 43% of the initial total phosphorus concentration in the soil was represented by the metal-bound phosphorus fraction (0.115 mg/g of soil), a fraction that is not directly available for the plants but that can release bioavailable phosphorus under anoxic conditions. The Olsen phosphorus

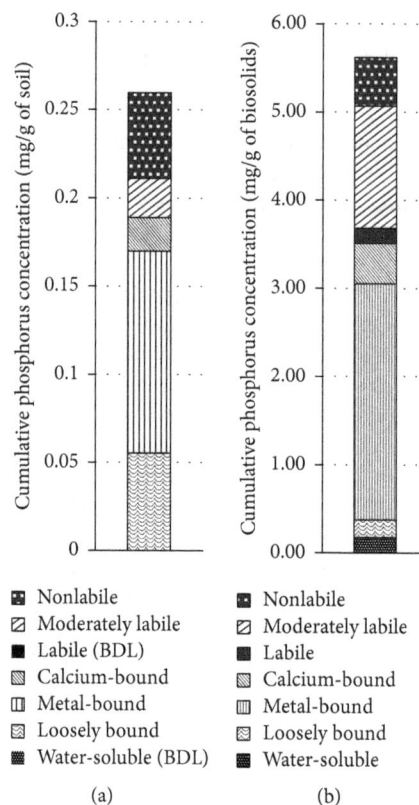

FIGURE 1: Cumulative initial phosphorus fractions in soil (a) and biosolids (b). Note different concentration scales between the two graphs.

concentration (0.060 mg/g) suggested that the soil used in the experiment was relatively rich in phosphorus and would be classified as a "no response" soil [32] meaning that the soil is capable of producing high-yielding crops with little or no additional phosphorus fertilizer. However, such soil might increase the potential of phosphorus migration to surrounding water bodies [33].

Biosolids analysis (Figure 1(b)) revealed that the total phosphorus concentration (5.617 mg/g of biosolids) in biosolids was 21 times higher than the total phosphorus concentration in initial soil. Individual phosphorus fractions were also greatly exceeding those in initial soil. This indicates that these biosolids have the potential to fertilize the initial soil. The distribution of the different phosphorus fractions contributed to the total phosphorus concentration, however, was similar to the distribution observed for soil prior to amendment. Such a distribution was unexpected as the organic fractions of phosphorus were expected to be larger (relative to total P) in the biosolids compared to the soil. An advanced level of mineralization of the biosolids may explain the relatively low organic phosphorus content. The cumulative contribution of relatively unavailable phosphorus fractions, such as inorganic calcium-bound (0.454 mg/g of biosolids) and organic nonlabile (0.550 mg/g of biosolids) fractions, was 17%. The metal-bound phosphorus fraction (2.677 mg/g of biosolids) represented the biggest phosphorus

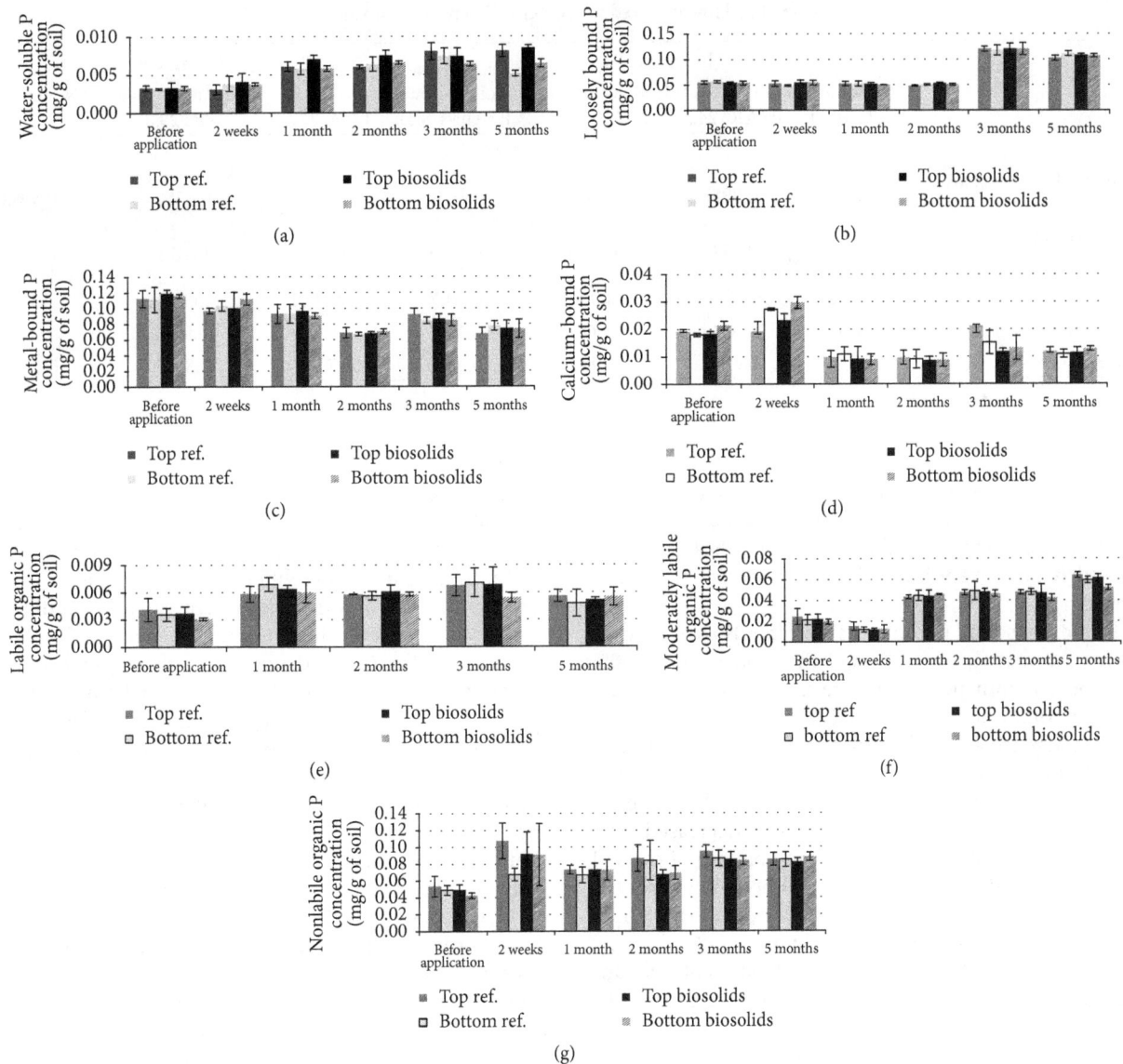

FIGURE 2: Phosphorus (P) fractions in soil measured in the reference and biosolids treated soil columns: a) inorganic water-soluble P, b) inorganic loosely bound P, c) inorganic metal-bound P, d) inorganic calcium-bound P, e) organic labile P, f) organic moderately labile P, and g) organic nonlabile P. The error bars represent standard deviation among replicate samples.

pool in biosolids and contributed 47% to the total phosphorus concentration. High metal-bound phosphorus percentage in biosolids has already been reported by other studies [34, 35]. This is consistent with the wastewater treatment procedure for phosphorus removal. Readily available water-soluble (0.174 mg/g of biosolids), inorganic loosely bound (0.201 mg/g of biosolids), and organic labile (0.175 mg/g of biosolids) fractions in biosolids represented minor pools of phosphorus. Their cumulative contribution to the total phosphorus concentration was 10%. The greatest difference between biosolids and the soil in the relative contribution to the total phosphorus was found for the moderately labile organic fraction: 1.385 mg/g of biosolids (25% of total phosphorus) versus 0.022 mg/g of soil (8% of total phosphorus).

This high value of moderately labile organic fraction was expected as it is relatively common in the biosolids [35].

The results of soil samples analysis after biosolids application are illustrated in Figure 2. When analyzing the statistical significance of changes in phosphorus fractions concentrations, the family of mixed models that consistently resulted in best fit (based on AIC and residuals distribution) were those that fitted a log normal distribution in the response variables (P pools) and that modeled a compound symmetry residuals covariance structure. All reported statistical results are based upon models of this form (Table 1).

As a general trend, concentrations of all P fractions varied over time (Table 1). The water-soluble (Figure 2(a)) and the organic labile (Figure 2(e)) phosphorus fractions increased

TABLE 1: Generalized linear mixed model (GLMM) results for various P pools.

	Inorganic P				Organic P		
	Water-soluble P (AIC 48.3)	Loosely bound P (AIC 24.5)	Metal-bound P (AIC 42.2)	Calcium-bound P (AIC 118.5)	Labile P (AIC 27.7)	Moderately labile P (AIC 29.8)	Nonlabile P (AIC 32.4)
				p-value			
Treatment (biosol. versus ref.)	0.3213	0.8956	0.4507	0.7984	0.5585	0.6280	0.5083
Location (top versus bottom)	0.5762	0.7523	0.4934	0.5819	0.3401	0.5174	0.0551
Time	<0.0001	<0.0001	<0.0001	0.0017	<0.0001	<0.0001	<0.0001
Time × treatment	0.8539	0.8745	0.5661	0.5791	0.8268	0.9211	0.9942
Time × location	0.0248	0.6792	0.8801	0.6515	0.7214	0.8461	0.2205

over time, showing some potential for long term phosphorus remobilisation. Although these increases were statistically significant, the maximum amount by which these fractions increased remains very low at less than 0.005 mg/g of soil (Figure 2(a)).

The loosely bound phosphorus concentrations were constant for the first two months and then increased by 50% at the three-month period (Figure 2(b)). As the increase in loosely bound phosphorus concentrations was observed for both reference and biosolids-amended soils, it could be explained in part by a 30% decrease in the metal-bound phosphorus fraction (Figure 2(c)) rather than by biosolids application. The metal-bound phosphorus decreased over time for both treatments in both top and bottom of the soil column. The transformation of the metal-bound phosphorus to the loosely bound phosphorus could have occurred under anoxic conditions inside of the soil columns.

Over a five-month period, the concentration of calcium-bound phosphorus decreased by 35% (Figure 2(d)). The calcium-bound fraction constitutes mainly phosphates from the composition of hydroxyapatite mineral, which makes it a relatively stable fraction [27]. The reduction over time of the calcium-bound fraction is hard to explain, and further investigation may be required. In general, the hydroxyapatite phosphorus constitutes a fraction that leads to an overestimation of phosphorus impact on ecosystems in studies where only the total phosphorus is measured. The dissolution of this form in soil is usually limited to a substantial drop in pH. The sandy loam soil used in the present study has little buffering capacity and as a consequence, accumulation of organic acids may lead to some local dissolution of the hydroxyapatite fraction. The organic moderately labile (Figure 2(f)) and nonlabile (Figure 2(g)) phosphorus concentrations also increased over the five-month period. No explanation for the changes in organic moderately labile and organic nonlabile phosphorus concentrations was found.

One of the main objectives of the present study was to establish in a laboratory setting if biosolids may bring a significant contribution to different phosphorus fractions in agricultural soils. Surprisingly, the chemical analysis of the pools of phosphorus fractions showed that no statistically significant difference was induced by the application of biosolids. No statistically significant difference between biosolids-amended soils and reference soils was observed for any phosphorus pool, nor did changes in phosphorus concentrations over time differ between treatments (time * treatment effect). While some forms of phosphorus did show a time-related change, this was due to chemical and biological processes in the soil and not linked to the application of biosolids. Same trends were observed in both the treated and the reference experimental columns. Therefore, it could be concluded that no measureable increase occurred in any phosphorus fraction as a result of biosolids application (Figure 2). Only water-soluble P differed over time between the top and bottom (time * location effect). While water-soluble P increased over time in both the top and the bottom of the soil column, the increase was greater in the top portion of the soil than in the deeper soil.

According to the model (Table 2), an increase in almost all phosphorus fractions was expected in the top 5 cm of the soil columns immediately following biosolids application (67% increase for water-soluble, 4% increase for loosely bound phosphorus, 29% increase for metal-bound phosphorus, 27% increase for calcium-bound phosphorus, 60% increase for organic labile phosphorus, 77% increase for organic moderately labile phosphorus, and 24% increase for organic nonlabile phosphorus). However, the expected increase in most pools was near or below the nominal limit of detection for phosphorus analysis. When phosphorus enrichment to the entire column was considered, the relative increase in each fraction was smaller still, with absolute increases below the detection limit for all pools except metal-bound phosphorus. Therefore, this model predicted that application of biosolids with the moderately high total phosphorus concentration, at recommended rates in Ontario, to a fertile soil would result in no measureable increase in phosphorus pools for the integrated soil column, and only marginally discernable increases in the top 5 cm, where the material is most concentrated. The interesting aspect of land applications of biosolids is that despite their relatively large phosphorus concentrations of both bioavailable and unbioavailable forms when compared to the soil concentrations, their impact on the final concentrations in amended soils, due to limits imposed on the application rates, is rather limited. Percentage

TABLE 2: Quantitative phosphorus concentration increases expected in the entire length of the soil columns within 1 week from biosolids application.

	Inorganic fractions				Organic fractions		
	Water-soluble P	Loosely bound P	Metal-bound P	Calcium-bound P	Labile P	Moderately labile P	Nonlabile P
Expected increase in P fractions (%) Top 5 cm	67.0	4.5	28.7	26.5	59.9	76.8	23.5
Expected increase in P fractions (%) Entire column	8.4	0.6	3.6	3.3	7.5	9.6	2.9

change calculations from biosolids concentrations, soil concentrations, and application rates translate to relatively small increases in the nominal concentrations of the phosphorus forms in the soil. The detection limit allows accurate measurement of ≥ 0.1 mg P/kg soil. The changes in concentration are below this value, so soil phosphorus remained virtually unaffected after biosolids application.

The experimental design accounted for both temporal changes and potential vertical migration of the phosphorus in the soil columns. The narrow diameter of the columns was meant to avoid lateral escapes of the phosphorus, limiting the potential migration of the phosphorus fractions to the vertical direction. The data obtained from the upper layer and from a depth of 35 cm in the columns showed no difference for any phosphorus fraction between the treated and untreated soils. This suggests that the vertical transport of phosphorus remained insignificant for the duration of the experiment, regardless of the phosphorus fraction considered. As mentioned, changes in concentrations of P at the top of the column were low to statistically insignificant. The possible migration of these nutrients towards the lower region of the column did not occur in the case of this study. Soil adsorption and biological fixation are both mechanisms that could explain this lack of mobility.

The soil and biosolids used in this experiment represented a kind of worst-case scenario, that is, phosphorus-rich anaerobically digested biosolids applied to soil that was also phosphorus-rich, providing conditions under which the phosphorus might be expected to migrate through the soil towards the drainage system rather than being immobilized [33]. This was not observed. Accordingly, it is unlikely that phosphorus vertical migration would be observed following application to more phosphorus-poor soils with higher adsorption capacity for phosphorus, or in soils receiving biosolids produced by further processing such as alkaline stabilization or drying which may decrease phosphorus mobility.

In order to account for the phosphorus movement in the soil columns, the laboratory-based experimental setup allowed the collection of water percolating down through the soil. These leachate samples collected from both biosolids-amended and reference soil columns were analyzed for soluble reactive phosphorus.

The concentrations of soluble reactive phosphorus in leachate samples collected from biosolids-amended soil columns were noted to be very close to the concentrations of soluble reactive phosphorus in leachate samples collected from reference soil columns (Figure 3). No significant difference was observed between samples ($t_{46} = 0.31$, $p = 0.758$). The absence of a significant difference in phosphorus concentrations between leachate from biosolids-amended columns and leachate from reference columns is in accord with the lack of statistical difference between the concentrations of various forms of phosphorus in biosolids-amended columns and reference columns (Figure 2). Within the five months of the research, leachate from both biosolids-amended soil columns and reference soil columns demonstrated phosphorus concentrations around 0.039–0.054 mg/L. These concentrations were much smaller than the concentrations of phosphorus measured in runoff from biosolids-amended soils [17, 20, 29, 36–38]. but still exceeded concentrations suggested as optimal for limiting eutrophication potential (below 0.025 mg/L in streams and 0.01 mg/L in lakes) [39]. Horizontal migration of P in runoff and erosion soon after land application would seem to be a much more important route of potential loading to surface water than vertical transport and groundwater transfers.

4. Conclusions

The results of the soil analysis revealed that concentrations of several phosphorus fractions changed significantly over time. The water-soluble, the organic labile phosphorus, and the loosely bound phosphorus concentrations increased during the five months of the experiment. However, this increase was observed for both reference and biosolids-amended soils and can be linked to a decrease in the metal-bound phosphorus fraction which could have occurred under anoxic conditions inside of the soil columns.

No statistically significant difference between biosolids-amended soils and reference soils was observed for any phosphorus pool, nor did changes in phosphorus concentrations over time differ between treatments (time * treatment effect). Therefore, it can be concluded that no measureable increase occurred in any phosphorus fraction as a result of biosolids application. Thus, phosphorus from the biosolids-amended surface does not measurably migrate through the soil profile towards underground tile drainage systems under the conditions simulated within the current research. This was confirmed by the lack of significant difference between

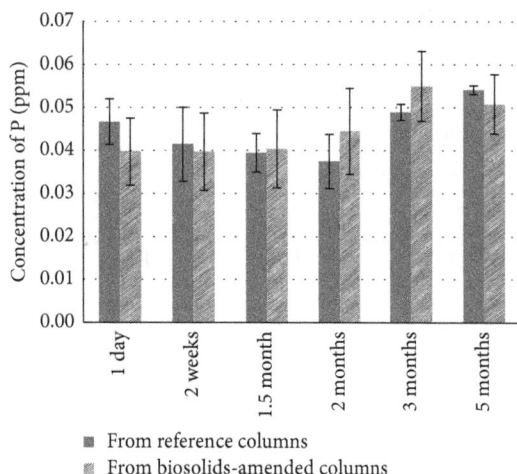

FIGURE 3: Concentrations of soluble reactive phosphorus in analyzed leachate. The error bars represent standard deviation among replicate samples.

concentrations of soluble reactive phosphorus in the leachate from various columns.

The current study provided laboratory-based evidence that anaerobically digested biosolids with phosphorus concentration of 5.6 mg/g of biosolids or below, when applied at the rate of 8 dry tons (dry weight) ha^{-1} to sandy loam soils with no or minimal slope, do not increase P leaching.

Additional Points

Core Ideas. We examined the impact of biosolids on soil phosphorus fractions in vertical profiles. Concentrations of several phosphorus fractions changed significantly over time. Mobile phosphorus fractions increased during the five months of the experiment. No difference between biosolids-amended and reference soils was observed. In specific conditions, biosolids can be applied without risk of increased P leaching.

Competing Interests

The authors declare that they have no competing interests.

Acknowledgments

This paper was possible due to the financial support provided by the Environmental Applied Science and Management graduate program from the Yeates School of Graduate Studies (YSGS) at Ryerson University.

References

[1] EPA, *Land Application of Biosolids: Process Design Manual*, Taylor & Francis, 1997.

[2] A. Atalay, C. Bronick, S. Pao et al., "Nutrient and microbial dynamics in biosolids amended soils following rainfall simulation," *Soil and Sediment Contamination*, vol. 16, no. 2, pp. 209–219, 2007.

[3] A. L. Shober and J. T. Sims, "Phosphorus restrictions for land application of biosolids: current status and future trends," *Journal of Environmental Quality*, vol. 32, no. 6, pp. 1955–1964, 2003.

[4] O. Oenema and C. W. J. Roest, "Nitrogen and phosphorus losses from agriculture into surface waters; the effects of policies and measures in the Netherlands," *Water Science and Technology*, vol. 37, no. 3, pp. 19–30, 1998.

[5] T. C. Daniel, A. N. Sharpley, and J. L. Lemunyon, "Agricultural phosphorus and eutrophication: a symposium overview," *Journal of Environmental Quality*, vol. 27, no. 2, pp. 251–257, 1998.

[6] B. Li and M. T. Brett, "The impact of alum based advanced nutrient removal processes on phosphorus bioavailability," *Water Research*, vol. 46, no. 3, pp. 837–844, 2012.

[7] V. Smil, "Phosphorus in the environment: natural flows and human interferences," *Annual Review of Energy and the Environment*, vol. 25, pp. 53–88, 2000.

[8] K. A. Anderson and J. A. Downing, "Dry and wet atmospheric deposition of nitrogen, phosphorus and silicon in an agricultural region," *Water, Air, and Soil Pollution*, vol. 176, no. 1–4, pp. 351–374, 2006.

[9] J. J. Hanway and J. M. Laflen, "Plant nutrient losses from tile outlet terraces," *Journal of Environmental Quality*, vol. 3, no. 4, pp. 351–356, 1974.

[10] G. W. Randall, T. K. Iragavarapu, and M. A. Schmitt, "Nutrient losses in subsurface drainage water from dairy manure and urea applied for corn," *Journal of Environmental Quality*, vol. 29, no. 4, pp. 1244–1252, 2000.

[11] R. Gächter, J. M. Ngatiah, and C. Stamm, "Transport of phosphate from soil to surface waters by preferential flow," *Environmental Science & Technology*, vol. 32, no. 13, pp. 1865–1869, 1998.

[12] D. W. Lucero, D. C. Martens, J. R. McKenna, and D. E. Starner, "Accumulation and movement of phosphorus from poultry litter application on a Starr clay loam," *Communications in Soil Science and Plant Analysis*, vol. 26, no. 11-12, pp. 1709–1718, 1995.

[13] J. Magid, M. B. Jensen, T. Mueller, and H. C. B. Hansen, "Phosphate leaching responses from unperturbed, anaerobic, or cattle manured mesotrophic sandy loam soils," *Journal of Environmental Quality*, vol. 28, no. 6, pp. 1796–1803, 1999.

[14] K. C. Makris, J. H. Grove, and C. J. Matocha, "Colloid-mediated vertical phosphorus transport in a waste-amended soil," *Geoderma*, vol. 136, no. 1-2, pp. 174–183, 2006.

[15] J. T. Sims, R. R. Simard, and B. C. Joern, "Phosphorus loss in agricultural drainage: historical perspective and current research," *Journal of Environmental Quality*, vol. 27, no. 2, pp. 277–293, 1998.

[16] B. L. Turner and P. M. Haygarth, "Phosphorus forms and concentrations in leachate under four grassland soil types," *Soil Science Society of America Journal*, vol. 64, no. 3, pp. 1090–1099, 2000.

[17] A. Hanief, D. Matiichine, A. E. Laursen, I. Vadim Bostan, and L. H. McCarthy, "Nitrogen and phosphorus loss potential from biosolids-amended soils and biotic response in the receiving water," *Journal of Environmental Quality*, vol. 44, no. 4, pp. 1293–1303, 2015.

[18] H. A. Elliott, R. C. Brandt, and G. A. O'Connor, "Runoff phosphorus losses from surface-applied biosolids," *Journal of Environmental Quality*, vol. 34, no. 5, pp. 1632–1639, 2005.

[19] C. J. Penn and J. T. Sims, "Phosphorus forms in biosolids-amended soils and losses in runoff: effects of wastewater

treatment process," *Journal of Environmental Quality*, vol. 31, no. 4, pp. 1349–1361, 2002.

[20] R. Quilbé, C. Serreau, S. Wicherek, C. Bernard, Y. Thomas, and J.-P. Oudinet, "Nutrient transfer by runoff from sewage sludge amended soil under simulated rainfall," *Environmental Monitoring and Assessment*, vol. 100, no. 1–3, pp. 177–190, 2005.

[21] N. C. Hansen, T. C. Daniel, A. N. Sharpley, and J. L. Lemunyon, "The fate and transport of phosphorus in agricultural systems," *Journal of Soil and Water Conservation*, vol. 57, no. 6, pp. 408–417, 2002.

[22] W. G. Harris, R. D. Rhue, G. Kidder, R. B. Brown, and R. Littell, "Phosphorus retention as related to morphology of sandy coastal plain soil materials," *Soil Science Society of America Journal*, vol. 60, no. 5, pp. 1513–1521, 1996.

[23] A. E. Hartemink, "Encyclopedia of soils in the environment (4 volumes), D. Hillel, J.L. Hatfield, D.S. Powlson, C. Rosenzweig, K.M. Scow, M.J. Singer, D.L. Sparks (Eds.), 2005, ISBN 0-12-348530-4, Elsevier Academic Press, Amsterdam, Hardbound, 2119 pp., US$1,095," *Encyclopedia of Soils in the Environment*, vol. 132, no. 1-2, pp. 240–246, 2006.

[24] Y. Markunas, *Vertical phosphorus migration in biosolids-amended soils: concentrations in soils and leachates [Ph.D. thesis]*, Ryerson University, Toronto, Canada, 2014.

[25] Ministry of Environment and Energy, Ministry of Agriculture, Food, and Rural Affairs, *Guidelines for the Utilization of Biosolids and Other Wastes on Agricultural Land*, Ministry of Environment and Energy, Ministry of Agriculture, Food and Rural Affairs, Toronto, Canada, 1996.

[26] N. Gottschall, M. Edwards, E. Topp et al., "Nitrogen, phosphorus, and bacteria tile and groundwater quality following direct injection of dewatered municipal biosolids into soil," *Journal of Environmental Quality*, vol. 38, no. 3, pp. 1066–1075, 2009.

[27] J. Kovar and G. Pierzynski, *Methods of Phosphorus Analysis for Soils, Sediments, Residuals, and Waters*, Virginia Tech University, Blacksburg, Va, USA, 2nd edition, 2009.

[28] J. Murphy and J. P. Riley, "A modified single solution method for the determination of phosphate in natural waters," *Analytica Chimica Acta*, vol. 27, no. C, pp. 31–36, 1962.

[29] T. W. Andraski and L. G. Bundy, "Relationships between phosphorus levels in soil and in runoff from corn production systems," *Journal of Environmental Quality*, vol. 32, no. 1, pp. 310–316, 2003.

[30] J. Liu, H. Aronsson, L. Bergström, and A. Sharpley, "Phosphorus leaching from loamy sand and clay loam topsoils after application of pig slurry," *SpringerPlus*, vol. 1, no. 1, article 53, pp. 1–10, 2012.

[31] P. P. Motavalli and R. J. Miles, "Soil phosphorus fractions after 111 years of animal manure and fertilizer applications," *Biology and Fertility of Soils*, vol. 36, no. 1, pp. 35–42, 2002.

[32] OMAFRA, *Macro and Secondary Nutrients—Soil Diagnostics*, Edited by Q. S. P. F. Ontario, Soil Diagnostics, Government of Ontario, 2009.

[33] J. Legg, *Topsoil Report Ranges*, Agri-Food Laboratories, Guelph, Canada, 2013.

[34] Z. He, H. Zhang, G. S. Toor et al., "Phosphorus distribution in sequentially extracted fractions of biosolids, poultry litter, and granulated products," *Soil Science*, vol. 175, no. 4, pp. 154–161, 2010.

[35] X.-L. Huang, Y. Chen, and M. Shenker, "Chemical fractionation of phosphorus in stabilized biosolids," *Journal of Environmental Quality*, vol. 37, no. 5, pp. 1949–1958, 2008.

[36] F. R. Cox and S. E. Hendricks, "Soil test phosphorus and clay content effects on runoff water quality," *Journal of Environmental Quality*, vol. 29, no. 5, pp. 1582–1586, 2000.

[37] A. N. Sharpley, W. W. Troeger, and S. J. Smith, "Water quality: the measurement of bioavailable phosphorus in agricultural runoff," *Journal of Environmental Quality*, vol. 20, no. 1, pp. 235–238, 1991.

[38] J. W. White, F. J. Coale, J. T. Sims, and A. L. Shober, "Phosphorus runoff from waste water treatment biosolids and poultry litter applied to agricultural soils," *Journal of Environmental Quality*, vol. 39, no. 1, pp. 314–323, 2010.

[39] V. H. Smith, G. D. Tilman, and J. C. Nekola, "Eutrophication: impacts of excess nutrient inputs on freshwater, marine, and terrestrial ecosystems," *Environmental Pollution*, vol. 100, no. 1–3, pp. 179–196, 1998.

Integrated Crop-Livestock Management Effects on Soil Quality Dynamics in a Semiarid Region: A Typology of Soil Change Over Time

J. Ryschawy,[1] M. A. Liebig,[2] S. L. Kronberg,[2] D. W. Archer,[2] and J. R. Hendrickson[2]

[1]Université de Toulouse, INRA, INP-ENSAT, UMR 1248 AGIR, 31324 Castanet-Tolosan, France
[2]USDA-ARS, Northern Great Plains Research Laboratory, P.O. Box 459, Mandan, ND 58554-0459, USA

Correspondence should be addressed to M. A. Liebig; mark.liebig@ars.usda.gov

Academic Editor: Teodoro M. Miano

Integrated crop-livestock systems can have subtle effects on soil quality over time, particularly in semiarid regions where soil responses to management occur slowly. We tested if analyzing temporal trajectories of soils could detect trends in soil quality data which were not detected using traditional statistical and index approaches. Principal component and cluster analyses were used to assess the evolution in ten soil properties at three sampling times within two production systems (annually cropped, perennial grass). Principal component 1 explained 33% of the total variance of the complete dataset and corresponded to gradients in extractable N, available P, and C:N ratio. Principal component 2 explained 25.4% of the variability and corresponded to gradients of soil pH, soil organic C, and total N. While previous analyses found no differences in Soil Quality Index (SQI) scores between production systems, annually cropped treatments and perennial grasslands were clearly distinguished by cluster analysis. Cluster analysis also identified greater dispersion between plots over time, suggesting an evolution in soil condition in response to management. Accordingly, multivariate statistical techniques serve as a valuable tool for analyzing data where responses to management are subtle or anticipated to occur slowly.

1. Introduction

Integrated crop-livestock systems (ICLS) are recognized globally for their contributions to improve agricultural sustainability [1–4]. Recent emphasis on conservation agriculture, climate-smart agriculture, and sustainable intensification has underscored the potential role of ICLS to create more productive and resilient agricultural systems [5, 6]. An inherent emphasis on multiple enterprises makes ICLS well-suited for future growing conditions, where production synergies between enterprises can serve to enhance adaptability to increasingly variable weather and market conditions, while concurrently minimizing input costs [7, 8].

Integrated crop-livestock systems can be more management and labor intensive than single-enterprise production systems [4]. Accordingly, producers need to know if investments in implementing ICLS translate into improvements in key response metrics. Relevant information would utilize metrics characterizing agroecosystem sustainability and, ideally, do so quickly following adoption of management practices to assess trajectory over time [9]. Such information could serve to justify investments in ICLS or, if necessary, provide quantifiable guidance for adjusting management to more effectively meet production, economic, and/or environmental goals.

Soil quality serves as an important response metric to agroecosystem management given its foundational role in affecting agricultural and environmental outcomes through impacts on ecosystem services [10]. Formally defined, soil quality refers to the capacity of soil to function [11]. Accordingly, soil quality contributes to numerous ecosystem services within agricultural landscapes (e.g., water retention/filtration, climate regulation, and biodiversity conservation) [12].

The central role of soil to deliver key ecosystem services has served as a focal point for the development of numerous assessment methods. Over the past 30 years, soil quality assessments have evolved from the evaluation of a minimum dataset of soil properties [13] to elaborate indices involving scoring functions based on management goals across multiple scales [14–16]. As agricultural practices are intensified to meet anticipated production and energy needs [17], soil quality assessments will be essential to ensure deployed practices are sustainable.

Previous soil quality evaluations of ICLS have focused on treatment effects on single soil physical, chemical, or biological properties or indices of multiple properties [4, 18–21]. Outcomes from these evaluations have helped understand soil responses to ICLS, as results are typically easy to interpret and convey, thereby improving understanding by clientele. Despite this benefit, there are unique attributes associated with ICLS that may make traditional assessment of soil property dynamics, whether by single properties or an index, problematic. Soil properties can change slowly in ICLS, particularly in semiarid regions where soil responses to management occur slowly [22]. Accordingly, long-term research may be necessary to assess ICLS effects on soil quality, yet short-term funding cycles may make it difficult to conduct such research. Secondly, traditional Soil Quality Index methods "merge" outcomes from multiple soil properties to provide an aggregated numerical rating [11, 16]. In doing so, traditional indexing methods may mask nuances inherent to the data and, thus, miss important spatiotemporal trends.

Given this context, we sought to analyze near-surface soil quality indicator data collected over a six-year period from an ICLS in a semiarid region using principal component analysis (PCA). These data were previously analyzed using traditional statistical and index approaches to evaluate soil property responses to residue management, frequency of hoof traffic, season, and production system [18, 23]. We hypothesized a reanalysis of the same data with multivariate statistical techniques would identify trends in soil property dynamics not previously observed.

2. Materials and Methods

2.1. Site and Treatment Description. The ICLS experiment was located at the USDA-ARS Northern Great Plains Research Laboratory near Mandan, ND, USA (46°46′ W, 100°54 N). The experimental site is within the temperate steppe ecoregion of the USA, with a semiarid climate characterized by long, cold winters and short, hot summers. Mean annual precipitation and temperature at the site are 414 mm and 4°C, respectively, and the average frost-free period is 131 days. Gently rolling uplands (0 to 3% slope) characterize the topography of the experimental site, and predominant soil types are Temvik-Wilton silt loams (fine-silty, mixed, superactive, frigid Typic, and Pachic Haplustolls) [24].

A thorough treatment description can be found elsewhere [18]. Briefly, two 6.0 ha crested wheatgrass [*Agropyron desertorum* (Fisch. ex. Link) Schult.] pastures were

sprayed with glyphosate [N-(phosphonomethyl) glycine; 0.7 kg a.i. ha^{-1}] twice in mid-May and converted to an annual cropping sequence of oat/pea (*Avena sativa* L./*Pisum sativum* L.), triticale/sweet clover (*Triticum aestivum x Secale cereale*/*Melilotus officinalis* L.), and corn (*Zea mays* L.). Beginning in 2007, the crop sequence was changed to oat/alfalfa (*Medicago* spp.)/hairy vetch (*Vicia villosa* Roth)/red clover (*Trifolium pratense* L.), brown midrib sorghum-sudangrass (*Sorghum bicolor* L. Moench)/sweet clover/red clover, and corn also using no-till planting techniques. Each phase of the three-year cropping sequence was present in both pastures, which were used as replicates. Management decisions related to seeding, fertilizer, and weed control by herbicides followed recommended practices by area producers.

The oat and triticale/sorghum crop mixtures were harvested for grain from mid-August to early-September with the straw spreader removed from the combine, which created a swath of crop residue for winter grazing. The corn was swathed for forage in mid- to late-September. Each crop strip was split into three residue management treatments that included no residue removal (Control; 0.05 ha), residue removal with a baler (Hayed; 0.05 ha), and residue removal by grazing with livestock (Grazed; 1.69 ha). The Control and Hayed treatments were randomly assigned within each crop strip. For the Grazed treatment, the swathed crop residues from the cropping sequence represented winter forage for ten 4–6-year-old nonlactating bred Hereford cows, due to calve in late March. Plot areas within a crop phase were 27 m × 23 m (0.06 ha) for the Control and Hayed treatments and 54 m × 312 m (1.68 ha) for the Grazed treatment. Stocking rate within a crop phase for the Grazed treatment was 0.2 ha cow^{-1} (5.9 cows ha^{-1}), corresponding to winter grazing practices used by local producers.

Grazing commenced in mid-November and ended in mid-February, with the oat crop mixture grazed first, triticale/sorghum crop mixture second, and corn last. Access to crop swaths was controlled using electric fences oriented at right angles to the swaths. Fences were moved daily to provide access to fresh forage. A shelter and "frost-free" water fountain were located at the end of each pasture within the Grazed treatment.

Two perennial grass pastures were used for comparison to the annually cropped treatments. The pastures, each 6.0 ha, were composed of a mixture of native and introduced cool-season perennial grasses. Similar to Grazed treatments, swathed grass from the perennial pastures was used as winter forage from 1999 through 2002 for the same number of cows managed in the annually cropped treatments. Within each perennial pasture a nongrazed strip was split into Hayed and Control treatments as outlined above. Drought-induced limitations in forage in 2002 and 2003 resulted in the perennial grass treatments being hayed but not grazed in 2003. From 2004 through 2008, the perennial grass treatment was not swathed but lightly grazed with ten Hereford or Angus cows from mid-October to mid-January.

2.2. Soil Sample Collection, Processing, and Analyses. Soil samples were collected in April of 2002, 2005, and 2008 after the swathed crops had been grazed but not replanted. Within

TABLE 1: Description of soil properties used for data analysis.

Variable	Acronym	Units	Minimum value	Maximum value
Soil bulk density	SBD	$Mg\,m^{-3}$	0.80	1.61
Electrical conductivity	EC	$dS\,m^{-1}$	0.10	0.31
Soil pH	PH	$-\log[H^+]$	5.48	7.65
Soil nitrate	NO3N	$mg\,kg^{-1}$	0	44
Soil ammonium	NH4N	$mg\,kg^{-1}$	1	26
Available phosphorus	P	$mg\,kg^{-1}$	1	29
Potentially mineralizable N	PMN	$mg\,kg^{-1}$	19	166
Soil organic carbon	SOC	$g\,kg^{-1}$	13.2	41.9
Total nitrogen	TN	$g\,kg^{-1}$	1.2	3.5
Soil carbon to nitrogen ratio	CNRATIO		10.3	12.5

the annually cropped pastures, samples were collected from nine subplots (three per crop phase) in the Control and Hayed treatments, oriented randomly in each treatment but between crop rows. Samples from the Grazed treatment were collected from two transects differing in frequency of hoof traffic, also between crop rows. Nine subplots (three per crop phase) were established in each transect perpendicular to crop swaths approximately 100 m (representing high-traffic; HT) and 200 m (representing low-traffic; LT) from the shelter and water source. Sampling protocol for the western wheatgrass pastures followed that in the annually cropped pastures, with the exception of fewer subplots in each treatment (three in the Control, Hayed, and Grazed hoof traffic transects). Within each subplot, six soil cores were collected from the 0 to 7.5 cm depth using a 35 mm (i.d.) step-down probe and composited. Each sample was saved in a double-lined plastic bag, placed in cold storage at 5°C, and analyzed within 6 wk of collection.

Soil samples were dried at 35°C for 4 d and then ground by hand to pass a number 10 (2.0 mm) sieve. Identifiable plant material (>2.0 mm diameter, >10 mm length) was removed during sieving. Electrical conductivity and pH were estimated from a 1:1 soil-water mixture [25, 26]. Soil NO_3-N and NH_4-N were determined from 1:10 soil-KCl (2 M) extracts using cadmium reduction followed by a modified Griess-Ilosvay method and indophenol blue reaction [27]. Plant-available soil P was estimated by bicarbonate extraction [28]. Potentially mineralizable N was estimated from the NH_4-N accumulated after a 7 d anaerobic incubation at 40°C [29]. Total soil C and N were determined by dry combustion [30], and as soil pH was <7.2 within the surface 7.5 cm depth, total soil C was considered equivalent to SOC. All data were expressed on an oven-dry basis prior to data analyses.

2.3. Data Organization. Prior to statistical analyses, data were organized to facilitate characterization of all sampled locations as well as sampled locations specific to the annually cropped treatments. Data group 1 (S1) was composed of 10 soil properties from 40 plot locations in both annually cropped and perennial grass treatments (Table 1). Within S1, there were 24 locations in the annually cropped treatments and 16 locations within the perennial grass treatments. Data group 2 (S2) focused on the 24 locations within the annually cropped treatments, with an equivalent number of grazed and

ungrazed locations (12 each). For both S1 and S2 data groups, data were included from each sampling time: 2002, 2005, and 2008.

2.4. Statistical Analyses. Individual plot dynamics were quantified by combining a series of multivariate analyses using a method developed by Dolédec and Chessel [31] and adapted by García-Martínez et al. [32] and Ryschawy et al. [33]. The method distinguished impacts of external (the system environment) and internal (system structure and functioning) factors, while considering the evolution of individual plots over time. We applied this statistical methodology to S1 to analyse plot trajectories of near-surface soil quality indicators (Table 1).

To analyse plot trajectories, we used a particular type of PCA (Principal Component Analysis) called Within-Class Analysis (WCA) to consider evolution of individuals (plots) while not considering the general date effect on different plots. We thus considered the deviation of each soil property in a plot from the average across plots, considering time period effect. The first step of the WCA consisted in standardizing the data. For each soil property (column), data were centered by subtracting the mean of the column to the values and scaled by dividing the values of each column by their standard deviation (with a column mean of 0 and a standard deviation of 1). Within-Class Analysis used an R statistical software package dedicated to Analysis of Environmental Data (Exploratory and Euclidean Method; ade4 package) [34]. Within-Class Analysis used orthogonal PCAs where factors were used as covariables to partition rows (time was used as the main factor to partition rows considering each date separately for plots). The analysis characterized differences between plot locations once a general common trend over time was eliminated. We considered the three sampling times (2002, 2005, and 2008) as the factor partitioning rows.

Main factors within data group S1 were summarized by PCA. We then performed a WCA resulting in a data frame W1 containing transformed data for each of the 40 plot locations at the three different dates considered, as the deviation between individual and date-class mean coordinates on within-analysis axes. The results of the WCA therefore represented interplot location trajectories.

TABLE 2: Principal component scores of measured variables and proportion of variability explained within each axis for data group 1 (S1).

	Axis 1	Axis 2	Axis 3	Axis 4
Soil bulk density	0.59	0.42	0.27	0.17
Electrical conductivity	−0.23	−0.32	0.62	0.53
Soil pH	−0.34	0.74	0.42	0.08
Soil nitrate	0.71	−0.53	0.16	0.30
Soil ammonium	0.56	−0.15	0.50	0.54
Available P	0.65	−0.30	0.33	−0.41
Potentially mineralizable N	−0.41	0.48	0.59	0.20
Soil organic carbon	−0.58	−0.65	0.46	−0.04
Total nitrogen	−0.45	−0.70	−0.50	−0.09
C : N ratio	−0.78	0.12	−0.04	0.20
Variability explained	*33.0%*	*25.4.%*	*13.6%*	*10.3%*

To summarize soil trajectories for each individual plot location for all dates, we performed a final PCA on the WCA individuals considered at each date. Per the Kaiser criterion, we thus considered the main factors with eigenvalues >1 as summarizing observed changes. A Hierarchical Cluster Analysis (HCA) (with squared Euclidean distance and Ward's aggregation method) was carried out on the four main factors of the PCA to establish a typology of plot locations according to their change over time.

To analyse changes in plot locations elucidated by HCA, we calculated means and standard deviation of variables for each measured variable. All statistical analyses were performed using R 3.1.0 software [34]. We repeated this statistical methodology on data group S2 to evaluate soil property dynamics based on grazed (Grazed) and ungrazed (Control, Hayed) areas within the 24 plot locations assigned to annually cropped treatments.

3. Results

3.1. Analysis of All Sampled Locations. The first four principal components accounted for 82.3% of the variability based on WCA of the whole sample (data group S1). Associations between soil properties were evident following the WCA, as three distinct groupings were observed graphically (Figure 1). Soil nitrate (NO3N), ammonium (NH4N), and available phosphorus (P) appeared linked, as did soil carbon to nitrogen ratio (CNRATIO), soil organic C (SOC), total nitrogen (TN), and electrical conductivity (EC). Soil pH (PH) and potentially mineralizable N were also linked. Only soil bulk density (SBD) was not grouped with another soil property.

Contributions of measured soil properties to the four principal components are provided in Table 2. Principal component 1 explained 33% of the variability and corresponded to gradients of available N and P (NO3N, P) and CN ratio levels. Principal component 2 explained 25.4% of the variability and corresponded to gradients of PH, SOC, and TN. Principal component 3 explained 13.6% of the variability and corresponded gradients in EC and PMN. Principal component 4 explained 10.3% of the variability and was primarily influenced by EC and NH4N.

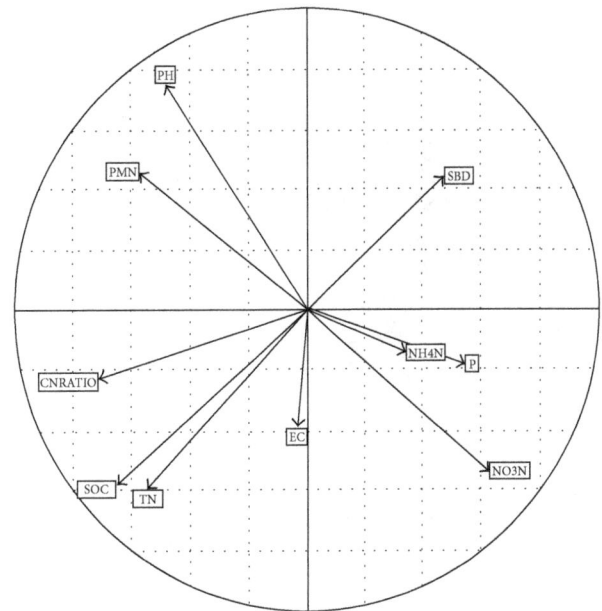

FIGURE 1: Spatial representation of measured variables on the first factorial map of the principal component analysis (PCA): SBD, soil bulk density; EC, electrical conductivity; PH, soil pH; NO3N, soil nitrate; NH4N, soil ammonium; P, available P; PMN, potentially mineralizable N; SOC, soil organic carbon; TN, total nitrogen; CNRATIO, soil carbon to nitrogen ratio.

Dispersion between plots showed a trend toward greater dissimilarity in soil properties over time (Figure 2). Based on standard deviation of mean values, plot differences were closer in composition in 2002 and 2005 than in 2008, with the latter showing substantial dispersion at the end of the 6 yr study period (Table 3).

3.2. Analysis of Production Systems. Cluster analysis found distinct groupings based on production system, as annually cropped treatments and perennial grasslands were clearly distinguished graphically (Figure 3). Such trends were evident by comparing soil properties in each treatment between 2002 and 2008 (Table 4). Plots under annual cropping had a greater

TABLE 3: Mean values (±1 standard deviation) for soil properties in 2002, 2005, and 2008.

Variable	2002	2005	2008
Soil bulk density ($Mg\,m^{-3}$)	1.05 ± 0.08	1.09 ± 0.10	1.12 ± 0.11
Electrical conductivity ($dS\,m^{-1}$)	0.24 ± 0.04	0.30 ± 0.06	0.41 ± 0.10
Soil pH ($-\log[H^+]$)	6.12 ± 0.38	6.23 ± 0.38	6.32 ± 0.55
Soil nitrate ($mg\,kg^{-1}$)	4 ± 3	5 ± 4	8 ± 9
Soil ammonium ($mg\,kg^{-1}$)	2 ± 1	2 ± 0	5 ± 3
Available phosphorus ($mg\,kg^{-1}$)	8 ± 3	5 ± 4	10 ± 6
Potentially mineralizable N ($mg\,kg^{-1}$)	74 ± 22	86 ± 24	86 ± 34
Soil organic carbon ($g\,kg^{-1}$)	30.1 ± 3.9	26.6 ± 3.6	29.6 ± 5.7
Total nitrogen ($g\,kg^{-1}$)	2.7 ± 0.3	2.3 ± 0.3	2.7 ± 0.5
C : N ratio	11.2 ± 0.4	11.4 ± 0.4	11.1 ± 0.4

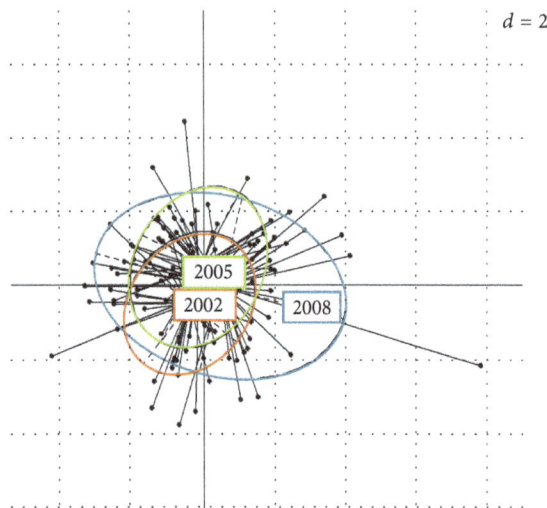

FIGURE 2: Spatial representation of three groups of measured variables based on sampling year (2002, 2005, and 2008).

accumulation of mineral nutrients (NO3N, NH4N, and P) compared to perennial grassland, whereas perennial grasslands had greater accumulation of organic matter (SOC, TN), increased mineralizable N (PMN), and a lower dispersion between plots compared to annually cropped treatments.

3.3. Analysis of Grazing Effects under Annual Cropping. In the analysis of grazing effects (data group S2), 83.4% of the variability was explained based on WCA. Principal component 1 explained 29.8% of the variability and corresponded to gradients in total carbon and nitrogen accrual rates (SOC, TN, and CNRATIO), whereas principal component 2 explained 24.7% of the variability and corresponded to gradients of mineral nutrient accrual rates (NO3N, NH4N, and P) (data not shown).

Hierarchical cluster analysis revealed four clusters that were differentiated by grazing and treatment replication (Figure 4). Treatment replications were distinguished graphically along the horizontal plane, with one replication (Field A) left of the vertical plane and the other replication (Field B) right

of the vertical plane. Differences in soil properties between treatments were limited to Field B, where soil pH was lower and available P was higher under grazing (Table 5).

4. Discussion

Traditional statistical and index-based approaches for assessing soil quality may provide a limited characterization of soil property dynamics [35]. Multivariate statistical analyses such as PCA provide a useful method for screening a diverse collection of functionally relevant soil properties to identify potential data trends [36]. These techniques provide an objective approach to extract and weight information in complex datasets [37], thereby providing a potentially valuable tool for researchers seeking to link the status of soil properties, soil function, and agroecosystem management.

Analysis of whole sample dynamics reflected the predominant influence of soil chemical properties on total variance in the dataset. Soil properties most influenced by additions of N and P through fertilization (e.g., NO3N and P) accounted for substantial variance between plots. Annual change in nutrient-related properties would be expected to vary based on weather-driven factors affecting plant growth and nutrient loss [38]. Electrical conductivity, a relative measure of the total quantity of ions in soil solution, also accounted for an appreciable amount of variance in the dataset.

Previous PCA evaluations characterizing soil quality have found organic matter-related properties to account for a large amount of variance in data [35, 36]. A similar outcome was observed in this evaluation, as SOC, TN, and CNRATIO accounted for a moderate amount of variation among soil properties across principal components. The limited measurement timespan (6 yr), coupled with the inherent difficulty in detecting change in properties with a large background signal due to previous management as a perennial grassland [39], may have compromised the usefulness of organic matter-related properties in this evaluation. In general, soil properties associated with organic matter are slow to change in semiarid agroecosystems, sometimes requiring a decade or longer for management effects to be detected [22].

Despite limited change in some soil properties in this evaluation, the aggregate visual assessment of data dispersion

TABLE 4: Mean values of near-surface soil properties within annually cropped and perennial grassland integrated crop-livestock systems in 2002, 2005, and 2008 [18].

Year/change	Annual crop	Perennial grassland
	Soil bulk density ($Mg\,m^{-3}$)	
2002	1.07 ± 0.07^{a}	1.02 ± 0.09
2005	1.07 ± 0.12	1.11 ± 0.04
2008	1.16 ± 0.12	1.07 ± 0.06
ΔSBD^{b}	*+0.09*	*+0.05*
	Electrical conductivity ($dS\,m^{-1}$)	
2002	0.24 ± 0.03	0.25 ± 0.04
2005	0.30 ± 0.06	0.29 ± 0.05
2008	0.43 ± 0.12	0.38 ± 0.07
ΔEC	*+0.19*	*+0.13*
	Soil pH ($-\log[H^{+}]$)	
2002	5.89 ± 0.21	6.46 ± 0.34
2005	6.00 ± 0.33	6.58 ± 0.27
2008	5.95 ± 0.33	6.88 ± 0.24
ΔPH	*+0.05*	*+0.43*
	Soil nitrate ($mg\,kg^{-1}$)	
2002	6 ± 3	1 ± 1
2005	7 ± 3	1 ± 1
2008	13 ± 8	1 ± 1
$\Delta NO3N$	*+6*	*0*
	Soil ammonium ($mg\,kg^{-1}$)	
2002	2 ± 1	2 ± 1
2005	2 ± 1	3 ± 1
2008	6 ± 4	4 ± 1
$\Delta NH4N$	*+4*	*+2*
	Available P ($mg\,kg^{-1}$)	
2002	7 ± 2	9 ± 4
2005	8 ± 4	2 ± 2
2008	14 ± 5	5 ± 1
ΔP	*+7*	*−4*
	Potentially mineralizable N ($mg\,kg^{-1}$)	
2002	62 ± 14	92 ± 20
2005	70 ± 14	109 ± 12
2008	66 ± 24	117 ± 21
ΔPMN	*+4*	*+25*
	Soil organic C ($g\,kg^{-1}$)	
2002	29.5 ± 3.8	30.9 ± 4.2
2005	27.1 ± 3.5	26.0 ± 3.8
2008	28.1 ± 6.0	31.9 ± 4.4
ΔSOC	*−1.4*	*+1.0*
	Total N ($g\,kg^{-1}$)	
2002	2.6 ± 0.3	2.7 ± 0.3
2005	2.4 ± 0.3	2.2 ± 0.3
2008	2.6 ± 0.5	2.8 ± 0.3
ΔTN	*0.0*	*+0.1*
	C : N ratio	
2002	11.2 ± 0.4	11.4 ± 0.5
2005	11.3 ± 0.4	11.7 ± 0.4
2008	10.9 ± 0.3	11.3 ± 0.3
ΔCN	*−0.3*	*−0.1*

[a]Mean ± standard deviations are given for each variable in 2002, 2005, and 2008. [b]Change reflects the difference in soil property between 2008 and 2002.

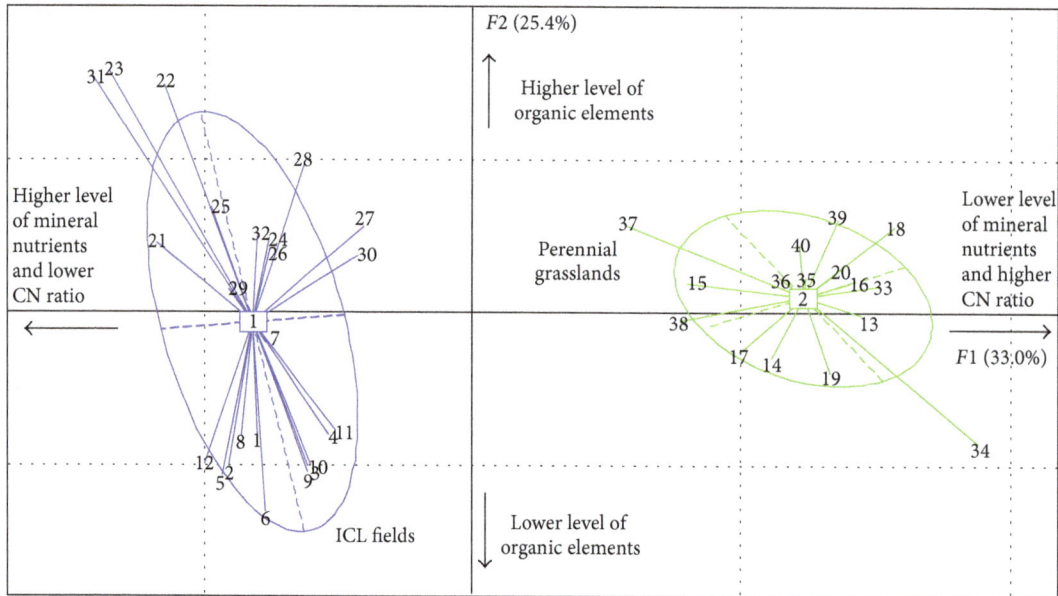

FIGURE 3: Spatial representation of two groups of soil trajectory types following principal component analysis of the complete dataset.

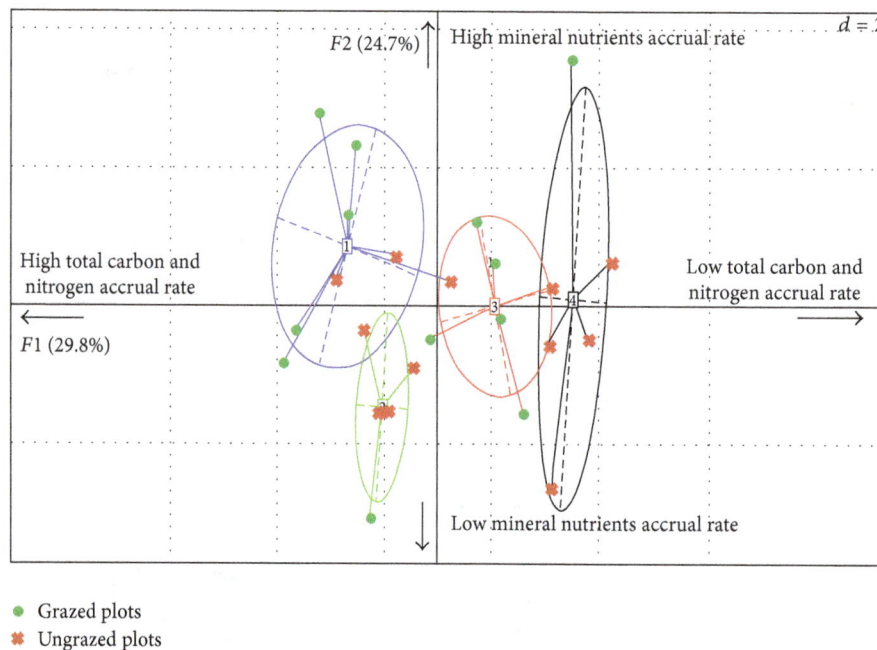

- ● Grazed plots
- ✴ Ungrazed plots

FIGURE 4: Spatial representation of four groups of soil trajectory types following principal component analysis of annually cropped treatments.

proved useful in detecting a general evolution of the dataset. Such dispersion was confirmed upon inspection of standard deviation values for soil properties within each year, as mean values of all soil properties, except CNRATIO, had the greatest standard deviation in the last year of the evaluation. Coupled graphical and tabular data characterizations seem particularly useful for datasets like the one analyzed in this evaluation, as visual assessments infer possible data trends generally, while tabular assessments provide confirmation specifically.

Outcomes from the cluster analysis of the complete dataset showed data groupings based on production system, with perennial grass and annually cropped treatments separated visually. Such separation was evident with cluster analysis despite limited observed differences in mean values of soil properties between production systems (Table 4) [18]. Moreover, multivariate analysis and graphical representation of these previously analyzed data clearly showed limitations associated with the aggregated Soil Quality Index (SQI) approach to assessing soil quality, as differences in the overall

TABLE 5: Mean values of near-surface soil properties within annually cropped fields according to grazing treatment.

Variable	Field A			Field B		
	Grazed	Ungrazed	Pvalue	Grazed	Ungrazed	Pvalue
Soil bulk density (Mg m^{-3})	1.03 ± 0.11[a]	1.04 ± 0.05	0.68[b]	1.18 ± 0.12	1.15 ± 0.06	0.46
Electrical conductivity (dS m^{-1})	0.22 ± 0.05	0.24 ± 0.04	0.26	0.20 ± 0.04	0.22 ± 0.05	0.40
Soil pH (−log[H$^+$])	5.76 ±0.16	5.78 ± 0.15	0.70	5.99 ± 0.13	6.27 ± 0.24	<0.01
Soil nitrate (mg kg^{-1})	9 ± 4	9 ± 4	0.64	10 ± 10	6 ± 4	0.13
Soil ammonium (mg kg^{-1})	3 ± 1	3 ± 2	0.67	4 ± 6	3 ± 1	0.31
Available phosphorus (mg kg^{-1})	9 ± 4	7 ± 4	0.29	12 ± 5	9 ± 4	0.04
Potentially mineralizable N (mg kg^{-1})	61 ± 16	62 ± 13	0.83	75 ± 20	65 ± 20	0.16
Soil organic carbon (g kg^{-1})	30.4 ± 5.5	31.2 ± 2.0	0.54	26.5 ± 3.1	24.9 ± 4.0	0.18
Total nitrogen (g kg^{-1})	2.7 ± 0.5	2.8 ± 0.2	0.34	2.4 ± 0.3	2.3 ± 0.3	0.15
C : N ratio	11.4 ± 0.5	11.2 ± 0.3	0.37	10.9 ± 0.2	10.9 ± 0.3	0.99

[a]Mean ± 1 standard deviation; [b]P value associated with mean comparison within a field between grazed and ungrazed treatments. Field A corresponds to clusters 1 and 2, while field B corresponds to clusters 3 and 4 (Figure 4).

SQI between production systems were not observed in any year [18]. From a soil management perspective, outcomes from the cluster analysis underscored the value of perennial grass systems to enhance soil quality through increased accrual of soil carbon and nitrogen, while minimizing levels of nutrients susceptible to loss. Such attributes are important features of sustainable agricultural systems [40] and are central to agroecosystem resilience to climate-induced stressors and improvements in environmental quality from agricultural production [41, 42].

In contrast to whole production system analyses, outcomes were less distinct when data were restricted to annually cropped fields with and without grazing. Cluster analysis revealed an overarching effect of replication on data groupings, of which grazed and ungrazed plots lacked homogeneity. It is possible that the limited number of data points, coupled with a need for greater time-in-treatment to resolve management effects on soil properties, resulted in indeterminate responses to PCA. It is important to note that no differences in soil properties were found between grazed and ungrazed systems in any year when the same data were analyzed using traditional statistical analyses [18]. Future evaluations of grazed and ungrazed treatments in both annually cropped and perennial grassland systems are warranted given the important role of livestock to affect biomass productivity through influences on soil condition [43].

Collectively, statistical methods used in this evaluation contribute to an ongoing evolution of soil quality assessment. As the interpretation of individual soil properties can be complicated by multiple response functions [16], use of aggregate approaches, such as PCA and HCA, potentially offer novel insights into how soil responds to management [14]. Use of statistical approaches capable of resolving changes in soil condition in the near-term can be particularly useful in regions where alterations in individual soil properties require a decade or more to be detected [22]. Similarly, these approaches may be valuable in detecting subtle changes in soil induced by systemic drivers (i.e., climate). The capacity to detect shifts in soil condition quickly under the context of anticipated climate change will be increasingly valuable, particularly for dryland agriculture [44]. Timely management interventions will be essential in the future to mitigate soil degradation and maintain soil function.

5. Conclusions

Use of multivariate statistical analyses revealed trends in data not previously detected using traditional statistical and index approaches for assessment of soil quality. Principal component and hierarchical cluster analyses provided a helpful means to visually discriminate between annually cropped and perennial grass treatments, while concurrently identifying a clear trend toward greater dispersion in the data over time. Based on the typology used in this evaluation, apparent multivariate statistical techniques represent a valuable tool for analysis of data collected at different time periods where changes in response variables are anticipated to occur slowly.

Application of these techniques to datasets in other semiarid regions may contribute toward identifying trajectories in soil responses to applied treatments in the near-term. Such information could justify research expenditures, and therefore continuation of experimental treatments until soil responses can be verified with traditional statistical approaches and/or indices. Conversely, these techniques could be used to inform possible changes in applied treatments if visual discrimination is not detected.

Abbreviations

CNRATIO: Soil carbon to nitrogen ratio
EC: Electrical conductivity
ICLS: Integrated crop-livestock systems
NH4N: Soil ammonium
NO3N: Soil nitrate
P: Available P
PCA: Principal component analysis
PH: Soil pH
PMN: Potentially mineralizable N
SBD: Soil bulk density

SOC: Soil organic carbon
TN: Total nitrogen
WCA: Within-Class Analysis.

Disclosure

The US Department of Agriculture (USDA) prohibits discrimination in all its programs and activities on the basis of race, color, national origin, age, disability, and, where applicable, sex, marital status, family status, parental status, religion, sexual orientation, genetic information, political beliefs, and reprisal, or because all or part of an individual's income is derived from any public assistance program (not all prohibited bases apply to all programs.). USDA is an equal opportunity provider and employer. Mention of commercial products and organizations in this manuscript is solely to provide specific information. It does not constitute endorsement by USDA-ARS over other products and organizations not mentioned.

Acknowledgments

The authors gratefully acknowledge the contributions of Justin Hartel, Marvin Hatzenbuhler, and Robert Kolberg for planting and swathing crops, Clay Erickson, Gordy Jensen, and Curt Klein for managing cattle, and John Bullinger, Jason Gross, and Holly Johnson for assisting with soil sampling, processing, and analyses.

References

[1] G. Lemaire, A. Franzluebbers, P. C. D. F. Carvalho, and B. Dedieu, "Integrated crop-livestock systems: Strategies to achieve synergy between agricultural production and environmental quality," *Agriculture, Ecosystems and Environment*, vol. 190, pp. 4–8, 2014.

[2] J. Ryschawy, A. Joannon, J. P. Choisis, A. Gibon, and P. Y. Le Gal, "Participative assessment of innovative technical scenarios for enhancing sustainability of French mixed crop-livestock farms," *Agricultural Systems*, vol. 129, pp. 1–8, 2014.

[3] J. R. Hendrickson, J. D. Hanson, D. L. Tanaka, and G. Sassenrath, "Principles of integrated agricultural systems: Introduction to processes and definition," *Renewable Agriculture and Food Systems*, vol. 23, no. 4, pp. 265–271, 2008.

[4] M. P. Russelle, M. H. Entz, and A. J. Franzluebbers, "Reconsidering integrated crop-livestock systems in North America," *Agronomy Journal*, vol. 99, no. 2, pp. 325–334, 2007.

[5] K. L. Steenwerth, A. K. Hodson, A. J. Bloom et al., "Climate-smart agriculture global research agenda: scientific basis for action," *Agriculture & Food Security*, vol. 3, no. 11, pp. 1–39, 2014.

[6] J. A. Foley, N. Ramankutty, K. A. Brauman et al., "Solutions for a cultivated planet," *Nature*, vol. 478, no. 7369, pp. 337–342, 2011.

[7] R. J. Wilkins, "Eco-efficient approaches to land management: A case for increased integration of crop and animal production systems," *Philosophical Transactions of the Royal Society B: Biological Sciences*, vol. 363, no. 1491, pp. 517–525, 2008.

[8] J. D. Hanson, M. A. Liebig, S. D. Merrill, D. L. Tanaka, J. M. Krupinsky, and D. E. Stott, "Dynamic cropping systems: Increasing adaptability amid an uncertain future," *Agronomy Journal*, vol. 99, no. 4, pp. 939–943, 2007.

[9] C. Bockstaller, L. Guichard, O. Keichinger, P. Girardin, M.-B. Galan, and G. Gaillard, "Comparison of methods to assess the sustainability of agricultural systems. A review," *Agronomy for Sustainable Development*, vol. 29, no. 1, pp. 223–235, 2009.

[10] D. A. Robinson, N. Hockley, E. Dominati et al., "Natural capital, ecosystem services, and soil change: why soil science must embrace an ecosystems approach," *Vadose Zone Journal*, vol. 11, no. 1, 2012.

[11] D. L. Karlen, M. J. Mausbach, J. W. Doran, R. G. Cline, R. F. Harris, and G. E. Schuman, "Soil quality: a concept, definition, and framework for evaluation," *Soil Science Society of America Journal*, vol. 61, no. 1, pp. 4–10, 1997.

[12] K. R. Olson, M. Al-Kaisi, R. Lal, and L. W. Morton, "Soil ecosystem services and intensified cropping systems," *Journal of Soil and Water Conservation*, vol. 72, no. 3, pp. 64A–69A, 2017.

[13] W. E. Larson and F. J. Pierce, "Conservation and enhancement of soil quality," in *Evaluation for Sustainable Land Management in the Developing World*, vol. 2 of *IBSRAM Proc. 12(2)*, pp. 175–203, Int. Board for Soil Res. and Management, Bangkok, Thailand, 1991.

[14] G. Renzi, L. Canfora, L. Salvati, and A. Benedetti, "Validation of the soil Biological Fertility Index (BFI) using a multidimensional statistical approach: a country-scale exercise," *CATENA*, vol. 149, pp. 294–299, 2017.

[15] A. Ferrara, L. Salvati, A. Sabbi, and A. Colantoni, "Soil resources, land cover changes and rural areas: Towards a spatial mismatch?" *Science of the Total Environment*, vol. 478, pp. 116–122, 2014.

[16] S. S. Andrews, D. L. Karlen, and C. A. Cambardella, "The soil management assessment framework: a quantitative soil quality evaluation method," *Soil Science Society of America Journal*, vol. 68, no. 6, pp. 1945–1962, 2004.

[17] J. L. Hatfield and C. L. Walthall, "Meeting global food needs: Realizing the potential via genetics × environment × management interactions," *Agronomy Journal*, vol. 107, no. 4, pp. 1215–1226, 2015.

[18] M. A. Liebig, D. L. Tanaka, S. L. Kronberg, E. J. Scholljegerdes, and J. F. Karn, "Integrated crops and livestock in central North Dakota, USA: Agroecosystem management to buffer soil change," *Renewable Agriculture and Food Systems*, vol. 27, no. 2, pp. 115–124, 2012.

[19] V. Acosta-Martínez, C. W. Bell, B. E. L. Morris, J. Zak, and V. G. Allen, "Long-term soil microbial community and enzyme activity responses to an integrated cropping-livestock system in a semi-arid region," *Agriculture, Ecosystems and Environment*, vol. 137, no. 3-4, pp. 231–240, 2010.

[20] M. W. Maughan, J. P. C. Flores, I. Anghinoni, G. Bollero, F. G. Fernández, and B. F. Tracy, "Soil quality and corn yield under crop-livestock integration in Illinois," *Agronomy Journal*, vol. 101, no. 6, pp. 1503–1510, 2009.

[21] A. J. Franzluebbers and J. A. Stuedemann, "Early response of soil organic fractions to tillage and integrated crop-livestock production," *Soil Science Society of America Journal*, vol. 72, no. 3, pp. 613–625, 2008.

[22] M. M. Mikha, M. F. Vigil, M. A. Liebig et al., "Cropping system influences on soil chemical properties and soil quality in the Great Plains," *Renewable Agriculture and Food Systems*, vol. 21, no. 1, pp. 26–35, 2006.

[23] M. A. Liebig, D. L. Tanaka, S. L. Kronberg, E. J. Scholljegerdes, and J. F. Karn, "Soil hydrological attributes of an integrated crop-livestock agroecosystem: increased adaptation through resistance to soil change," *Applied and Environmental Soil Science*, vol. 2011, Article ID 464827, 6 pages, 2011.

[24] USDA, "Natural Resources Conservation Service, United States Department of Agriculture, Web Soil Survey," 2017, http://web-soilsurvey.nrcs.usda.gov.

[25] M. E. Watson and J. R. Brown, "pH and lime requirement," in *Recommended Chemical Soil Test Procedures for the North Central Region*, J. R. Brown, Ed., NCR Publ. 221 (revised), pp. 13–16, Missouri Agric. Exp. Stn. Bull. SB1001, Columbia, Mo, USA, 1998.

[26] D. A. Whitney, "Soil salinity," in *Recommended Chemical Soil Test Procedures for the North Central Region*, J. R. Brown, Ed., NCR Publ. 221 (revised), pp. 59-60, Missouri Agric. Exp. Stn. Bull. SB1001, Columbia, Mo, USA, 1998.

[27] R. L. Mulvaney, "Nitrogen—inorganic forms," in *Methods of Soil Analysis. Part 3. Chemical Methods*, D. L. Sparks, Ed., SSSA Book Series no. 5, pp. 1123–1184, Soil Science Society of America and American Society of Agronomy, Madison, Wis, USA, 1996.

[28] S. Kuo, "Phosphorus," in *Methods of Soil Analysis. Part 3. Chemical Methods*, D. L. Sparks, Ed., SSSA Book Series no. 5, pp. 869–919, Soil Science Society of America and American Society of Agronomy, Madison, wis, USA, 1996.

[29] L. G. Bundy and J. J. Meisinger, "Nitrogen availability indices," in *Methods of Soil Analysis. Part 2—Microbiological and Biochemical Properties*, R. W. Weaver, S. Angle, and P. Bottomley, Eds., SSSA Book Series no. 5, pp. 951–984, Soil Science Society of America, Madison, Wis, USA, 1994.

[30] D. W. Nelson and L. E. Sommers, "Total carbon, organic carbon, and organic matter," in *Methods of Soil Analysis. Part 3. Chemical Methods*, D. L. Sparks, Ed., SSSA Book Series no. 5, pp. 961–1010, Soil Science Society of America and American Society of Agronomy, Madison, Wis, USA, 1996.

[31] S. Dolédec and D. Chessel, "Rythmes saisonniers et composantes stationnelles en milieu aquatique. I- Description d'un plan d'observations complet par projection de variables," *Acta Oecologica, Oecologia Generalis*, vol. 8, no. 3, pp. 403–426, 1987.

[32] A. García-Martínez, A. Olaizola, and A. Bernués, "Trajectories of evolution and drivers of change in European mountain cattle farming systems," *Animal*, vol. 3, no. 1, pp. 152–165, 2009.

[33] J. Ryschawy, N. Choisis, J. P. Choisis, and A. Gibon, "Paths to last in mixed crop-livestock farming: Lessons from an assessment of farm trajectories of change," *Animal*, vol. 7, no. 4, pp. 673–681, 2013.

[34] R Development Core Team, *R: A Language and Environment for Statistical Computing*, R Foundation for Statistical Computing, Vienna, Austria, 2011, http://www.R-project.org.

[35] M. M. Wander and G. A. Bollero, "Soil quality assessment of tillage impacts in Illinois," *Soil Science Society of America Journal*, vol. 63, no. 4, pp. 961–971, 1999.

[36] S. S. Andrews, D. L. Karlen, and J. P. Mitchell, "A comparison of soil quality indexing methods for vegetable production systems in Northern California," *Agriculture, Ecosystems and Environment*, vol. 90, no. 1, pp. 25–45, 2002.

[37] S. A. Mulaik, "Objectivity and Multivariate Statistics," *Multivariate Behavioral Research*, vol. 28, no. 2, pp. 171–203, 1993.

[38] R. F. Daubenmire, *Plants and Environment*, John Wiley & Sons, New York, NY, USA, 1974.

[39] M. A. Liebig, J. R. Hendrickson, J. D. Berdahl, and J. F. Karn, "Soil resistance under grazed intermediate wheatgrass," *Canadian Journal of Soil Science*, vol. 88, no. 5, pp. 833–836, 2008.

[40] A. J. Franzluebbers, J. Sawchik, and M. A. Taboada, "Agronomic and environmental impacts of pasture-crop rotations in temperate North and South America," *Agriculture, Ecosystems and Environment*, vol. 190, pp. 18–26, 2014.

[41] L. W. Bell, R. C. Hayes, K. G. Pembleton, and C. M. Waters, "Opportunities and challenges in Australian grasslands: Pathways to achieve future sustainability and productivity imperatives," *Crop and Pasture Science*, vol. 65, no. 6, pp. 489–507, 2014.

[42] R. Lal, J. A. Delgado, P. M. Groffman, N. Millar, C. Dell, and A. Rotz, "Management to mitigate and adapt to climate change," *Journal of Soil and Water Conservation*, vol. 66, no. 4, pp. 276–282, 2011.

[43] R. M. Sulc and A. J. Franzluebbers, "Exploring integrated crop-livestock systems in different ecoregions of the United States," *European Journal of Agronomy*, vol. 57, pp. 21–30, 2014.

[44] J. Huang, H. Yu, X. Guan, G. Wang, and R. Guo, "Accelerated dryland expansion under climate change," *Nature Climate Change*, vol. 6, no. 2, pp. 166–171, 2016.

7

Spatial Variability and Relationship of Mangrove Soil Organic Matter to Organic Carbon

Pasicha Chaikaew[1] and Suchana Chavanich[2]

[1]Department of Environmental Science, Faculty of Science, Chulalongkorn University, Bangkok, Thailand
[2]Department of Marine Science, Faculty of Science, Chulalongkorn University, Bangkok, Thailand

Correspondence should be addressed to Pasicha Chaikaew; pasicha.c@chula.ac.th

Academic Editor: Teodoro M. Miano

Degradation and destruction of mangrove forests in many regions have resulted in the alteration of carbon cycling. Objectives of this study were established to answer the question regarding how much soil organic carbon (SOC) is stored in wetland soils in part of the upper northeastern Gulf of Thailand and to what extent SOC is related to organic matter (OM). A total of 29 soil samples were collected in October 2015. Soil physiochemical analyses followed the standard protocol. Spatial distributions were estimated by a kriging method. Linear regression and coefficient were used to determine the suitable conversion factor for mangrove soils. The results showed that surface soil (0–5 cm) contained higher SOC content as compared to subsurface soil (5–10 cm). Considering a depth of 10 cm, this area had a high potential to sequester carbon with a mean ± standard deviation of 5.59 ± 2.24%. The spatial variability of OM and SOC revealed that organic matter and carbon decreased with the distance from upstream areas toward the gulf. Based on the assumption that OM is 50% SOC, the conversion factor of 2 is recommended for more accuracy rather than the conventional factor of 1.724.

1. Introduction

Since the Industrial Revolution, there has been a speedy increase in atmospheric carbon dioxide (CO_2) concentration. Global levels of CO_2 already passed above 400 ppm mark in 2015 [1]. The natural movement of carbon across the atmosphere, vegetation, soils, and the oceans is the key to mitigate climate change due to elevated CO_2. Globally, soils store more carbon, about 3.3 times, than the atmosphere and 4.5 times the bionic pool [2]. Among different ecosystems, wetlands often represent the largest carbon pool because of their anoxic conditions and thus play a vital role in carbon cycles. Of the 1,500 Pg C stored in the Earth's soils, peat-forming wetlands are estimated to contain about 300–600 Pg C [3]. Southeast Asian peatlands account for the most important carbon sink, representing 68.5 Pg C or 77% of global tropical peatlands [4]. Despite accounting for the importance of wetlands to ecosystems and humans, one-third of global wetlands have been lost due to land use change, aquaculture inversion, and wetland degradation. Carbon emissions from mangrove loss are uncertain, but it is estimated that about 10% of all carbon emissions are released by deforestation worldwide [5]. Thailand alone is estimated to have lost approximately 82% of its wetlands [6]. Chonburi Province, Thailand, also has experienced wetland loss as a result of urban sprawl, industrial development, and deforestation. Between 1961 and 2007, its wetland coverage was reduced from 38.2 km^2 to 7.8 km^2, or by about 80% [7]. Wetland conversion directly affects the size of the soil carbon pool.

Soil organic carbon (SOC) is part of the carbon stored in soil organic matter (OM). The estimation of SOC has followed various techniques from the oldest to the simplest method. The assumption that OM contains 58% carbon was first established by Sprengel in 1826 [8]. Since then, the conversion factor of 1.724 has been repeatedly used as a rough estimate of organic carbon content for universal applications. Other early studies published in German scientific literature assumed that OM holds 58% carbon (conversion factor of 1.724) [9, 10] and 60% carbon (conversion factor of 1.667) [11, 12]. However, these numbers are considered too low in many areas. The application of 1.9 and 2.5 was recommended to convert OM to SOC for soils in the top and subsurface layers, respectively, worldwide [13]. Soil type and environment result in SOC-to-OM variations. In wetland ecosystems, a factor of 1.842 has been determined for most peats [14], 1.88 (advocated 2) for peat in Wales [15], and 2.07 and 2.34 for wetlands in England [16, 17]. A recent study has claimed that a factor of 2, based on the assumption that OM is 50% carbon, is more accurate in almost all cases than the conventional factor of 1.724 [18].

Soil management can manipulate the balance of carbon sink and source processes. When OM is broken down, some of the SOC can be quickly mineralized and converted to CO_2. Carbon thus disappears from the soil. The high amount of soil organic matter tends to be limited to the soil surface, probably at a depth of 5 to 10 cm [19]. Organic matter often binds to fine particles, particularly clay, and the binding processes further prevent soil carbon loss [20]. The study of SOC and its relationship to OM is important; yet there are research gaps to be filled regarding mangrove soils in Thailand. It is still questionable how much SOC is stored in the wetland soils and to what extent SOC is related to OM. The aim of this study is to spatially investigate the SOC content in mangrove areas at surface and subsurface layers, as well as the association of SOC-to-OM conversion. The result of the conversion factor from this study is useful for estimating either SOC or OM in similar environments. The specific objectives of this study are threefold:

(1) Investigating the concentrations of SOC in surface soil (0–5 cm) and subsurface soil (5–10 cm).

(2) Assessing spatially explicit OM and SOC contents (0–10 cm).

(3) Identifying the conversional function between OM and SOC in mangrove soils (0–10 cm).

2. Materials and Methods

2.1. Site Description. Estuarine mangrove forests along the coast of the Gulf of Thailand are influenced by the open-coastal environment during high tide and fresh water flow during low tide. The study area is part of the Nature Education Center for Mangrove Conservation and Ecotourism located in Chonburi Province (13°20′37.05″N, 100°56′34.83″E) on the eastern coast of the Gulf of Thailand. The center was established in 2001 for mangrove conservation and education purposes. The conservation area covers approximately 480,000 m^2, but the focus area of interest is limited to the area within and around the wooden walkway due to the constraints of area accessibility. The area within the walkway covers 113,240 m^2 and extends to 220,000 m^2 for the study area. The water channel receives runoff from residential communities and industries upstream and then drains through the mangrove conservation area before being discharged into the Gulf of Thailand. Chonburi has a tropical savanna climate [21] and temperatures rise in April, with the average daily maximum at 35.2°C (95.4°F). The monsoon season, May through October, has heavy rain and somewhat cooler temperatures during the daytime but nights remain warm. Rich variations

of fauna such as mud skipper, flying fox, fiddler crab, sesarma mederi, oyster, cockle, banana shrimp, and snake are commonly found. The main mangrove dominance is *Avicennia alba* Bl. due to their adaptive characteristics in tolerating a saline environment. Other plant species found across the conservation area include *Rhizophora apiculata* Blume, *Rhizophora mucronata*, *Sonneratia griffithii*, *Sonneratia alba*, *Xylocarpus granatum*, and *Xylocarpus moluccensis*. The presence of poor mangrove growth was near the entrance of the walkway and appears in the southeast center of the study area.

2.2. Sampling Design. The sampling site was marked by a handheld GPS (Garmin GPSMAP® 62sc model) on site. The boundary of the area was mapped using Google Earth imaginary layout [22]. The geographic coordinate system used was WGS84 before being projected to the UTM_Zone_47N system for consistent data analysis. The spatial sampling sites were planned on the grid-based sampling design with a cell size of 80 m × 80 m. From an initial grid design providing 35 soil samples covering the entire area, a total of 29 soil samples from the surface (0–10 cm) across the study area were available for analysis due to limited accessibility (Figure 1). Soil samples were collected in October 2015. All the samples were collected during low tides; thus a hand collection technique was implemented. To collect samples, a 10 × 5 cm^2 soil core was used from each designated location within a square grid. The collected sediments were cut into two layers: surface (0–5 cm) and subsurface (5–10 cm). All samples were kept at 4°C at the sampling moisture content for laboratory processing.

2.3. Analysis of Soil Characteristics. The physiochemical sediment properties, that is, texture, pH, salinity, electrical conductivity (EC), total dissolved solid (TDS), and soil texture, were analyzed. After air-dried samples were prepared, the percentages of sand, silt, and clay in the sediments were determined by hydrometer testing [23] and then compared with a soil texture triangle to identify soil texture [24]. The ratio of soil to water was 1 : 2.5 as suggested by Jackson [25]. The measurement of pH was determined by a pH meter (Vetus, pH-009 (III), USA). Salinity, EC, and TDS were analyzed by a portable handheld EC meter (Hach, sensION 156, USA) [25].

2.4. Assessment of Organic Matter. The loss on ignition (LOI) method was employed to analyze organic matter content at a 10 cm depth. The wet soil samples were completely oven-dried at a temperature of 105°C until the weight of dried soils remained constant. The dried soil samples were then burned in a furnace at 550°C for 4 hours before being weighed. Organic matter content was calculated from the soil sample weight before and after burning by using the following equation:

Organic matter content (%)

$$= \frac{\text{Oven-dried soil}\,(g) - \text{Burned soil}\,(g)}{\text{Oven-dried soil}\,(g)} \times 100. \tag{1}$$

2.5. Assessment of Soil Organic Carbon. The soil samples were freeze-dried overnight in a freeze dry/shell system

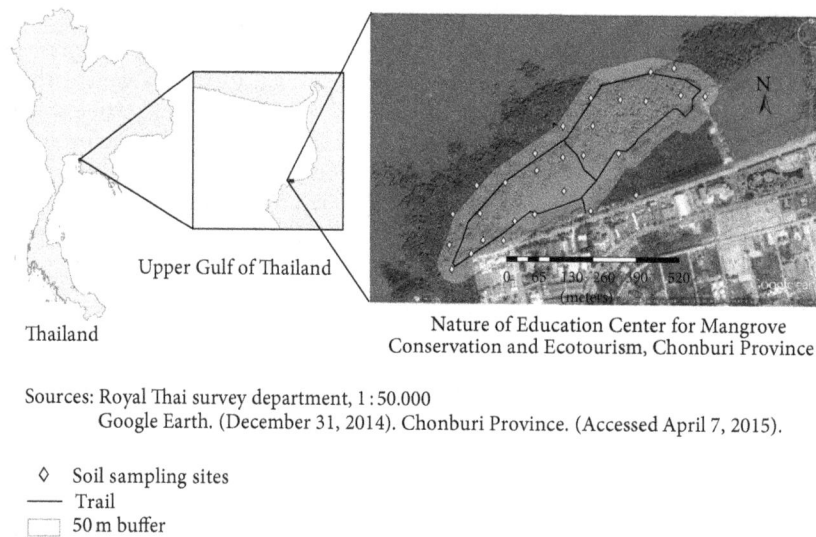

Sources: Royal Thai survey department, 1 : 50.000
Google Earth. (December 31, 2014). Chonburi Province. (Accessed April 7, 2015).

◇ Soil sampling sites
—— Trail
☐ 50 m buffer

FIGURE 1: The Nature Education Center for Mangrove Conservation and Ecotourism located in coastal western Chonburi Province, Thailand.

(FreeZone6, Labconco, USA) and then ground to a fine powder. The fractions passing a 2-mm sieve were stored in a polypropylene centrifuge tube. Total carbon (TC) and soil inorganic carbon (SIC) were measured by the Shimadzu gas analyzer (TOC-V CPH) using separate gas analysis procedures. The CO_2 evolution using tiny ball-milled samples with gas combustion was applied at different temperatures. The temperature was set at 900°C for the combustion process to measure TC, while adding phosphoric acid (H_3PO) reaction and setting the temperature at 200°C to measure IC [26, 27]. The organic carbon can then be derived by subtracting SIC from TC.

2.6. Statistical and Geostatistical Analyses. Descriptive statistics were performed to establish physiochemical properties information using SPSS 22.0. An ordinary kriging approach was used to estimate the spatial SOC distributions in both surface and subsurface layers. Three main steps in the kriging included (1) exploring measured data to characterize spatial continuity, (2) developing a semivariogram model for investigating spatial autocorrelation, and (3) estimating unknown values based on optimized weighted averages of defined neighboring locations [28]. The relationship between OM and SOC was carried out by a Pearson's correlation coefficient and a simple linear regression in *R* statistical software (version 3.2.0).

3. Results

3.1. Soil Organic Carbon and Physiochemical Characteristics at 0–5 cm and 5–10 cm Depths. Sandy clay loam covered the majority of the surface soil, followed by sandy loam and loam. On the other hand, clay texture was dominant in the subsurface soil, followed by sandy clay loam and clay loam. Low salinity with mean values of 2.04 ppt in surface soil and 2.22 ppt in subsurface soil was influenced by strong

water inflow from the estuary to the seaside because the data collected in this study was taken during low tides. These values were consistent with average EC values of 3.66 mS/cm and 3.96 mS/cm which indicated moderately saline condition and sea water was diluted by the fresh water. Its salinity, the mixture of saline and fresh water, ranged from 1.00 to 3.52 ppt. According to the salinity classification of Levinton [29], mangrove soil in this area was considered under brackish and slightly alkaline condition with an average pH above 7. Descriptive statistics of SOC content and related parameters are summarized in Table 1.

The measured values of SOC were equally alike to the amount of TC, indicating very low inorganic carbon forms in this area. The surface soil exhibited a higher SOC content than the subsurface layer. The results show that, vertically, SOC content decreased with increasing depth. The mean observation with standard deviation SOC was 3.08 ± 1.25% and 2.51 ± 1.47% for the surface and subsurface soils, respectively. Thus surface soils with higher productive mangroves in this area retain a greater amount of SOC input.

Local topography and soil drainage are considered important drivers in determining carbon storage. Since the mangrove area in this study was generally flat, the erosion process to transport carbon from topsoil to the bottom of the slopes was difficult. The water-logged condition in the mangrove forest could promote depositional rates and inhibit decomposition rates [30]. It can be observed that the disturbance of soil surface from tillage, dredging, or other aquaculture activities was extremely low in the conservation area compared to agricultural space. Less soil disturbance contributed to a net positive SOC storage of estuarine wetlands [31]. The different amount of carbon deposits along depths was largely influenced by vegetation. In an area with similar decomposition rates, carbon inputs greatly depend on primary productivity. This may be explained by the size of fraction SOC inputs and decomposition rates. Since a large

TABLE 1: Descriptive statistics of total carbon, soil organic carbon, soil inorganic carbon contents, and soil characteristics in surface and subsurface soils.

Parameters	Surface soil (0–5 cm) ($n = 29$)				Subsurface soil (5–10 cm) ($n = 29$)			
	Max	Min	Mean	SD	Max	Min	Mean	SD
Total carbon (%)	6.20	0.02	3.09	1.25	5.08	0.02	2.53	1.49
Soil organic carbon (%)	6.20	0.02	3.08	1.25	5.01	0.02	2.51	1.47
Soil inorganic carbon (%)	0.06	0.00	0.01	0.02	0.23	0.00	0.03	0.05
pH	8.01	6.95	7.38	—	8.12	7.35	7.81	—
Salinity (ppt)	3.04	1.00	2.04	0.47	3.52	1.20	2.22	0.52
EC (mS/cm)	5.28	1.88	3.66	0.78	6.06	2.24	3.96	0.86
TDS (g/L)	2.64	0.94	1.83	0.39	3.03	1.12	1.98	0.43

ppt = parts per thousand; mS/cm = microsiemens/centimeter; EC = electrical conductivity; TDS = total dissolved solids; g/L = grams/liter.

TABLE 2: Descriptive statistics of organic matter (OM) obtained by loss on ignition (LOI) and soil organic carbon (SOC) obtained by gas analyzation for the 0–10 cm layer.

Variable ($n = 29$)	Min	Max	Mean	Median	SD	Skewness
OM (%)	6.14	17.34	11.12	11.04	2.23	0.59
SOC (%)	2.26	9.80	5.59	5.98	2.24	0.15

fraction comes from root turnover, the variability of root inputs is an influential factor in carbon storage [32].

3.2. Carbon Storage (0–10 cm). At a depth of 10 cm, SOC varied across vegetation species and areas. The Nature Education Center for Mangrove Conservation and Ecotourism in this study, heavily dominated by Avicennia alba Bl., stored SOC at about 5.59% and OM 11.12%. Observed organic matter ranged from 6.14 to 17.34% with a mean ± standard deviation of 11.12 ± 2.23%. Soil organic matter ranged from 2.26 to 9.80% with a mean ± standard deviation of 5.59 ± 2.24% (Table 2).

3.3. Estimated Spatial Distribution of Organic Matter and Soil Organic Carbon (0–10 cm). The close values between mean and median with low skewness described the normal central tendency of the data. Log transformation was not required to perform krigged estimates. The total spatial distribution of OM showed a minimum of 9.19%, maximum of 12.47%, and mean ± standard deviation of 11.08 ± 0.75%. The SOC variability demonstrated a minimum of 3.32%, maximum of 7.91%, and mean ± standard deviation of 5.48 ± 1.16%. The estimated OM and SOC derived by ordinary kriging are shown in Figure 2. A spherical model was applied to fit the experimental semivariograms of OM and SOC with a cell size of $3 \times 3\,\text{m}^2$. The semivariograms of the data were found to vary between OM and SOC. Organic matter had longer correlation range (945.84 m) than SOC (322.49 m). The nugget and sill values of OM were 3.65 and 6.30, whereas nugget and sill of SOC were 2.61 and 6.34. Moderate spatial autocorrelation was found for both OM (0.58) and SOC (0.41) as indicated by the nugget-to-sill ratio. A small ratio in SOC informs the stronger spatial dependency as compared to OM. The large spatial autocorrelation range of 945.8 m of

OM explained the broader homogenous scale compared to a shorter range of 322.5 m of SOC. Organic matter and SOC showed uniform spatial patterns, with high content near the land and low content toward the Gulf of Thailand. The tidal export of organic matter tended to be impeded by the mangrove root system and forest structure. The findings designated that plant debris and riverine influx of dissolved and particulate organic matter upstream supply a substantial degree of organic loading near the upper intertidal area.

3.4. SOC-to-OM Conversion. The average SOC content of OM in soils at a 10-cm depth was approximately 50%. The relationship between SOC and OM is illustrated by a scatter plot with a linear conversion in Figure 3, the regression formula being

$$\hat{y} = 0.05 + 0.50\,(x), \tag{2}$$

where $\hat{y}$ represents dried-weight soil organic carbon (%) measured by gas combustion analysis and x represents dried-weight soil organic matter (%) measured by loss on ignition method.

In Figure 3(a), the relationship between OM and SOC of soils in this study area was weak but statistically significant ($R^2 = 0.25$, p value < 0.001) and Pearson's correlation coefficient was 0.50 for all the samples. When considering 17 samples near the ocean (Figure 3(b)), a model was moderately fitted with statistical significance ($R^2 = 0.48$, p value < 0.001), and Pearson's correlation coefficient was 0.68. When considering 12 samples near the land, the regression model showed a very weak relationship and was not statistically significant ($R^2 = 0.01$, p value = 0.74) and Pearson's correlation coefficient was as low as 0.10. The different relationships of OM and SOC detected between samples close to the ocean and far from the ocean were probably due to the natural settings such as inputs of OM, anaerobic environment, and influence of tides.

An average ratio of spatial OM : SOC was 2.01 ± 0.35 across the entire study area (Figure 4). Higher ratios were observed in the direction toward the ocean and findings from spatial analysis were compatible with results from our point-based data. Mangrove soils comprised of organic matter holding 50% carbon identified a conversion factor of 2. Even

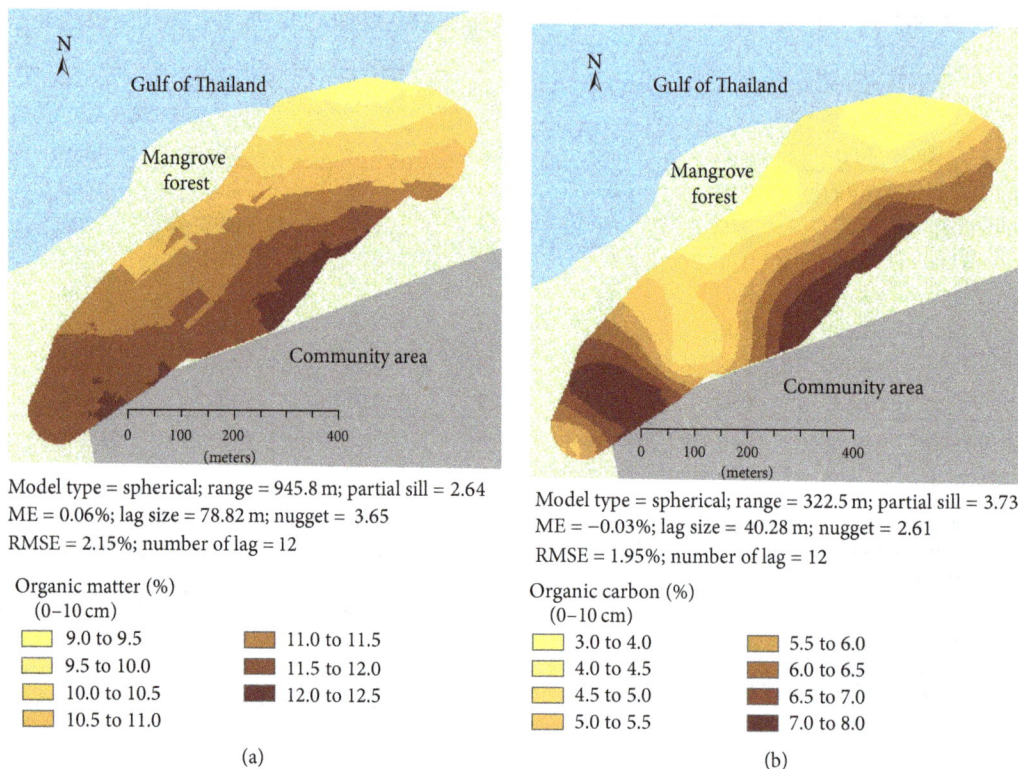

Model type = spherical; range = 945.8 m; partial sill = 2.64
ME = 0.06%; lag size = 78.82 m; nugget = 3.65
RMSE = 2.15%; number of lag = 12

Organic matter (%)
(0–10 cm)
☐ 9.0 to 9.5   ▧ 11.0 to 11.5
☐ 9.5 to 10.0   ▧ 11.5 to 12.0
☐ 10.0 to 10.5 ▧ 12.0 to 12.5
☐ 10.5 to 11.0

(a)

Model type = spherical; range = 322.5 m; partial sill = 3.73
ME = −0.03%; lag size = 40.28 m; nugget = 2.61
RMSE = 1.95%; number of lag = 12

Organic carbon (%)
(0–10 cm)
☐ 3.0 to 4.0   ▧ 5.5 to 6.0
☐ 4.0 to 4.5   ▧ 6.0 to 6.5
☐ 4.5 to 5.0   ▧ 6.5 to 7.0
☐ 5.0 to 5.5   ▧ 7.0 to 8.0

(b)

FIGURE 2: Estimated predictions of (a) soil organic matter content (%) and (b) soil organic carbon (%) content based on an ordinary kriging using 29 soil samples. ME is mean error and RMSE is root mean square error.

though the practical use of the 1.724 factor found at the beginning of nineteenth century has been established through authority and has been adopted in many publications, the results from this study were similar to those of other investigators who directly and indirectly indicated the value of 1.724 to be too low to estimate the amount of OM of mangrove and/or wetland soils [15–18].

4. Discussion and Comparisons

Looking back at Table 2 we can see that the results were comparatively equal as compared to carbon content under dominance of *Rhizophora mangle* (mean SOC 2.80 + 2.70 = 5.50%) but were lower when compared to dominance of *Avicennia schaueriana* (mean SOC 6.10 + 3.80 = 9.90%) in Brazil (Table 3) [33]. In Indonesia, soils under *Avicennia* forest and *Ceriops* forest showed mean ± SD SOC contents of 3.96 ± 0.18% and 11.40 ± 0.64% at a depth of 20 cm, respectively. An overview of SOC content studies at different depths and mangrove forests are shown in Table 3.

Of note also is Howard's study [16], where a regression model was expressed across various soil types as OM (%) = 1.997 + 1.872 ∗ SOC (%). However, the altered factors and amount of SOC in OM relied on type of soils being from 1.77 to 1.93 (51.92–56.39%) in moors, 1.90 to 1.95 (51.29–52.50%) in acid peats, 1.92 (52.19–52.20%) in fen peats, 1.81–1.83 (48.36–50.87%) in alluvial soils, and 1.97–2.07 in mulls. Published results early in the twentieth century showed that 1.8 was a suitable OM-to-SOC factor for the marine sediment

[37]. Later, the simulated model of OM and nutrients in mangrove wetland soils located in the Shark River estuary of south Florida showed OM : SOC ratios from 1.81 to 2.10 with a mean value of 1.98 based on four mangrove sites [38]. Robinson et al. [15], Ponomareva and Plotnikova [39], and Pribyl [18] advocated 2 as the recommended factor for accurate conversion. The SOC-to-OM conversion factor of 2 is, thus, considered to provide better accuracy for estimating the carbon content in mangrove environments or similar types of wetland. Our results coincided with the study from Everglades National Park, Florida [38], in which OM concentrations decreased from 82% to 30% in upstream locations down to marine sites.

The recent study mentioned two sources of variability in any estimated factor: first, the different methods used to measure OM and SOC and, second, the diversity in soil composition [18]. In this study, the methods used to estimate OM and SOC were consistent; thus natural factors played a dominant role. The natural setting of the coastal area can reflect differences in carbon content of OM that, in this study, showed higher conversion factors of SOC to OM in the mangrove environment further from shore. In nature, mixed origins of organic matter in mangrove soil can be very diverse and complex. The near-ocean area showed a high density of mangrove plants that can gain soil OM from fallen detritus decomposition. On the contrary, near-shore areas potentially receive OM sources from land via human activity as it is closer to the community. Therefore, the rate and source of organic matter input affect microbial

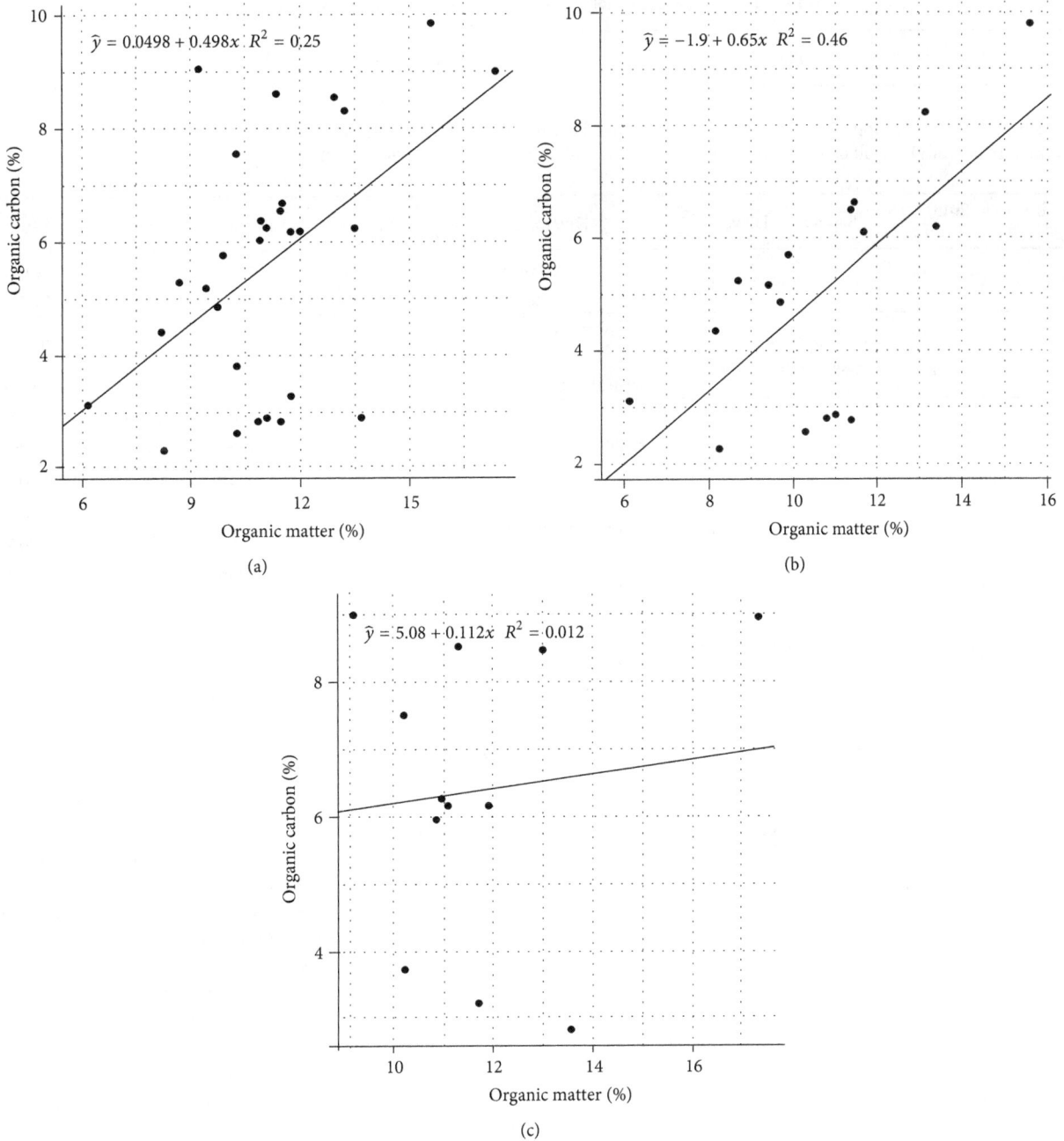

$\hat{y} = 0.0498 + 0.498x \quad R^2 = 0.25$

(a)

$\hat{y} = -1.9 + 0.65x \quad R^2 = 0.46$

(b)

$\hat{y} = 5.08 + 0.112x \quad R^2 = 0.012$

(c)

FIGURE 3: Linear regression of soil organic matter (OM) measured by loss on ignition (LOI) to soil organic carbon (SOC) measured by gas combustion analysis for (a) all soil samples ($n = 29$), (b) soil samples near the ocean ($n = 17$), and (c) soil samples near the land ($n = 12$). The delineation of soil sample groups for (b) and (c) was divided from the center area parallel to the shoreline along the study area.

reactions as well as nutrient availability [40]. Continued production and slow decomposition can lead to very large OM content in soil with long periods of water saturation, but raised OM decomposition occurs in more aerated or less anoxic conditions. However, the speed and ease of carbon mineralization depend on the OM fraction in which it resides; for example, humus-carbon mineralizes slowly when compared to plant residue, particulate OM, and soil microbial biomass [41]. The tidal process is one factor driving a great

influx of SOC content and turnover in mangrove forests [42]. Coastal mangroves act as a barrier to any fresh water and tidal flows; this buffer zone often develops suspended matter including OM and SOC in the directions parallel to the coast. Areas further from shore with a high density of mangroves have a vertical mixing of water caused by strong waves and tides which tend to circulate suspended matter; thus OM and SOC usually remain in this zone. Estimates of OM : SOC in soil vary from site to site and may be affected by other factors

TABLE 3: Overview of available soil carbon content in mangrove forests.

Authors	Place	Site characteristics	Depth	Reported statistics	Carbon content
Chaikaew and Chavanich (this study)	Upper eastern coast of the Gulf of Thailand, Thailand	Mixed natural and planted mangrove forests with dominated *Avicennia alba* Bl.	0–5 cm	Mean ± SD	3.08 ± 1.25%
			5–10 cm	Mean ± SD	2.51 ± 1.47%
Chandra et al. (2015) [34]	Sarawak, Malaysia	Diverse mangrove species	40 cm	Range	1.73–6.24%
Donato et al. (2011) [5]	Indo-Pacific region	Oceanic mangrove	2 m	Mean	14.6%
		Estuarine mangrove	2 m	Mean	7.69%
Ray et al. (2011) [35]	Northeast coast of the Bay of Bengal, India	Natural mangrove forest (before monsoon)	30 cm	Mean	0.51%
		Natural mangrove forest (after monsoon)	30 cm	Mean	0.65%
Lacerda et al. (1995) [33]	Sepetiba Bay, Brazil	Experimental *Rhizophora mangle* forest	1.5 cm	Mean	2.80%
			5–10 cm	Mean	2.70%
			10–15 cm	Mean	2.70%
		Experimental *Avicennia schaueriana* forest	1.5 cm	Mean	6.10%
			5–10 cm	Mean	3.80%
			10–15 cm	Mean	3.80%
Sukardjo (1994) [36]	East Kalimantan, Indonesia	Soils dominated by *Avicennia* forest	20 cm	Mean ± SD	3.96 ± 0.18%
		Soils dominated by *Ceriops* forest	20 cm	Mean ± SD	11.40 ± 0.64%

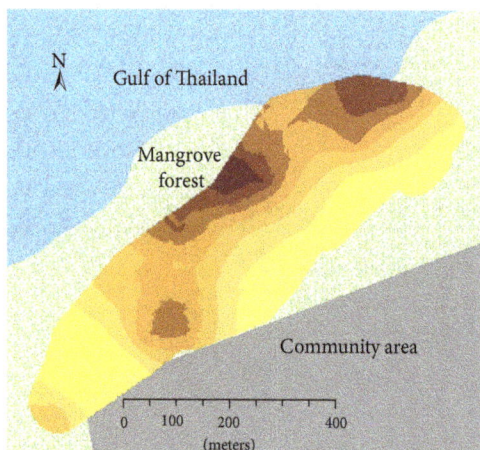

Ratios of organic matter to organic carbon (0–10 cm)

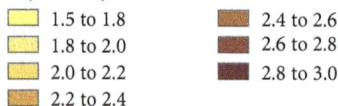

- 1.5 to 1.8
- 1.8 to 2.0
- 2.0 to 2.2
- 2.2 to 2.4
- 2.4 to 2.6
- 2.6 to 2.8
- 2.8 to 3.0

FIGURE 4: Ratios of soil organic matter obtained by loss on ignition (LOI) and organic carbon measured by gas analyzer based on the ordinary krigged estimated (Figure 2(a) divided by Figure 2(b)).

such as types of vegetation cover, clay in the soil, degree of decomposition, and soil composition, which are out of the study scope.

5. Conclusions

The Nature Education Center for Mangrove Conservation and Ecotourism under dominance of mixed natural and planted *Avicennia alba* Bl. showed a great potential to sequester carbon in soils with high amounts of SOC stored in the top 10 cm. Surface soil (0–5 cm) contained an average ± standard deviation in SOC of 3.08 ± 1.25% which is greater when compared to subsurface soil (5–10 cm) with 2.51 ± 1.47%. Considering the ecosystem in the study area, the origins of the carbon inputs were derived from primary productivity such as leaf debris and plant roots. The spatial variability of OM and SOC demonstrated the decreasing pattern from land to sea or upper to lower tidal zones. Besides the influence of the tides, we summarized that a substantial load of dissolved and particulate organic matter comes from canals and the community upstream. In terms of the strength of the spatial autocorrelation, the nugget-to-sill ratio of 0.41 for SOC was stronger than 0.58 for OM yet indicated moderate spatial dependency for both variables. Organic matter showed homogenous spatial variability due to the long spatial autocorrelation range. This study contradicts the assumption that has been repeatedly used for more than a century that OM is 58% organic carbon and that the conversion factor is 1.724 (van Bemmelen factor). The use of a low factor can cause sequential errors in estimating OM when multiplying carbon by 1.724. Organic matter in mangrove soils in this study contained about 50% organic carbon with the conversion factor of 2. The use of the factor of 2 in different types of wetlands was also recommended by considerable supporting evidence. However, this number is not for universal purposes but rather for rough estimation. The conversion factor may vary depending on vegetation cover, temperature, soil type, soil depth, and the technique used to analyze OM and SOC.

Competing Interests

The authors declare that there is no conflict of interests regarding the publication of this paper.

Acknowledgments

Funding research was provided by Grants for Development of New Faculty Staff, Ratchadaphiseksomphot Endowment Fund, Chulalongkorn University.

References

[1] E. Dlugokencky and P. Tans, "Trends in atmospheric carbon dioxide," 2015, http://www.esrl.noaa.gov/gmd/ccgg/trends/.

[2] R. Lal, "Soil carbon sequestration impacts on global climate change and food security," *Science*, vol. 304, no. 5677, pp. 1623–1627, 2004.

[3] M. Apps, C. Cerri, T. Fujimori et al., Technological and economic potential of options to enhance, maintain, and manage biological carbon reservoirs and geo-engineering, 2001, http://www.grida.no/climate/ipcc_tar/wg3/index.htm.

[4] S. Page, R. Wüst, and C. Banks, "Past and present carbon accumulation and loss in Southeast Asian peatlands," *PAGES News*, vol. 18, pp. 25–27, 2010.

[5] D. C. Donato, J. B. Kauffman, D. Murdiyarso, S. Kurnianto, M. Stidham, and M. Kanninen, "Mangroves among the most carbon-rich forests in the tropics," *Nature Geoscience*, vol. 4, no. 5, pp. 293–297, 2011.

[6] C. P. Immirzi, E. Maltby, and R. S. Clymo, "The global status of peatlands and their role in carbon cycling," 1992.

[7] N. Pumijumnong, "Mangrove forests in Thailand," in *Mangrove Ecosystems of Asia: Status, Challenges and Management Strategies*, I. Faridah-Hanum, A. Latiff, K. R. Hakeem, and M. Ozturk, Eds., p. 470, Springer, New York, NY, USA, 2014.

[8] C. Sprengel, "Ueber Pflanzenhumus, Humussäure und humussaure Salze," *Archiv für die Gesammte Naturlehre*, vol. 8, pp. 145–220, 1826.

[9] E. Wolff, "Entwurf zur Bodenanalyse," *Zeitschrift für Analytische Chemie*, vol. 3, no. 1, pp. 85–115, 1864.

[10] G. Loges, "Mittheilungen aus dem agriculturchemischen Laboratorium der Versuchs-Station Kiel: III. Ueber die Bestimmung des Humus in Ackererden," *Die Landwirthschaftlichen Versuchs-Stationen*, vol. 28, pp. 229–245, 1883.

[11] F. Schulze, "Anleitung zur Untersuchung der Ackererden auf ihre wichtigsten physikalischen Eigenschaften und Bestandtheile," *Journal für Praktische Chemie*, vol. 47, no. 1, pp. 241–335, 1849.

[12] W. Detmer, "Mittheilungen aus dem agriculturchemischen laboratorium der Universität Leipzig. VI. Die natürlichen humuskörper des bodens und ihre landwirthschaftliche bedeutung," *Die Landwirthschaftlichen Versuchs-Stationen*, vol. 14, pp. 248–296, 1871.

[13] F. E. Broadbent, "The soil organic fraction," *Advances in Agronomy*, vol. 5, pp. 153–183, 1953.

[14] R. A. Gortner, "The organic matter of the soil: 1. Some data on humus, humus carbon and humus nitrogen," *Soil Science*, vol. 2, no. 5, pp. 395–442, 1916.

[15] G. W. Robinson, W. McLean, and R. Williams, "The determination of organic carbon in soils," *The Journal of Agricultural Science*, vol. 19, no. 2, pp. 315–324, 1929.

[16] P. J. A. Howard, "The carbon-organic matter factor in various soil types," *Oikos*, vol. 15, no. 2, pp. 229–236, 1965.

[17] P. J. A. Howard and D. M. Howard, "Use of organic carbon and loss-on-ignition to estimate soil organic matter in different soil types and horizons," *Biology and Fertility of Soils*, vol. 9, no. 4, pp. 306–310, 1990.

[18] D. W. Pribyl, "A critical review of the conventional SOC to SOM conversion factor," *Geoderma*, vol. 156, no. 3-4, pp. 75–83, 2010.

[19] B. Murphy, "Key soil functional properties affected by soil organic matter—evidence from published literature," *IOP Conference Series: Earth and Environmental Science*, vol. 25, Article ID 12008, 2015.

[20] M. V. Lützow, I. Kögel-Knabner, K. Ekschmitt et al., "Stabilization of organic matter in temperate soils: Mechanisms and their relevance under different soil conditions—a review," *European Journal of Soil Science*, vol. 57, no. 4, pp. 426–445, 2006.

[21] M. C. Peel, B. L. Finlayson, and T. A. McMahon, "Updated world map of the Köppen-Geiger climate classification," *Hydrology and Earth System Sciences*, vol. 11, no. 5, pp. 1633–1644, 2007.

[22] Google Earth, "Chonburi Province: 13°20′37.05″N, 100°56′34.83″E, Eye alt 1.01 km," 2014, http://www.google.com/earth/index.html.

[23] G. J. Bouyoucos, "Hydrometer method improved for making particle size analyses of soils," *Agronomy Journal*, vol. 54, no. 5, pp. 464–465, 1961.

[24] USDA, *USDA Textural Soil Classification*, 1987, http://www.wcc.nrcs.usda.gov/ftpref/wntsc/H&H/training/soilsOther/soil-USDA-textural-class.pdf.

[25] M. L. Jackson, *Soil Chemical Analysis*, Prentice Hall of India, New Delhi, India, 1967.

[26] C. W. Ross, S. Grunwald, and D. B. Myers, "Spatiotemporal modeling of soil organic carbon stocks across a subtropical region," *Science of the Total Environment*, vol. 461-462, pp. 149–157, 2013.

[27] Shimadzu, *TOC-V CPH/CPN Total Organic Carbon Analyzer User's Mannual*, Corporation Process and Environmental Instrumentation Division, Kyoto, Japan, 2001.

[28] P. Goovaerts, *Geostatistics for Natural Resources Evaluation*, Oxford University Press, New York, NY, USA, 1997.

[29] J. S. Levinton, *Marine Biology: Function, Biodiversity, Ecology*, Oxford University Press, New York, NY, USA, 1995.

[30] A. A. Berhe, J. Harte, J. W. Harden, and M. S. Torn, "The significance of the erosion-induced terrestrial carbon sink," *BioScience*, vol. 57, no. 4, pp. 337–346, 2007.

[31] P. D. Howe, E. M. Markowitz, T. M. Lee, C.-Y. Ko, and A. Leiserowitz, "Global perceptions of local temperature change," *Nature Climate Change*, vol. 3, no. 4, pp. 352–356, 2013.

[32] J. A. Bird and M. S. Torn, "Fine roots vs. needles: a comparison of 13C and 15N dynamics in a ponderosa pine forest soil," *Biogeochemistry*, vol. 79, no. 3, pp. 361–382, 2006.

[33] L. D. Lacerda, V. Ittekkot, and S. R. Patchineelam, "Biogeochemistry of mangrove soil organic matter: a comparison between *Rhizophora* and *Avicennia* soils in South-Eastern Brazil," *Estuarine, Coastal and Shelf Science*, vol. 40, no. 6, pp. 713–720, 1995.

[34] I. A. Chandra, G. Seca, R. Noraini et al., "Soil carbon storage in dominant species of Mangrove Forest of Sarawak, Malaysia," *International Journal of Physical Sciences*, vol. 10, no. 6, pp. 210–214, 2015.

[35] R. Ray, D. Ganguly, C. Chowdhury et al., "Carbon sequestration and annual increase of carbon stock in a mangrove forest," *Atmospheric Environment*, vol. 45, no. 28, pp. 5016–5024, 2011.

[36] S. Sukardjo, "Soils in the mangrove forests of the Apar nature reserve, Tanah Grogot, East Kalimantan, Indonesia," *Southeast Asian Studies (Kyoto)*, vol. 32, no. 3, pp. 385–398, 1994.

[37] P. D. Trask, "Organic content of recent marine sediments," in *Recent Marine Sediments*, P. D. Trask, Ed., pp. 428–453, Dover Publications, New York, NY, USA, 1938, http://archives.data-pages.com/data/sepm_sp/SP4/Organic_Content_of_Recent.htm.

[38] R. Chen and R. R. Twilley, "A simulation model of organic matter and nutrient accumulation in mangrove wetland soils," *Biogeochemistry*, vol. 44, no. 1, pp. 93–118, 1999.

[39] V. V. Ponomareva and T. A. Plotnikova, "Data on the degree of intramolecular oxidation of humus in various soil groups (problem of the carbon to humus conversion factor)," *Soviet Soil Science*, vol. 7, pp. 924–933, 1967.

[40] D. E. Canfield, E. Kristensen, and B. Thamdrup, "Aquatic geomicrobiology," in *Advances in Marine Biology*, vol. 48, pp. 517–599, Elsevier, Amsterdam, Netherlands, 2005.

[41] NRCS, *Soil Quality Indicators*, 2009, https://www.nrcs.usda.gov/Internet/FSE_DOCUMENTS/nrcs143_019177.pdf.

[42] J. P. Zhang, W. X. Yi, C. D. Shen et al., "Quantification of sedimentary organic carbon storage and turnover of tidal mangrove stands in southern China based on carbon isotopic measurements," *Radiocarbon*, vol. 55, no. 2-3, pp. 1665–1674, 2013.

8

Geotechnologies and Soil Mapping for Delimitation of Management Zones as an Approach to Precision Viticulture

José Maria Filippini Alba,[1] Carlos Alberto Flores,[1] and Alberto Miele[2]

[1]Empresa Brasileira de Pesquisa Agropecuária, Centro de Pesquisa Agropecuária de Clima Temperado, P.O. Box 403, 96010-971 Pelotas, RS, Brazil
[2]Empresa Brasileira de Pesquisa Agropecuária, Centro Nacional de Pesquisa de Uva e Vinho, P.O. Box 130, 95701-008 Bento Gonçalves, RS, Brazil

Correspondence should be addressed to José Maria Filippini Alba; jose.filippini@embrapa.br

Academic Editor: Claudio Cocozza

Data of the physical and chemical properties of soils from three vineyards located in Vale dos Vinhedos, Bento Gonçalves, Rio Grande do Sul state, in southern Brazil, were processed. Soil mapping was performed by means of four profiles and the digital elevation model in detailed scale. Then, superficial soils (0–20 cm) were sampled according to a grid pattern. Analysis of variance (ANOVA), kriging, and unsupervised classification methods were applied on physical and chemical data of superficial soils sampled according to grid pattern. This study aimed to compare both methods, the conventional soil mapping and the map produced with superficial soil sampling, about their potential for definition of the management zones, as an approach for precision agriculture. Maps elaborated by conventional soil mapping overlapped partially with the maps derived from superficial sampling, probably due to the specific methodological differences of each case. Anyway, both methods are complementary because of the focus on vertical variability and horizontal variability, respectively. In that sense, slope appears as significant edaphic parameter, due to its control on water circulation in the profile of soil.

1. Introduction

Notable advances in pedological research were reached in the 1990s, after a period of stagnancy, when a significant search for a rational use of natural resources and the equilibrium of biogeochemical cycles took place. In this way, monitoring and evaluation of soil resources began a new age, due to the quality of information derived from new technologies as geographical information systems (GIS) and remote sensing. Tayari et al. [1] discussed the relation among GPS, GIS, and precision agriculture (PA).

These technologies also contributed to improve precision viticulture, that is, "precision farming (or PA) applied to optimize vineyard performance, in particular maximizing grape yield and quality while minimizing environmental impacts and risk" [2]. According to McBratney et al. [3] "the definition of precision agriculture is still evolving as technology changes and our understanding of what is achievable grows. Over the years the emphasis has changed from simply 'farming by soil', through variable rate technologies, to vehicle guidance systems and will evolve to product quality and environmental management". Actually, in a more contemporaneous definition, "PA is a whole-farm management approach using information technology, satellite positioning data, remote sensing and proximal data gathering. These technologies seek improving returns on inputs while potentially reducing environmental impacts. The state-of-the-art of PA on arable land, permanent crops and within dairy farming are reviewed, mainly in the European context, altogether with some economic aspects of the adoption of PA" [4].

SCORPAN model for soil mapping considers the pedological parameters as a mathematical function evolving several factors: soil, climate, organisms (microorganisms, vegetation, or land use), relief, parent material, age, and spatial position [5, 6]. McBratney et al. [5] mentioned three main scales in pedometry:

(1) national context with resolution larger than 2 km;

(2) drainage basins and landscape need resolution between 20 m and 2 km;

(3) local context with resolution lesser than 20 m (the order 0 of USDA survey evolves pixel size lesser than 5 m × 5 m for applications in precision agriculture).

Soils are mainly components of "terroir" in viticulture [7]; however their survey is expensive and there is no direct relation to the type of wine. Cheaper methods of survey would be provided by the "geological model" or the "geomorphological model." Anyway, the mapped units do not agree with the classes of soil occurring in a vineyard due to scale problems, although there are agronomic variations intraunits of soil, according to small organic matter content changes or oscillations of horizon thickness. By this reason, new technologies must be used, as for instance digital elevation models, GIS, electric resistivity measures or remote sensing, aiming an efficient survey for the management of viticulture.

From a geospatial perspective, Flores et al. [8] considered each class of soil a management zone. However, Filippini-Alba et al. [9] integrated the classes of soil from the three vineyards by similarity, location and practical reasons, in five management zones. In that sense, Filippini-Alba et al. [10] implemented a microzoning in one vineyard, based on four variables: (1) content of clay; (2) organic matter level; (3) saturation of bases; (4) stoniness. Preferential class of aptitude was 150–350 g·kg^{-1}, <2.5%, 20–49%, and <0.5% for (1) to (4), respectively, and locates mainly on CXve 2 and CXve 3 units, northern part of the vineyard. Coincidence within management zones is not exactly but is very similar.

Several papers discuss the effects of phenolic compounds in wine [11], but the amount of papers is significantly reduced if the influence of soil on phenolic compounds is considered. A diversity of taxonomic classes is associated with the vineyards of Vale dos Vinhedos, which change radically in a tiny scale occasionally. These changes affect the content of phenolic compounds in the wine produced [11, 12]. Grenache noire vines planted with water or nitrogen restriction are related to grapes enriched with sugars and anthocyanins levels [13]. Moreover, the authors concluded that the type of soil does not affect the quality of grapes; however, this property would be associated with soil depth.

The previous developing suggest several questions related to the management of the vineyard and the use of geospatial modeling from a precision agriculture (PA) perspective or the application of a more realistic approach, including management zones defined by conventional soil mapping or GIS modeling. On the one hand we have the precision of geotechnologies and the horizontal geospatial variability; on the other hand we have the recognized pedological model, where the analysis of the vertical variability and the associated cycle of water have provided so many benefits for agriculture. These approaches are discussed in some way in this paper considering three vineyards located on Vale dos Vinhedos, Bento Gonçalves, Rio Grande do Sul state, in southern Brazil. Some more specific objectives/questions are the following. (1) Soil mapping is an expensive procedure, costly, and time consuming; is there some technique with the same level of efficiency and cheaper in the perspective of PA? (2) Geospatial analysis and GIS procedures are related to PA, but standard procedures for interpolation and fusion data are generally absent from bibliographies. A specific method in that sense is presented and discussed here. (3) How are soil units related to viticulture variables?

2. Material and Methods

The three vineyards were established with Merlot grapevine, clone 347, grafted on the 1103 Paulsen rootstock. Two of them were established in 2005 (vineyards 1 and 3) and the remaining vineyard was set up in 2006 (vineyard 2). Grapevines were vertical trellised and spur-pruned. The total area covered by vines was 2.42 ha.

The vine rows and 249 plants inside the vineyards were registered with a Sokkia SET 610 total station and a Sokkia GSR 2600 GPS receiver, in the way that maps of altimetry and slope were elaborated. Then, a regular network with cells of 10 m × 10 m was also delineated, which was used as reference for the pedological survey and superficial soil sampling. Coordinates of four strategic points were performed with the GPS receiver, because, in this case the reception time is too long for measure of full positions. Then, the total station was used for measure distances and angles, thus, the other coordinates and altitudes were calculated. WGS 84 was used as a reference system.

Four trenches were opened for sampling soil profiles, including all horizons up to 150 cm deep. The soils were classified [8] according to the Brazilian System of Soil Classification [14].

Superficial soil samples were collected with a shovel cutting, near the marked vines for the Ap horizon in January, 2011: 28 samples in vineyard 1; 54 samples in vineyard 2; and 26 samples in vineyard 3.

Physical and chemical analyses were performed on the Soil Laboratory at *Universidade Federal do Rio Grande do Sul* (UFRGS). The following variables were evaluated: pH; Al, Ca, Na, K, and P (exchangeable) contents; cationic exchangeable capacity (CEC); coarse fraction (pebbly and gravel); granulometric composition (coarse sand, fine sand, silt, clay, and flocculation degree); organic carbon; and nitrogen. Methods of soil analysis included physical separation, soft extractors for exchangeable elements, dichromate in acid media for organic C and Kjeldahl method for total N. All these methods are detailed by Embrapa [15].

Data were organized and integrated on the software ArcGIS [16] as information lawyers. Analysis of variance (ANOVA) and related procedures were processed in SPSS software [17], considering groups defined by the superficial samples of soil inside each class of soil. The semivariograms of each nutrient and the granulometric classes were processed with GS+ [18]. The spherical model adjusted the semivariograms of most of the variables, with reach of 129 m prevailing. pH and Al and P content showed nugget effect. Then, the spatial parameters were inserted on ArcGIS and kriging was applied for each variable, except for Al, P, and pH, when the inverse distance weighted interpolation was used.

TABLE 1: Properties of the units of soil [8]. Symbols based on Brazilian System of Soil Science (SiBCS) and correspondence SiBCS and soil taxonomy.

Mapping unit	SiBCS	Soil taxonomy	Texture/relief/stoniness	Declivity	Area
PBACal 1	*Argissolo*	Ultisol	Moderately clayey	3–8%	0.13 ha
PBACal 2	*Argissolo*	Ultisol	Medium clayey to clayey, moderately wavy	8–13%	0.12 ha
PBACal 3	*Argissolo*	Ultisol	Clayey loam to clayey, wavy	13–20%	0.38 ha
CXve 1	*Cambissolo*	Inceptisol	Clay loam to clayey	13–20%	0.19 ha
CXve 2	*Cambissolo*	Inceptisol	Clay loam to clayey	20–45%	0.47 ha
CXve 3	*Cambissolo*	Inceptisol	Clay loam to clayey, stony	20–45%	0.18 ha
RRh1	*Neossolo*	Entisol	Sandy clay loam to loam, stony	3–8%	0.29 ha
RRh 2	*Neossolo*	Entisol	Sandy clay loam to loam, stony	8–13%	0.10 ha
RRh 3	*Neossolo*	Entisol	Sandy clay loam to loam, stony	13–20%	0.21 ha
RRh 4	*Neossolo*	Entisol	Sandy clay loam to loam, stony	20–45%	0.35 ha

The six nutrients (Ca, C, K, Mg, N, and P content) and the granulometric variables (clay, silt, fine sand, and coarse sand) were integrated in digital files of six and four information lawyers, respectively. Thus, unsupervised classification by maximum likelihood method was applied on each file by means of ER-Mapper [19]. Afterwards, data were edited and organized in accordance with the integrated level of enrichment of nutrients or fine fractions (clay, silt and fine sand), expressed in percentage.

A new map of management zones was elaborated by overlapping both maps mentioned before, by visual appreciation, taking into account the necessity of condensed information for industrial processes (winemaking).

3. Results and Discussion

3.1. Soils Survey. Soils legend discriminates three classes of soils in level of order that represent ten mapping units (Table 1). Correspondence between the Brazilian System of Soil Science (SiBCS) and Soil Taxonomy (US) is an adaptation, because *Argissolo*, *Cambissolo*, and *Neossolo* are considered as Entisol, Inceptisol, and Ultisol, respectively, but the terms do not have the same meaning exactly. Thus, Brazilian terms were used. The mapping units are different due to the fourth or fifth category level, mainly by Al saturation, bases saturation, texture, stony and phases of relief. These classes occupied 26.03%, 34.71%, and 39.26% of the total area of the three vineyards, respectively.

The spatial distribution of the units of soil is represented in Figure 1. The image of satellite in background show direction E-W for vines in the vineyards 1 and 2, but direction of vine lines is N-S for vineyard 3.

Mapping units PBACal refer to soils with textural B horizon, immediately below A horizon in this case, with high content of clay, grayish brown color, and Al enrichment in several horizons (Table 2), thus with alic condition. The profiles 2 and 4 (PBACal) were studied up to 1.5 m deep, with enrichment in Ap horizon for bases content, organic C, N, and P, as well as differences in declivity and texture (Table 1). BC horizon is intermediary between a B horizon and a C horizon, but with prevalent materials from B horizon. Abrupt condition of units PBACal 2 and PBACal 3 refers to the occurrence of high content of clay in B horizon, deriving on strong differentiation between horizons A and B.

Each class of soil showed different sequence of horizons, but Ap horizon, a superior A horizon plow or removed, occurred in all cases. Mapping unit RRh 4 of *Neossolo* is in accordance with Ap1 and Ap2. *Argissolo* (PBACal,) have a subdivision for B horizon in deep, with accumulation of clay (Bt1 and Bt2). *Cambissolo* (CXve 3,) has an intermediary horizon AB and in sequence, an incipient B horizon (Bi) and a BC horizon. PBACal 2 and 3 units have abruptic condition due to the occurrence of more clay in B horizon than in A horizon, with strong contrast between these two horizons.

From a physicochemical point of view, *Cambissolo* and *Argissolo* are different between them, mainly for Al saturation, bases content and clay content (Table 2). These soils had greater organic matter than *Neossolo*, with significant gravel in this case. About declivity, *Argissolo* ranged from 3% to 20%, *Cambissolo* ranged from 13% to 45%, and *Neossolo* ranged from 3% to 45%, this one with the greatest variation. This feature, associated with texture, affects significantly the dynamic of water circulation through the different edaphic horizons.

High contents of Cu, Mn, P, and Zn in the superficial horizons of the three classes of soils, when compared to the subsuperficial horizons, suggest a strong influence of the local viticulture management.

Water storage capacity (WSC) was different for each horizon and profile, suggesting strong dependence with climate. For instance, CXve3, PBACal2, and PBACal1 have good storage capacity for subsuperficial horizons, but RRha does not show a good storage capacity. So, vines production may be impaired for dry climate in the case of RRha; contrariwise, production may be improved for wet climate. An opposite situation can happen for the remained soils.

3.2. Superficial Soil Data Processing

3.2.1. Statistical Analysis of Superficial Soil Data. Variables with variation coefficients greater than 50% presented almost the total population near of the minimum values, but with some high extreme values in the case of pebbly and level of Al. However, when the variation coefficients ranged between

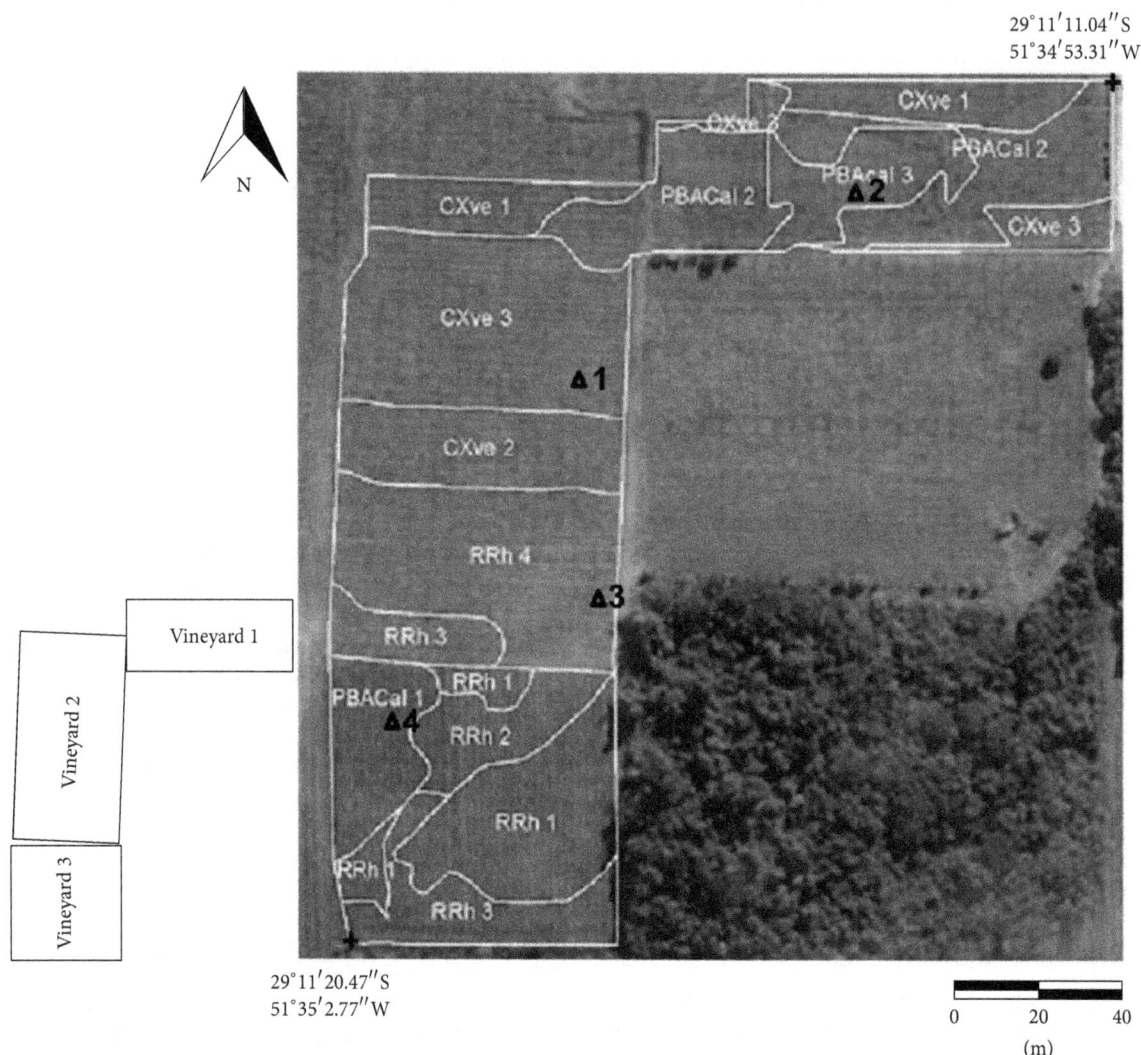

FIGURE 1: Map of soils of the study area on the orbital image as background (Google Earth®, 2013). Triangles indicate the location of the soil profiles.

30% and 35%, the mean located near the half of the range of variation (Table 3).

Statistics were processed for each class of soil including homogeneity and analysis of variance or ANOVA (Table 4). All variables showed significance lesser than 15% for variance homogeneity test with the exception of coarse sand and pH. Thus, precision of ANOVA test would be affected by both variables. Anyway, all variables were significant for ANOVA test at level of 5%, so, in general terms, at least one mean of the classes of soils is different to the other ones.

Variation of the means is represented by line graphs (Figure 2). Main variations occur among classes of soil (*Argissolo*, *Cambissolo*, and *Neossolo*); however softer variations happen within the classes, suggesting declivity differences or degree of stoniness because these two parameters affected differences among classes.

3.2.2. Geospatial Analysis of Superficial Soil Data. Semivari-ograms were adjusted mostly by the spherical model and the predominant reach was 129 m. The spatial distribution of the granulometric variables (clay, silt, fine sand and coarse sand) in the Ap horizon is shown in Figure 3. A sudden transition from vineyard 2 to vineyard 3 is observed for clay, silt and fine sand, thus, result of soil map is confirmed (Figure 1). Separation between *Cambissolo* and *Neossolo* in vineyard 2 also appears in Figure 3, as well as the high clay content related to *Argissolo*.

Spatial distribution of nutrients (Ca, C, K, Mg, N, and P) in the vineyards 1, 2, and 3 is showed in Figures 4, 5, and 6, respectively. P was constant in vineyard 1 and C and N were very similar in the three vineyards. The greatest variability for C, Ca, Mg, N, and P was associated with vineyard 3.

Observation of the means of each class of soil (Figure 2) in comparison to the geospatial distribution of granulometric fractions (Figure 3) and level of chemical elements (Figures 4, 5 and 6) suggest a coherent adjustment between both procedures of processing data.

TABLE 2: Physical and chemical parameters of the soil profiles for each horizon.

Horizonte	CXve 3 (profile 1)				PBACal 2 (profile 2)				PBACal 1 (profile 4)				RRha (profile 3)		
	Ap	AB	Bi	BC	Ap	Bt1	Bt2	BC	Ap	Bt1	Bt2	BC	Alp	A2	CR
Thickness (cm)	14	21	40	80	24	18	37	71	15	16	48	71	25	36	109
Ca ($cmol_c/kg^{-1}$)	13.6	6.1	8	6.4	6.3	3	1.9	1.8	10	3.7	2.9	1.5	11.1	12.9	3.1
Mg ($cmol_c/kg^{-1}$)	3.1	1.9	3.9	4.5	1.1	1.3	0.9	1	3.9	2	0.9	0.8	2.7	4.1	1
K ($cmol_c\cdot kg^{-1}$)	0.8	0.3	0.3	0.5	0.3	0.2	0.2	0.3	5	2	2	2	0.7	1.5	1.3
Organic C ($g\cdot kg^{-1}$)	27	7	6	3.7	16	13	8	5.8	23	12	10	5.2	16	15	4.9
N ($g\cdot Kg^{-1}$)	2.3	0.7	0.6	0.4	1.3	1.1	0.7	0.6	2.1	1.2	1	0.5	1.6	1.5	0.5
CEC ($cmol_c/kg$)	21	11	16	30	12	12	11	10	18	11	13	12	18	20	13
Bases saturation (%)	83	76	79	38	66	36	27	31	78	51	30	21	80	87	33
Sum of bases ($cmol_c/kg$)	17	8	13	11	8	4	3	3	14	6	4	2	15	17	4
P (mg/kg)	67	9	7	8	20	8	8	7	52	12	9	13	74	64	15
pH in water	6.0	6.0	5.8	5.0	5.4	5.0	4.8	4.8	5.5	5.3	4.8	4.7	6.3	6.0	4.7
Al exch. ($cmol_c/kg$)	0	0	0	11	0.2	3	4	3	0.4	1	5	7	0	0	7
Al saturation (%)	0	0	0	58	3	42	59	52	3	14	56	75	0	0	63
Gravel ($g\cdot kg^{-1}$)	10	0	0	0	0	0	10	0	0	0	20	60	200	230	250
Clay ($g\cdot kg^{-1}$)	400	470	520	610	360	610	590	570	410	460	570	420	260	220	220
Silt ($g\cdot kg^{-1}$)	250	200	190	60	310	210	230	220	370	360	290	220	250	310	190
Fine sand ($g\cdot kg^{-1}$)	100	90	110	140	90	50	50	60	70	70	50	110	140	150	110
Coarse sand ($g\cdot kg^{-1}$)	250	240	180	190	240	130	130	150	150	110	90	250	350	320	480
Soil density	1.15	1.1	1.1	1.12	1.32	1.3	1.4	1.4	1.1	1.2	1.2	1.1	1.0	1.1	nd
WSC (mm)	5.3	6.9	17.6	17.9	12.7	4.7	9.2	19.9	6.8	9.9	22	15.1	7.9	7.5	nd
S (mg/dm^3)	6	2	2	12	4	21	20	13	10	15	47	42	70	4	22
Zn (mg/dm^3)	41	3	0.4	1	13	1	1	1	27	3	1	2	14	15	0.4
Cu (mg/dm^3)	351	18	3	3	123	4	3	4	110	10	5	3	97	105	1
B (mg/dm^3)	0.5	0.3	0.3	0.2	0.4	0.1	0.2	0.1	0.7	1.3	1.4	0.3	0.4	0.6	0.4
Mn (mg/dm^3)	55	19	11	6	45	19	14	13	57	25	6	4	40	25	16

CEC = capacity for exchangeable cations; WSC = water storage capacity.

TABLE 3: Statistics for variables related to the 108 soil samples (Ap horizon).

Variable	Unit	Mean	Range	VC (%)
Pebbly (>20 mm)	$g\cdot kg^{-1}$	50	0–270	132
Gravel (2–20 mm)	$g\cdot kg^{-1}$	160	0–410	69
Fine land (<2 mm)	$g\cdot kg^{-1}$	790	480–1000	17
C (organic)	$g\cdot kg^{-1}$	20	8–46	32
N	$g\cdot kg^{-1}$	2	0.5–4.4	34
pH (in water)		6	5–7.4	7
Al	$cmolc\cdot kg^{-1}$	0.01	0–0.5	441
Ca	$cmolc\cdot kg^{-1}$	11	4–23	32
Mg	$cmolc\cdot kg^{-1}$	2.4	1.2–4.5	31
K	$cmolc\cdot kg^{-1}$	0.6	0.1–1.1	34
P	$cmolc\cdot kg^{-1}$	86	9–423	85
Bases saturation	%	76	29–97	17

VC = variation coefficient (standard deviation expressed as percentage of the mean).

TABLE 4: Results of ANOVA as a factor of soil classes. LS = Levene statistic; S_{HV} = significance of homogeneity of variances test, %; S_A = significance of F statistic, % (ANOVA). Number of samples for the classes of soils and total: CXve1, CXve2, CXve3, PBAcal1, PBAcal2, PBAcal3, RRh1, RRh2, RRh3, and RRh4 and total equal to 8, 10, 22, 6, 15, 7, 11, 4, 8, 17, and 108, respectively.

Variable	Units	Extreme means	LS	S_{HV}	F	S_A
Coarse sand	g·kg^{-1}	180–340	1.43	18.5	4.94	0.0
Fine sand	g·kg^{-1}	73–139	1.64	11.3	13.96	0.0
Silt	g·kg^{-1}	206–370	1.61	12.2	3.77	0.0
Clay	g·kg^{-1}	222–479	1.56	13.7	5.25	0.0
C organic	g·kg^{-1}	15–30	3.06	0.3	7.08	0.0
N	g·kg^{-1}	1,4–3	2.09	3.7	8.39	0.0
pH		5,7–6,6	1.14	34.4	5.09	0.0
Ca	cmol$_c$·kg^{-1}	7–15,4	7.25	0.0	10.37	0.0
Mg	cmol$_c$·kg^{-1}	1,5–3,5	2.99	0.3	14.00	0.0
K	cmol$_c$·kg^{-1}	0,36–0,74	1.61	12.2	7.03	0.0
P	cmol$_c$·kg^{-1}	35–127	2.16	3.2	2.89	0.5
CEC	cmol$_c$·kg^{-1}	9–20	7.00	0.0	12.10	0.0
Bases saturation	%	63–86	3.93	0.0	3.36	0.1

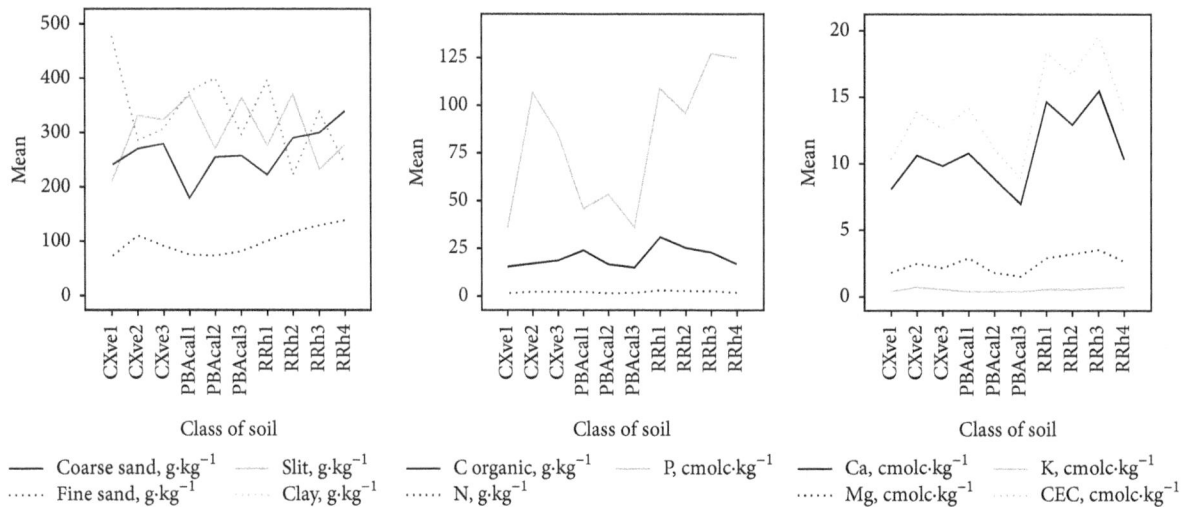

FIGURE 2: The linear graphs show the variation of the mean related to each class of soil for different variables.

3.3. Data Integration of Geospatial Data.

Map of integrated granulometric variables in Ap horizon overlapped partially conventional map of soils (Figure 7). Vineyard 1 was divided into two parts in direction E-W, with local variations, perhaps related to transition among edaphic units. A spot with content of clay + silt + fine sand from 700 to 750 g·kg^{-1} in vineyard 2 may be associated with *Cambissolos* (CXve1, CXve2, and CXve3) and the transition to *Neossolo* (RRha), with diagonal geometry suggesting the influence of slope, very strong in this zone. *Argissolo* was well characterized in vineyard 3 but the same did not happen to *Neossolos* (RRh1, RRh2, and RRh3).

The western part of vineyard 1 and the northern part of vineyard 2 showed similar spatial distribution of low values of nutrient content in Ap horizon (Figure 8). Then, a moderate zone in nutrient content was established, including the northern half of vineyard 3. The remaining half of vineyard 3 showed significant increase in nutrient content, mainly organic matter, Ca and Mg. According to the present land use, there is a woodland on the lateral part of the vineyard 3 (Figure 1), then perhaps part of the wood was removed before grapevines were cultivated. This may explain increase in organic matter but enrichment of alkaline earth elements seems to be associated with fertilization.

The map management zones (Figure 9) was elaborated by overlapping the maps of Figures 7 and 8. This map and previous maps [9, 10] were elaborated with different methods or based on different variables, but final result is similar. Conventional soil maps evaluate circulation of water and nutrients to all horizons, usually up to 1.5–2.0 m deep, evolving subsuperficial variability. On the other hand, the maps based on superficial sampling (Ap horizon) do not include that dimension but a significant analysis of horizontal variability.

3.4. Final Remarks.

Pedological maps and superficial soil maps overlap partially, perhaps due to their complementary

Coarse sand (g·kg⁻¹)	Thin sand (g·kg⁻¹)	Silt (g·kg⁻¹)	Clay (g·kg⁻¹)

The above represents the legend:

Coarse sand (g·kg^{-1}): 0–100, 100–200, 200–350, 350–400, 400–500

Thin sand (g·kg^{-1}): 0–40, 40–80, 80–120, 120–160, 160–200

Silt (g·kg^{-1}): 0–100, 100–200, 200–300, 300–400, 400–500

Clay (g·kg^{-1}): 0–150, 150–350, 350–450, 450–600, 600–1000

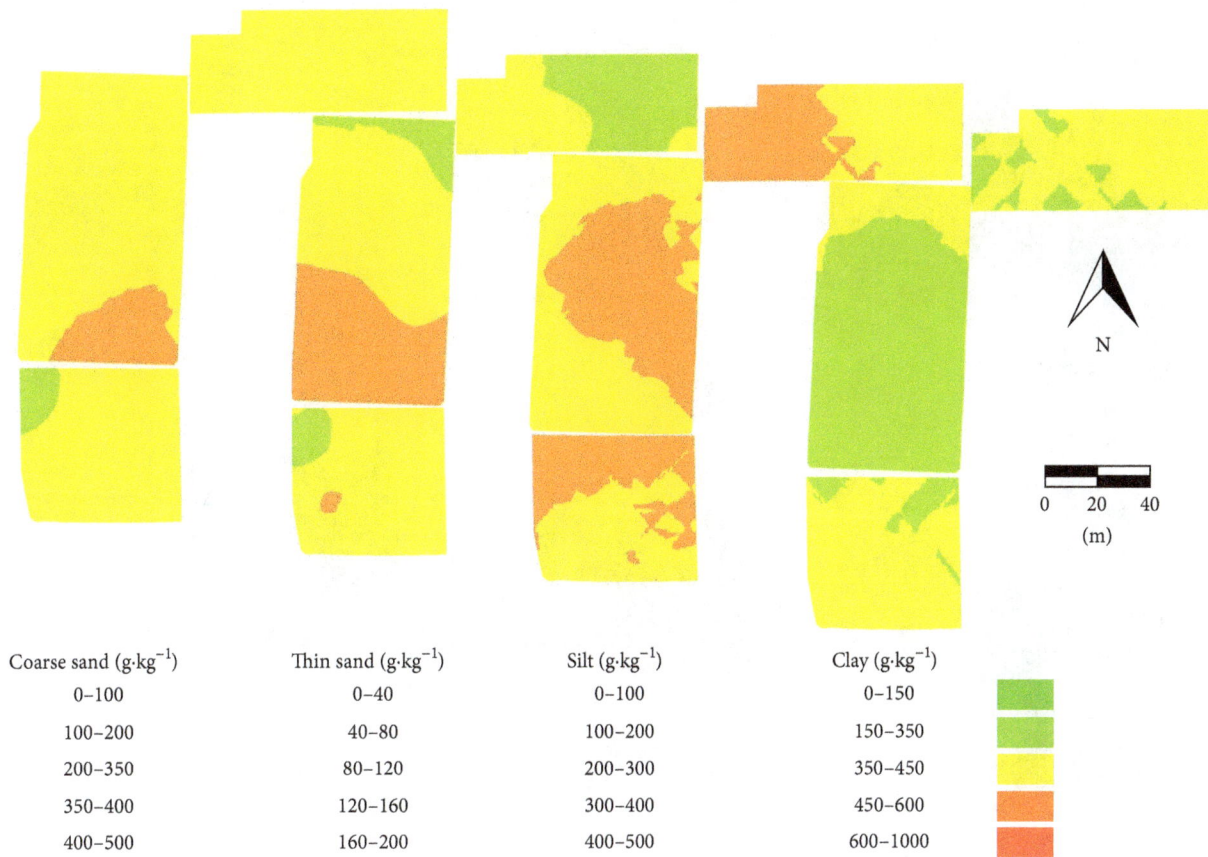

FIGURE 3: Spatial distribution of granulometric variables related to Ap horizon in the study area.

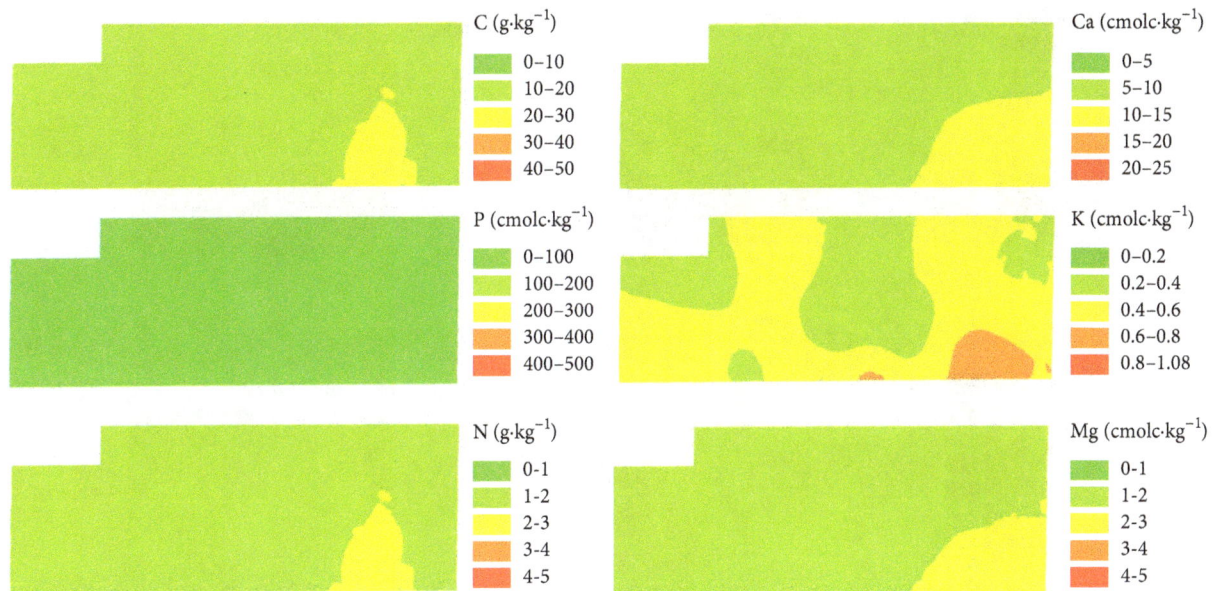

C (g·kg^{-1}): 0–10, 10–20, 20–30, 30–40, 40–50

Ca (cmolc·kg^{-1}): 0–5, 5–10, 10–15, 15–20, 20–25

P (cmolc·kg^{-1}): 0–100, 100–200, 200–300, 300–400, 400–500

K (cmolc·kg^{-1}): 0–0.2, 0.2–0.4, 0.4–0.6, 0.6–0.8, 0.8–1.08

N (g·kg^{-1}): 0-1, 1-2, 2-3, 3-4, 4-5

Mg (cmolc·kg^{-1}): 0-1, 1-2, 2-3, 3-4, 4-5

FIGURE 4: Spatial distribution of nutrients in the soils of vineyard 1 (horizon Ap).

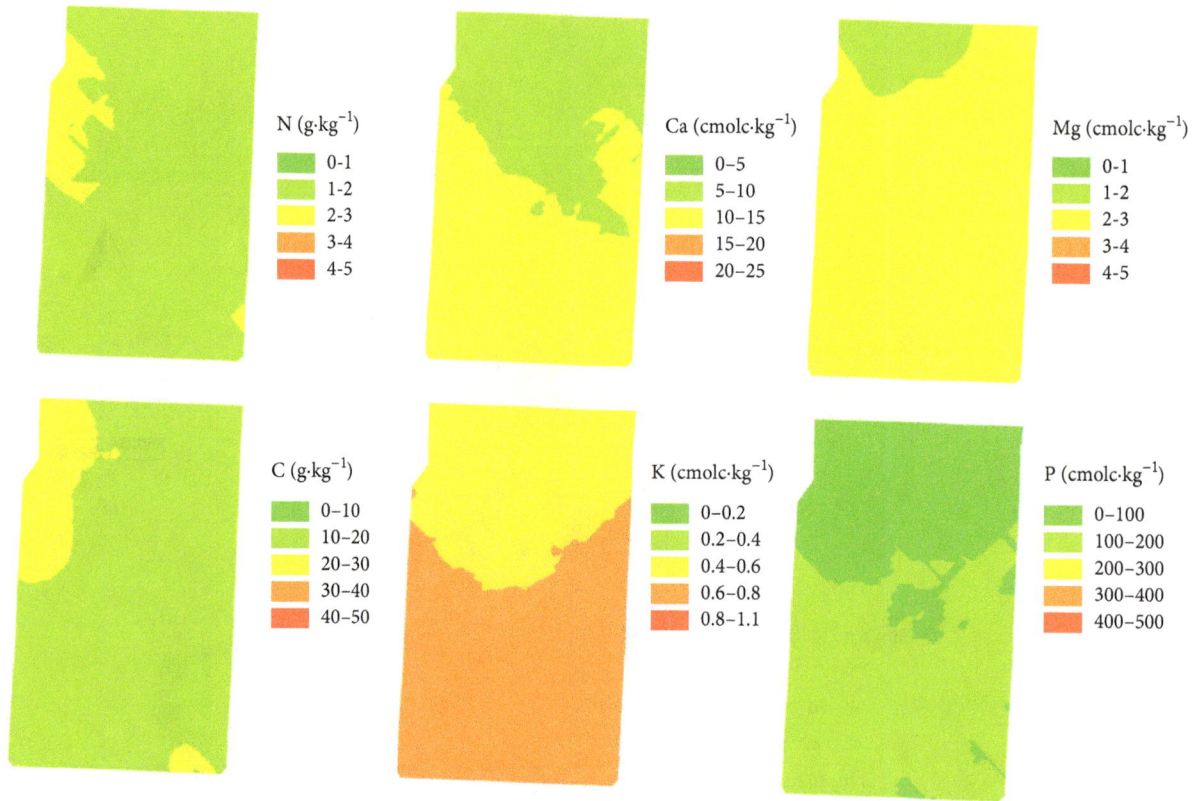

FIGURE 5: Spatial distribution of nutrients in the soils of vineyard 2 (horizon Ap).

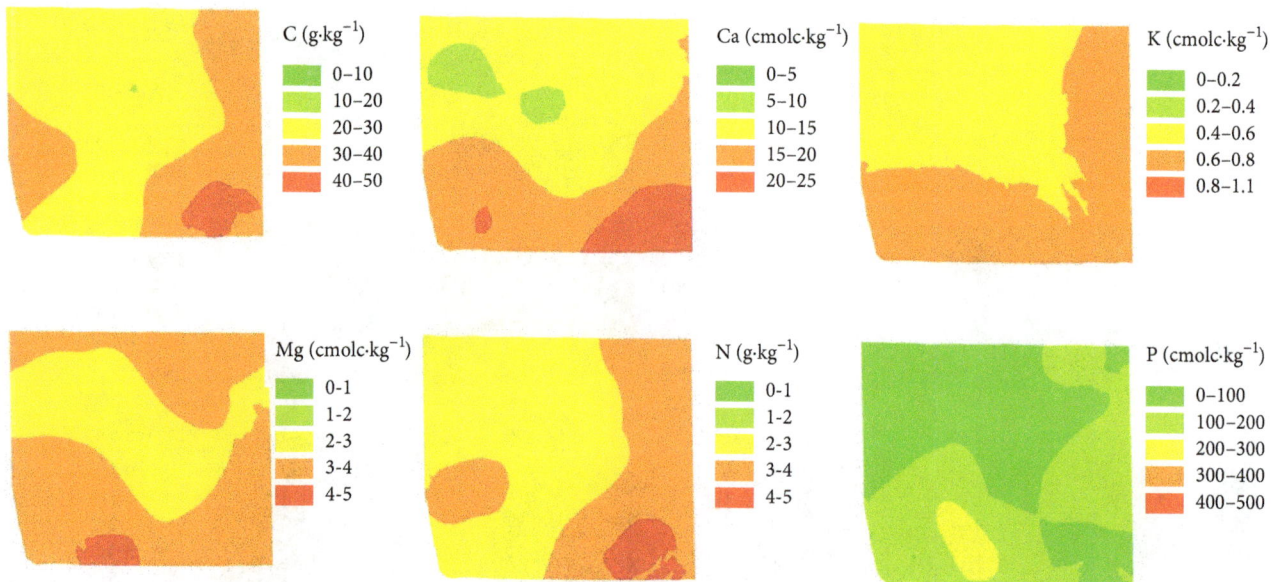

FIGURE 6: Spatial distribution of nutrients in the soils of vineyard 3 (horizon Ap).

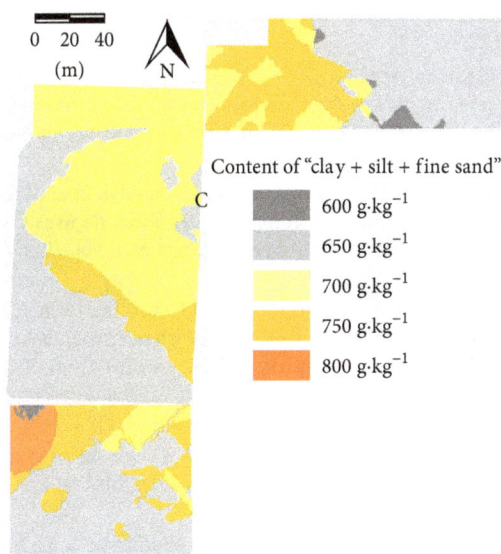

FIGURE 7: Spatial distribution of the content of fine minerals (clay, silt, and fine sand) in Ap horizon for the study area.

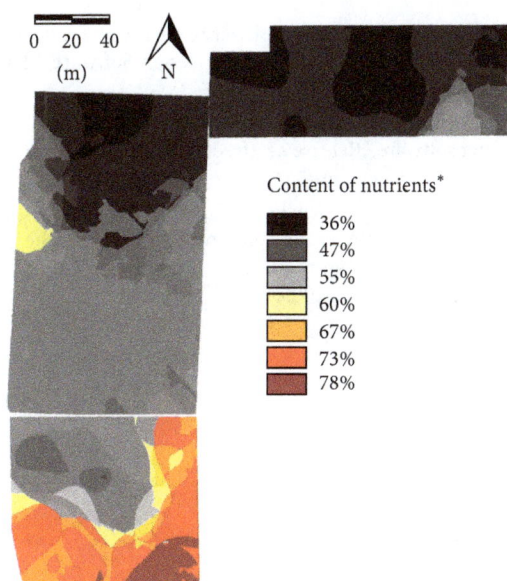

FIGURE 8: Spatial distribution of the content of nutrients (C, Ca, K, Mg, N, and P) in Ap horizon for the study area. *C + Ca + K + Mg + N + P.

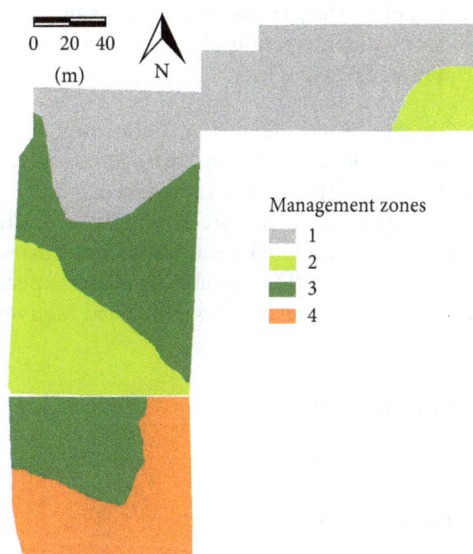

FIGURE 9: Management zones defined by integration of ten variables (granulometry and nutrients). 1: moderate CSS and low CN; 2: low CSS and moderate CN; 3: moderate CSS and moderate CN; 4: moderate CSS and high CN. CSS = content of clay + silt + fine sand; CN = content of nutrients.

(1) Soil Mapping Is an Expensive Procedure, Costly, and Time Consuming; Is There Some Other Technique with the Same Level of Efficiency and Cheaper? Several authors mentioned the influence of soil depth on grape quality; however the type of soil would not be an affecting factor in that sense. Some results of the present work suggest the influence of declivity that altogether with soil depth and other parameters control the water cycle of soil, therefore, the relation between soil and grapevines.

Hence, geotechnologies represent exceptional tools to support precision agriculture, improving the knowledge of the horizontal variability obtained from soil research, but a vertical component must be incorporated. The superficial sampling of soil must be complemented with a digital elevation model and the study of depth of soil and texture of subsuperficial horizons. In that sense, studies on apparent electrical conductivity represent a good option, since, data in two depths are supplied [20].

(2) Geospatial Analysis and GIS Procedures Are Related to PA, but Standard Procedures for Interpolation and Fusion Data Are Generally Absent of Bibliographies. A Specific Method in That Sense Is Presented and Discussed Here. A method of fusion data was used, including geostatistical processing and unsupervised classification of nutrient data and texture data. A final map of management zones was constructed by observation and synthesis. There was coincidence with previous management zones [9, 10], but local differences and variations from each methodology were observed. Correlation between the management zone map produced by this work and the conventional soil map [8] was weak in that sense. Management zones must be conferred according to grapes quality or derived wines, but in this work only the

nature but not substitutive. Management zones can be defined by means of the pedological map or when several information lawyers are integrated. Anyway, this and previous works [9, 10] pointed that different methods resulted in similar management zones but with local differences.

Perhaps, an integration of information lawyers, from an PA perspective, would be more efficient way to define management zones. However, the ideal set of variables must be found.

Addressing the specific questions mentioned in the introduction of this paper.

edaphic parameters were considered. Consequently, a new challenge will be the integration of these other kinds of data with information discussed here.

(3) How Are Soil Units Related to Viticulture Variables? Shallow soils provide few water and nitrogen to vines. Consequently, the yield of the vineyard is reduced and the quality of berries is improved. In the present study, this situation was corroborated by *Neossolo* RRh 4, which occurred with scarce depth (Table 2). A similar status would be expected for units RRh 1 and RRh 3; nevertheless, they had high levels of nutrients, as explained before.

Competing Interests

The authors declare that they have no competing interests.

Acknowledgments

The authors are thankful to Miolo Winery for making its vineyards available to carry out this research and to CNPq for providing part of resources.

References

[1] E. Tayari, A. R. Jamshid, and H. R. Goodarzi, "Role of GPS and GIS in precision agriculture," *Journal of Scientific Research and Development*, vol. 2, no. 3, pp. 157–162, 2015.

[2] T. Proffitt, R. Bramley, D. Lamb, and E. Winter, *Precision Viticulture: A New Era in Vineyard Management and Wine Production*, Winetitles, Adelaide, Australia, 2006.

[3] A. McBratney, B. Whelan, T. Ancev, and J. Bouma, "Future directions of precision agriculture," *Precision Agriculture*, vol. 6, no. 1, pp. 7–23, 2005.

[4] European Parliament, *Precision Agriculture: An Opportunity for EU Farmers—Potential Support with the Cap 2014–2020*, IP/B/AGRI/IC?2013_153, 2014.

[5] A. McBratney, M. Mendonça Santos, and B. Minasny, "On digital soil mapping," *Geoderma*, vol. 117, no. 1-2, pp. 3–52, 2003.

[6] B. Minasny and A. B. Bratney, "Methodologies for global soil mapping," in *Digital Soil Mapping: Bridging Research, Environmental Application, and Operation*, J. L. Boettinger, D. W. Howell, A. C. Moore, A. E. Hartemink, and S. Kienast-Brown, Eds., vol. 2 of *Progress in Soil Science*, pp. 429–436, Springer, Berlin, Germany, 2010.

[7] V. C. Leeuwen, J. P. Roby, D. Pernet, and B. Bois, "Methodology of soil based zoning for viticultural terrois," *Bulletin de l'OIV*, vol. 83, no. 947, pp. 13–29, 2010.

[8] C. A. Flores, J. M. Filippini-Alba, H. F. Levien, D. H. Zarnott, A. Miele, and C. Pavan, "Levantamento detalhado dos solos e a viticultura de precisão," in *Anais do XXXIII Congresso Brasileiro de Ciência do Solo*, Sociedade Brasileira de Ciência do Solo, Uberlândia, Brazil, 2011.

[9] J. M. Filippini-Alba, C. A. Flores, A. Miele, and L. M. Villani, "SIG para a gestão vitivinícola no Vale dos Vinhedos, RS," in *Agricultura de Precisão: Resultados de um Novo Olhar*, A. C. Bernardi, J. M. Naime, A. V. Resende, L. H. Bassoi, and R. Y. Inamasu, Eds., pp. 368–373, Embrapa, Brasília, Brazil, 2014.

[10] J. M. Filippini-Alba, C. A. Flores, and A. Miele, "Modelagem espacial do solo para apoio à viticultura de precisão: Vale dos Vinhedos, Serra Gaúcha, Rio Grande do Sul, Brasil," *Revista Brasileira de Viticultura e Enologia, Bento Gonçalves*, no. 4, pp. 8–17, 2012.

[11] A. Miele, C. A. Flores, and J. M. Filippini-Alba, "Efeito do tipo de solo nos compostos fenólicos e na atividade antioxidante do vinho," *Revista Brasileira de Viticultura e Enologia*, no. 6, pp. 40–47, 2014.

[12] G. Chavarria, H. Bergamaschi, L. C. da Silva et al., "Relações hídricas, rendimento e compostos fenólicos de uvas Cabernet Sauvignon em três tipos de solo," *Bragantia*, vol. 70, no. 3, pp. 481–487, 2011.

[13] J. Coipel, R. B. Lovelle, C. Sipp, and C. van Leeuwen, "'Terroir' effect, as a result of enviromental stess, depends more on soil depth than on soil type (*Vitis vinifera* L. cv. Grenache Noir, Côtes du Rhône, France, 2000)," *Journal International des Sciences de la Vigne et du Vin*, vol. 40, no. 4, pp. 177–185, 2006.

[14] H. G. Santos, P. K. T. Jacomine, L. H. C. dos Anjos et al., Eds., *Sistema Brasileiro de Classificação de Solos*, Embrapa Solos, Rio de Janeiro, Brazil, 2006.

[15] Embrapa, *Manual de Métodos de Análise de Solo*, Embrapa Solos, Rio de Janeiro, Brazil, 2nd edition, 1997.

[16] ESRI (Environmental Systems Research Institute), *ArcGIS Desktop*, Release 10, ESRI, Redlands, Calif, USA, 2011.

[17] IBM, *Statistical Package for Social Science (SPSS)*, Release 19.0.0, 2010.

[18] Gamma (Gamma Design Software), *GS+ User's Guide. 1998–2008, Version 9*, Gamma (Gamma Design Software, Plainwell, Mich, USA, 2008.

[19] ERDAS (Earth Resource Data Analysis System), "ERDAS-ER-Mapper software," Release 7.2. (free trial) http://erdas-er-mapper.software.informer.com/7.2/.

[20] K. A. Sudduth, N. R. Kitchen, W. J. Wiebold et al., "Apparent electrical conductivity to soil properties across the north-central USA," *Computers and Electronics in Agriculture*, vol. 46, no. 1—3, pp. 263–283, 2005.

9

Potential of Using Nanocarbons to Stabilize Weak Soils

Jamal M. A. Alsharef,[1] Mohd Raihan Taha,[1,2]
Ali Akbar Firoozi,[1] and Panbarasi Govindasamy[1]

[1]Department of Civil and Structural Engineering, Universiti Kebangsaan Malaysia (UKM), 43600 Bangi, Selangor, Malaysia
[2]Institute for Environment and Development (LESTARI), Universiti Kebangsaan Malaysia (UKM), 43600 Bangi, Selangor, Malaysia

Correspondence should be addressed to Jamal M. A. Alsharef; jamalshref@yahoo.com

Academic Editor: Teodoro M. Miano

Soil stabilization, using a variety of stabilizers, is a common method used by engineers and designers to enhance the properties of soil. The use of nanomaterials for soil stabilization is one of the most active research areas that also encompass a number of disciplines, including civil engineering and construction materials. Soils improved by nanomaterials could provide a novel, smart, and eco- and environment-friendly construction material for sustainability. In this case, carbon nanomaterials (CNMs) have become candidates for numerous applications in civil engineering. The main objective of this paper is to explore improvements in the physical properties of UKM residual soil using small amounts (0.05, 0.075, 0.1, and 0.2%) of nanocarbons, that is, carbon nanotube (multiwall carbon nanotube (MWCNTs)) and carbon nanofibers (CNFs). The parameters investigated in this study include Atterberg's limits, optimum water content, maximum dry density, specific gravity, pH, and hydraulic conductivity. Nanocarbons increased the pH values from 3.93 to 4.16. Furthermore, the hydraulic conductivity values of the stabilized fine-grained soil samples containing MWCNTs decreased from $2.16E-09$ m/s to $9.46E-10$ m/s and, in the reinforcement sample by CNFs, the hydraulic conductivity value decreased to $7.44E-10$ m/s. Small amount of nanocarbons (MWCNTs and CNFs) decreased the optimum moisture content, increased maximum dry density, reduced the plasticity index, and also had a significant effect on its hydraulic conductivity.

1. Introduction

For a long time, soil stabilization has been one of the best ways to improve the effectiveness of subgrade soils. Compacted cohesive soils with low hydraulic conductivity are commonly used as a landfill cover barrier and bottom liner material. Regarding the long-term performance of the cover barrier system, the desiccation cracking of compacted clay liners is the central relevancy as desiccation will cause cracks in the compacted soil liner and consequently reduce the sealing effect of the cover system dramatically [1, 2]. Additionally, hydraulic conductivity increases with about three orders of magnitude due to desiccation cracking [2]. A few have considered soil additives (lime, sand, and cement) to increase the soil strength and resistance to cracking [3–6]. However, based on the previous studies, the lime or cement additives did not sufficiently suppress desiccation cracking and the low permeability of clayey soils with high water contents. Additionally,

the effects of artificial cementation with quicklime on the microtexture and mineralogy of marine clays from eastern Canada were studied by Choquette et al. [7]. They determined that the addition of lime results in abrupt agglomeration of clay particles and that the flocculated structure is maintained with the growth of cementitious bonds. They also found that lime stabilization results in the production of platy minerals, which partition the interaggregate space within the soil material and increase the volume of the micropores (porous product). They developed a correlation between the amount of structural change due to the addition of lime and the measured strength of the treated clay.

The word nanotechnology incorporates an extensive variety of advances performed on a nanometer scale for far-reaching applications. The principle distinction between a nanomaterial and a bigger scale material is the altogether bigger particular surface region, which allows for an expansion in the substance reactivity and/or a change in the physical properties of the material [8]. The properties displayed by this new class of materials have been utilised in manufacturing for a considerable time; the huge and fast incremental use of nanotechnology is due to the logical meeting of science, material science, science, and design streams [9]. Nanotechnology has opened a new world in nanoscale while civil engineering infrastructure is focused on macroscale. Despite the fact that good pavement is constructed with existing materials, applications of nanotechnology can improve pavement performance significantly. Scientists anticipate that nanotechnology may offer great potential in the fields of material design, manufacturing, properties, monitoring, and modelling for advances in asphalt pavement technology [8–15].

Nanotechnology is defined as the understanding and control of matter at dimensions between 1 and 100 nm, where unique phenomena enable novel applications [16, 17]. The application of nanotechnology to the environment and agriculture was addressed by the United States Department of Agriculture in a document published in September 2003 [18], rapidly evolving and revolutionising agriculture. Nanotechnology can play a major role in pollution sensing through surface-enhanced Raman scattering, surface plasmon resonance, fluorescent detection, electrochemical and optical detection, treatment through adsorption, photocatalysis treatment of pollutants, and reduction by nanoparticles and bioremediation.

The implications of the nanotechnology research in the environment and agriculture are developed based on the identification of the nanoresearch thematic areas of relevance to the environmental and agricultural system. Nanomaterials such as nanoparticles, carbon nanotubes, fullerenes, and biosensors play a role in controlled delivery systems. Nanofiltration finds relevant applications in agri-food thematic areas, such as natural resources management, delivery mechanisms in plants and soils, and use of agricultural waste and biomass, and in food processing and food packaging. Risk assessment is also being evaluated [19–21].

The processes affected by nanoparticles presence in soils (the role of the nanosize fraction) have gained importance recently. Sorption capacity, interfacial electron transfers reactions, mobility, and diffusive mass transfer play a major role in soil properties. Sorption capacity of the NPs is important in topics such as assessment of sorption capacity of the NPs in soils, assessment of the NPs interactions with other minerals of the soil matrix and the resulting effects on contaminant and nutrient adsorption/desorption in soils, usage of NPs for groundwater clean-up and remediation purposes, and the evaluation and quantification of the controls or effects of different variables (physical, chemical, and biological) on these processes. Environmental studies were done by a few scientists about sparing raw materials, wastewater and debased soil treatment, vitality stockpiling, and risky waste administration [20, 21]. The OECD proposes that nanotechnology can help to resolve important environmental issues such as the provision of clean drinking water and the change and detoxification of an extensive variety of contaminants such as PCBs, heavy metals, organochlorine pesticides, and solvents [22]. Nanomaterials exhibit intriguing synthetic and physical properties that make them appropriate for many applications in a field as broad as ecology [23].

The soil nanoalumina mixtures caused beneficial changes to the engineering properties of soil (i.e., compaction characteristics, volumetric expansive strain, the crack intensity factor, and volumetric shrinkage strain). These changes were mostly due to the displacement and rearrangement of soil particles by the addition of nanoalumina [22].

Nanocarbon (NC) fibers are primarily used in industrial sectors such as electronics, automotive, aeronautics, sports, marine, and concrete. NC is also a promising advanced material in the construction industry. NC fibers, especially carbon nanotubes (CNTs) and carbon nanofibers (CNFs), have promising material properties such as high tensile strength, elastic modulus, hardness, and electrical properties [23–25]. The history of carbon nanofibers goes back more than a century. Vapour-grown carbon nanofibers (VCNFs) are a type of carbon nanomaterial which was first explored in 1889 by Hughes and Chambers. It is reported that carbon filaments were grown from carbon, and its hollow graphitic structure was first revealed in the early 1950s by Radushkevich and Lukyanovich [26, 27]. VCNFs can be extensively synthesised by catalytic chemically vapour deposit (CVD) of a hydrocarbon (such as natural gas, propane, acetylene, benzene, and ethylene) or carbon monoxide using metal (Fe, Ni, Co, and Au) or metal alloy (Ni-Cu, Fe-Ni) catalysts at a temperature of 500–1500°C [28, 29].

Carbon nanotubes, first discovered by Iijima in 1991 [30], are one of the most promising classes of new materials to emerge from nanotechnology to date. Nanotubes are members of the fullerene structural family. The name is derived from its long, hollow structure with walls formed by one-atom-thick sheets of carbon called graphene. Carbon nanotubes (CNTs) are allotropes of carbon with a cylindrical nanostructure. Carbon nanotubes were found in Damascus steel from the 17th century, which may account for the legendary strength of swords made from this steel [31–33]. Carbon nanotubes have been constructed with a length-to-diameter ratio of up to 132,000,000 : 1, significantly larger than for any other material. It has an ideal structure formed by carbon atoms with one dimension [33]. These cylindrical carbon molecules have unusual properties which are valuable for nanotechnology, electronics, optics, and other fields of materials science and technology [34]. Moreover, since CNTs exhibit great mechanical properties along with extremely high aspect ratios (length-to-diameter ratio), they are expected to produce significantly stronger and tougher composites than traditional reinforcing materials (e.g., glass fibers or carbon fibers) [35]. Research has been conducted for potential use of CNT in environmental protection application. CNT is used as a selective sorbent for organic/biological contaminants in water streams to remove contaminants such as carcinogenic cyanobacterial microcystins [36]. Moreover, the use of carbon nanotubes as an efficient source and storage of hydrogen is an important research area because of CNT's adsorption characteristics, which are due to its particularly high surface area of 50–1315 m^2/g [37–39].

The main objective of this study was to investigate the effect of nanocarbons (MWCNTs and CNFs) on the physical

TABLE 1: Basic properties of UKM soil.

Characteristics	
Specific gravity	2.6
Passing number 200 sieve (%)	45.39
Clay content (<1 μm) (%)	23
Unified soil classification system (USCS)	SC
Chemical composition	
SiO_2 (%)	62.07
Al_2O_3 (%)	29.46
Fe_2O_3 (%)	5.7
MgO (%)	0.58
CaO (%)	0.03
TiO_2 (%)	1.17
Na_2O (%)	—
K_2O (%)	0.76
Other	0.05
Heat loss	0.18
Organic matter (%)	4.2

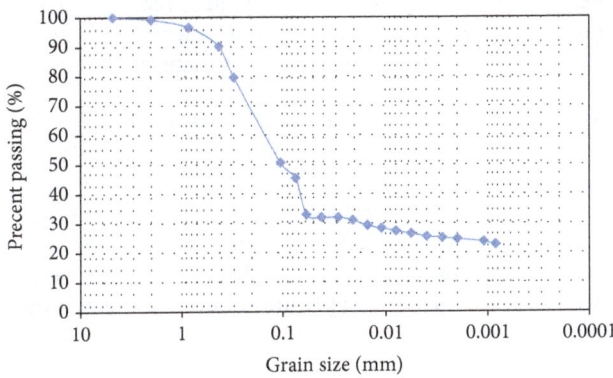

FIGURE 1: Grain size distributions for UKM soil.

FIGURE 2: UKM soil particles under SEM.

TABLE 2: Properties of CNT (Graphistrength C100).

Property	Value
Average diameter (nm)	10–15
Average length (μm)	1–10
Carbon purity, %	>95
Apparent density (kg/mc)	50–150
Relative density (g/mL) at 25°C	2,1
Aspect ratio	600–700
Applications	Reinforcements

TABLE 3: Properties of CNF-PR-19-XT-LHT.

Property	Value
Average diameter, nm (average)	200
Average length (μm)	50–200
Carbon purity, %	>98
Apparent density (kg/mc)	30–300
CVD carbon overcoat present on fiber	No
Nanofiber wall density, g/cc	2–2.1
Iron, ppm	12,466
Aspect ratio	1300–1500
Applications	Mechanical and electrical

properties of UKM soils. The parameters investigated in this study included Atterberg limits, optimum water content, maximum dry density, specific gravity, pH, and hydraulic conductivity.

2. Materials and Methods

A residual soil from the Universiti Kebangsaan Malaysia, Bangi, Malaysia campus, was chosen for the study and termed UKM soil. The soil was sampled from 0.5 to 1 m below the ground surface. The physical and chemical properties of the soil are shown in Table 1. The UKM soil was classified as clayey sand (SC) according to the Unified Soil Classification System. The UKM soil sieve analysis and particle size distribution are shown in Figure 1. Two types of nanocarbons, multiwall carbon nanotube (MWCNTs) and carbon nanofiber (CNFs) with the commercial names Graphistrength® C100 and PR-24-XT-LHT, were selected by 0.0, 0.5, 0.075, 0.1, and 0.2% content, respectively. The properties of CNTs and CNFs are listed in Tables 2 and 3, respectively. Distilled water was used to prepare solutions. Figures 2, 3, and 4 show

scanning electronic microscopes (SEM) and transmission electron microscopy (TEM) images of UKM soil, MWCNTs, and CNFs as used in this study.

2.1. Standard Compaction Test. The purpose of any moisture-density test, commonly called compaction test, is to determine the optimum moisture content (OMC) at which the maximum dry density (MDD) of the soil is attained. It is generally desirable to increase the density in the field to enhance the strength of the soil. The standard Proctor test was conducted for both the natural soils and the soil-nanocarbons mixtures to determine the compaction parameters. The BS 1377-4:1990 standard was followed as a base for these tests. In the standard Proctor test, the soil was compacted in a mould with a 1000 cm^3 volume, also referred to as the 1-liter mould. The soil used for compaction was dried in an oven and crushed with a rubber hammer until it passed the US number 4 sieve. The soils were moistened with tap water

FIGURE 3: Multiwall carbon nanotube (MWCNTs) particles under TEM.

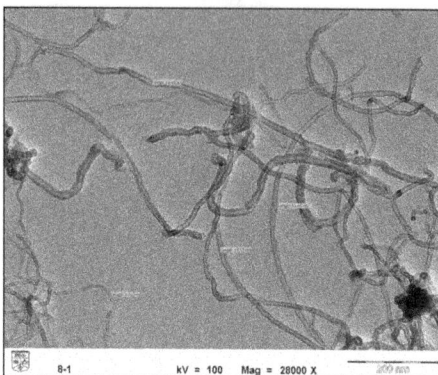

FIGURE 4: Carbon nanofiber (CNFs) particles under TEM.

using a spray bottle and stirred with a trowel during mixing to ensure an even distribution of water. Then, the soils were sealed in plastic bags and allowed to hydrate for at least 24 h prior to compaction. Three equal layers were compacted with 27 blows for each layer.

2.2. Atterberg Limits. The Atterberg limits (i.e., the liquid limit, the plastic limit, and the plasticity index) of each of the natural and treated soil samples were determined in accordance with BS 1377, part 2, (1990). The liquid limit tests were performed using a penetration cone assembly, which consisted of a stainless steel cone with a cone angle of $30° \pm 1°$. The plastic limit tests were performed using a manual method. Each sample was rolled at a sufficient pressure on a glass plate to form a thread with a uniform diameter of 3.2 mm along the full length of the sample. The plasticity index was calculated as the difference between the water contents at the liquid and plastic limits.

2.3. Specific Gravity. The specific gravity for a solid soil can be defined as a ratio of the weight of the solid at a specific volume to the weight of water at the same volume. It is actually the average value of all the mineral types found in each of the soil samples in the test. This test was conducted on control soil using a small pycnometer of 50 mL capacity. An average of three samples was taken to determine the specific gravity value of each type of soil used. This test was conducted in

accordance with the procedure suggested by BS: 1377, part 2:1990, Clause 8.3.

2.4. Hydraulic Conductivity. The soil was compacted according to the standard test method (BS1377: part 4:1990 Clause 3.4) for both the untreated soil and the soil containing nanomaterials. Cylindrical specimens with a diameter of 70 mm and a height of 35 mm were prepared from the mixtures compacted by standard compaction energy at optimum water content. Then, its hydraulic conductivity was determined following ASTM D5084, that is, using flexible membrane apparatus as shown in Figure 5. Porous stones and filter paper were placed against the ends of the samples to distribute the permeated deaired water across the entire end area of the sample. Once the sample had been prepared in the test cell, the cell was filled with water and the specimen was saturated by applying pressures gradually step by step from two directions, with back pressuring from the bottom and cell surrounding the sample to force water to enter the sample for saturation until the back pressure reached 215 kPa, providing a saturation degree of more than 98% [40, 41]. After the saturation had been completed, readings of inlet and outlet burettes were taken until the measured hydraulic conductivity reached a relatively steady-state condition.

2.5. pH Test. In this study, the electrometric method (BS 1377: part 3: 1990: Clause 9) was utilised to determine the pH value of soil suspension in water as this is considered to be the most accurate method. This method requires a soil-water ratio of 1 : 2.5 and can be obtained by passing 30 gr soil through a 2 mm sieve and diluted with 75 mL distilled water for at least 8 hours. Figure 6 shows the pH meter device which was used in this test.

3. Results and Discussion

The effects of multiwall carbon nanotube (MWCNTs) and carbon nanofiber (CNFs) on specific gravity, Atterberg limit's (liquid limit, plastic limit, and plasticity index), pH, and hydraulic conductivity for clayey sand soil (UKM soil) are shown in Figures 7–15.

3.1. Effect on Specific Gravity (G_s). The test result (Figure 7) shows that reinforcing soil with MWCNTs and CNFs decreased the specific gravity from 2.604 to 2.53 for MWCNTs and from 2.604 to 2.542 for soil reinforced by CNFs.

3.2. Effect on pH. The test results showed that the pH of the stabilized soil samples increased with an increase in nanomaterial percentage for both nanocarbon materials (Figure 8) [42, 43]. However, the increase in pH was insignificant for both nanocarbons.

3.3. Effect on Compaction Characteristics. The optimum water content and maximum dry density values of reinforced soils are shown in Figures 9, 10, 11, and 12, respectively. These figures indicate that increased nanocarbon contents (at the optimum nanomaterial ratio) were associated with decreased optimum water content and increased maximum

FIGURE 5: Sample preparation procedures for hydraulic conductivity test in flexible membrane apparatus.

dry density. Beyond a certain nanocarbon content, any increase in the nanocarbon percentage tended to reduce the dry density [44]. The decrease in water content is related to the tendency of nanocarbons to fill pores between soil particles. In Figure 11, containing information of a soil sample reinforced by MWCNTs, it can be seen that the maximum decrease in optimum water content was 13.6% at 0.075% MWCNTs content, and dry density increased from 1.86 to 1.89 (g/cm^3). For the soil sample reinforced by CNFs, the maximum decrease in optimum water content was 14% at 0.075% CNFs content. Furthermore, dry density increased from 1.86 to 1.88 (g/cm^3); see Figure 12.

3.4. *Effect on Atterberg's Limits.* Atterberg limits (plastic limit "PL," liquid limit "LL," and plasticity index "PI" = LL – PL) play an important role in soil identification and classification. These parameters indicate some of the geotechnical problems such as swell potential and workability. The results of the liquid limit test of the soil samples containing various percentages of MWCNTs and CNFs used are shown in Figures 13 and 14, respectively. The results indicate that, for reinforcement soil samples, the liquid limit decreased with increasing MWCNTs and CNFs contents, especially at the optimum nanomaterial content level. Reductions in the plasticity (Figures 13 and 14) indices are indicators of

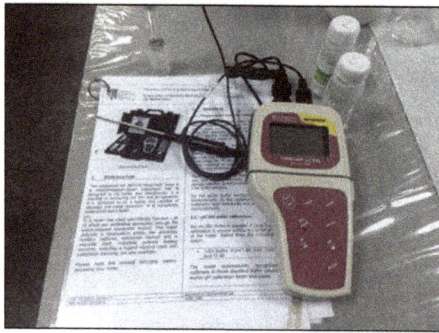

FIGURE 6: pH meter device used in this research.

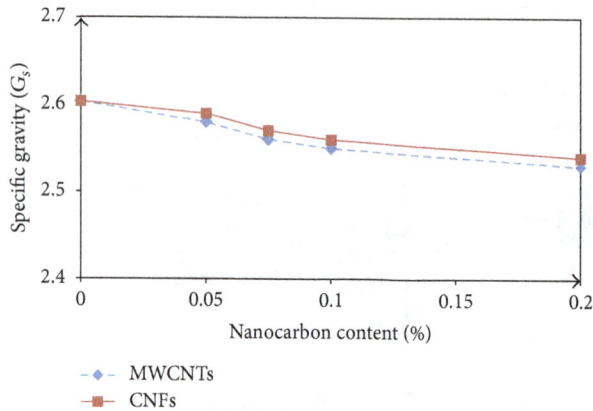

FIGURE 7: Effect of MWCNTs and CNFs on the specific gravity.

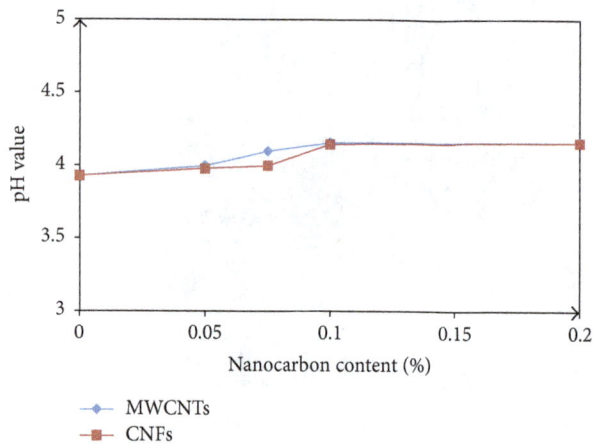

FIGURE 8: Effect of MWCNTs and CNFs on the pH.

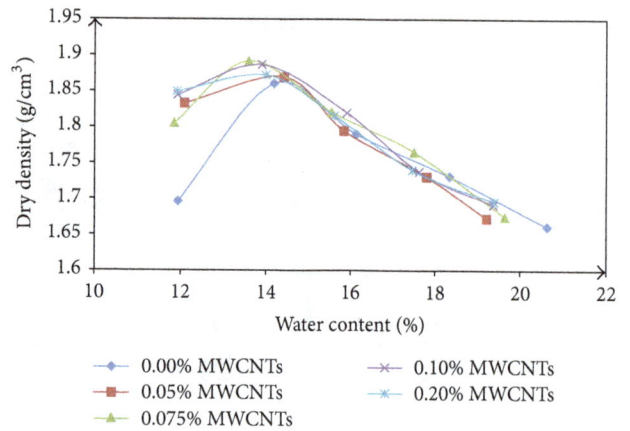

FIGURE 9: Compaction curves for treated UKM soil with MWCNTs.

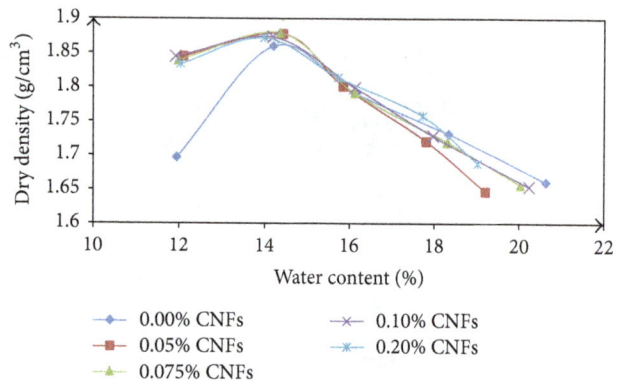

FIGURE 10: Compaction curves for treated UKM soil with CNFs.

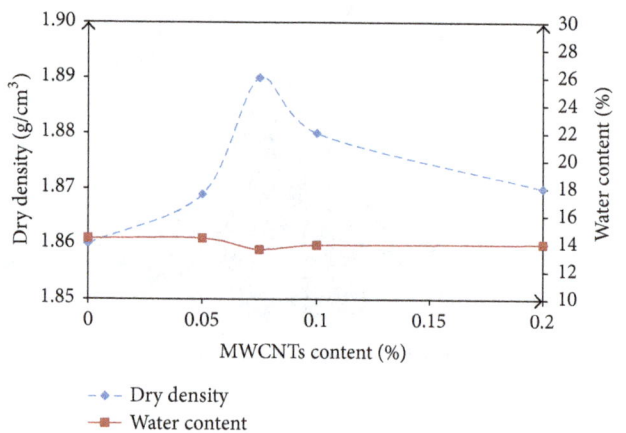

FIGURE 11: Effect of MWCNTs on the OMC and MDD.

soil improvement, because the plasticity index represents the range of water content over which a soil is plastic [45].

3.5. Effect on Soil Hydraulic Conductivity. The hydraulic conductivity (k) test result drawn in Figure 15 clearly shows a decrease in k value of the tested soils corresponding to an increase in the nanocarbon (MWCNTs and CNFs) percentage. Nanocarbon contents less than the optimum content significantly affected the k values. Compared with the natural fine-grained soil samples, k values of the stabilized fine-grained soil samples containing nanocarbons decreased

from $2.16E-09$ m/s to $9.46E-10$ m/s for soil samples reinforced by MWCNTs and $2.16E-09$ m/s to $7.44E-10$ m/s for soil samples reinforced by CNFs.

4. Conclusion

A study was conducted to evaluate the effects of adding a small amount (less than 0.2%) of multiwall carbon nanotube

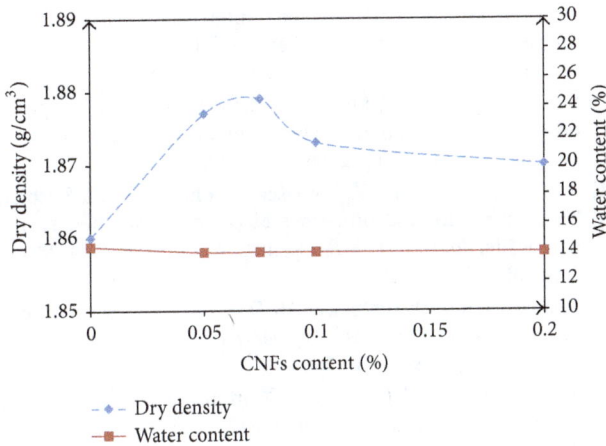

FIGURE 12: Effect of CNFs on the OMC and MDD.

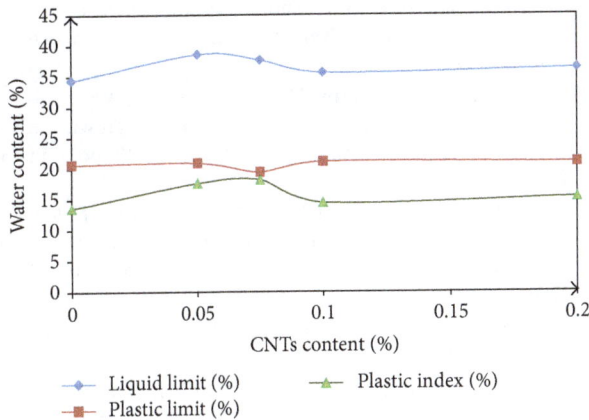

FIGURE 13: Effect of MWCNTs on LL, PL, and PI.

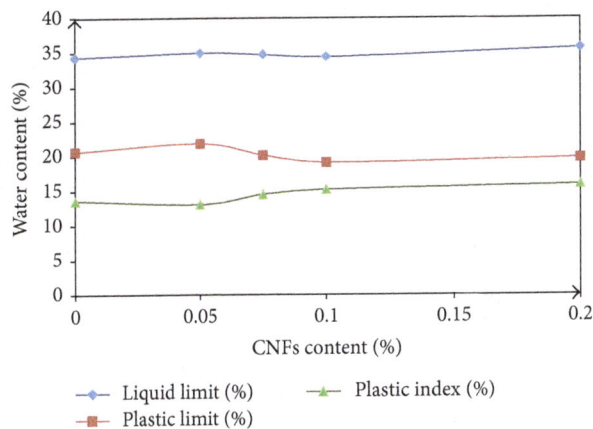

FIGURE 14: Effect of CNFs on LL, PL, and PI.

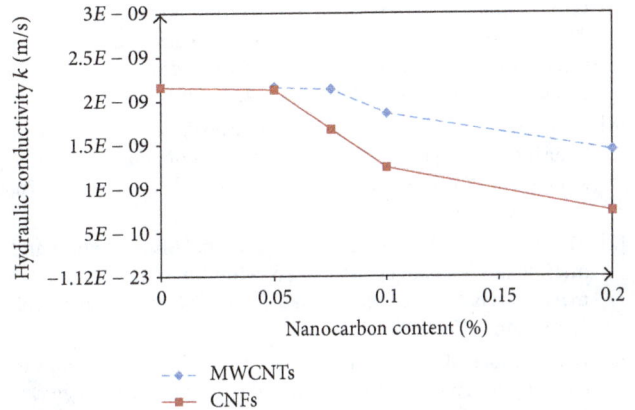

FIGURE 15: Effect of MWCNTs and CNFs on hydraulic conductivity.

reduced porosity by filling the space between soil particles and bonded the particles together. The results showed that decreased OMC and increased MDD were for treated soil by MWCNTs at 0.075% nanocarbon content. Lastly, the result of the hydraulic conductivity test showed that nanocarbons have a significant ability to reduce k value. A comparison between the results from the two different nanocarbons (MWCNTs and CNFs) showed that the soil samples treated with CNFs had a higher reduction in hydraulic conductivity. This differs from other treatment materials like polypropylene and other types of fiber, which reduces cracks in the soil but increases the hydraulic conductivity.

Competing Interests

The authors declare that they have no competing interests.

Acknowledgments

The authors acknowledge and appreciate the financial support and facilities provided by the Geotechnical Engineering Lab of Universiti Kebangsaan Malaysia (UKM) for this study and the Fuel Cell Institute for SEM and TEM tests.

References

[1] K. Witt and R. Zeh, "Cracks due to desiccation in cover lining systems phenomena and design strategy," in *Proceedings of the International Workshop LIRIGM*, Grenoble University, Grenoble, France, March 2005.

[2] S. Sadek, S. Ghanimeh, and M. El-Fadel, "Predicted performance of clay-barrier landfill covers in arid and semi-arid environments," *Waste Management*, vol. 27, no. 4, pp. 572–583, 2007.

[3] M. Leung and C. Vipulanandan, "Treating contaminated, cracked and permeable field clay with grouts," in *Proceedings of the Specialty Conference on Geotechnical Practice in Waste Disposal*, pp. 829–843, New Orleans, La, USA, February 1995.

[4] G. H. Omidi, J. C. Thomas, and K. W. Brown, "Effect of desiccation cracking on the hydraulic conductivity of a compacted clay liner," *Water, Air, and Soil Pollution*, vol. 89, no. 1-2, pp. 91–103, 1996.

(MWCNTs) and carbon nanofiber (CNFs) to clayey sand soil (UKM soil). The liquid limit and plastic limit increased with the amount of multiwall carbon nanotube (MWCNTs) and carbon nanofiber (CNFs) in the mixture. Moreover, the increase in the maximum dry density is due to the particle densities of nanocarbons which are greater than the particle density of natural soil. Furthermore, the nanocarbon fibers

[5] G. H. Omidi, T. V. Prasad, J. C. Thomas, and K. W. Brown, "The influence of amendments on the volumetric shrinkage and integrity of compacted clay soils used in landfill liners," *Water, Air, and Soil Pollution*, vol. 86, no. 1-4, pp. 263–274, 1996.

[6] A. A. Firoozi, M. R. Taha, A. A. Firoozi, and T. A. Khan, "The influence of freeze-thaw cycles on unconfined compressive strength of clay soils treated with lime," *Jurnal Teknologi*, vol. 76, no. 1, pp. 107–113, 2015.

[7] M. Choquette, M.-A. Bérubé, and J. Locat, "Mineralogical and microtextural changes associated with lime stabilization of marine clays from eastern Canada," *Applied Clay Science*, vol. 2, no. 3, pp. 215–232, 1987.

[8] A. A. Firoozi, M. R. Taha, and A. A. Firoozi, "Nanotechnology in civil engineering," *Electronic Journal of Geotechnical Engineering*, vol. 19, pp. 4673–4682, 2014.

[9] A. A. Firoozi, M. R. Taha, A. A. Firoozi, and T. A. Khan, "Effect of ultrasonic treatment on clay microfabric evaluation by atomic force microscopy," *Measurement: Journal of the International Measurement Confederation*, vol. 66, pp. 244–252, 2015.

[10] M. Diallo, S. Christie, P. Swaminathan et al., *Nanotechnology Applications for Clean Water: Solutions for Improving Water Quality*, William Andrew, 2009.

[11] K. Sobolev, "Modern developments related to nanotechnology and nanoengineering of concrete," *Frontiers of Structural and Civil Engineering*, vol. 10, no. 2, pp. 131–141, 2016.

[12] M. R. Taha, T. A. Khan, I. T. Jawad, A. A. Firoozi, and A. A. Firoozi, "Recent experimental studies in soil stabilization with bio-enzymes-a review," *Electronic Journal of Geotechnical Engineering*, vol. 18, pp. 3881–3894, 2013.

[13] M. S. Baghini, A. B. Ismail, M. R. B. Karim, F. Shokri, and A. A. Firoozi, "Effects on engineering properties of cement-treated road base with slow setting bitumen emulsion," *International Journal of Pavement Engineering*, vol. 9, no. 4, pp. 321–336, 2016.

[14] A. A. Firoozi, M. R. Taha, and A. A. Firoozi, "Analysis of the load bearing capacity of two and three-layered soil," *Electronic Journal of Geotechnical Engineering*, vol. 19, pp. 4683–4692, 2014.

[15] M. R. Taha, T. A. Khan, I. T. Jawad, A. A. Firoozi, and A. A. Firoozi, "Recent experimental studies in soil stabilization with bio-enzymes-a review," *Electronic Journal of Geotechnical Engineering*, vol. 18, pp. 3881–3894, 2013.

[16] C. Demetzos, "Introduction to nanotechnology," in *Pharmaceutical Nanotechnology*, pp. 3–15, Springer, Berlin, Germany, 2016.

[17] M. C. Roco, "Broader societal issues of nanotechnology," *Journal of Nanoparticle Research*, vol. 5, no. 3-4, pp. 181–189, 2003.

[18] N. Scott, H. Chen, and C. J. Rutzke, "Nanoscale science and engineering for agriculture and food systems: a report submitted to," in *A Report Submitted to Cooperative State Research, Education and Extension Service, the United States Department of Agriculture: National Planning Workshop*, USDA, Washington, DC, USA, November 2002.

[19] A. Al-Amoudi, "Nanofiltration membrane cleaning characterization," *Desalination and Water Treatment*, vol. 57, no. 1, pp. 323–334, 2016.

[20] A. A. Firoozi, G. Olgun, and S. Mobasser, "Carbon nanotube and civil engineering," *Saudi Journal of Engineering and Technology*, vol. 1, no. 1, pp. 1–4, 2016.

[21] H. Dai, E. T. Thostenson, and T. Schumacher, "Processing and characterization of a novel distributed strain sensor using carbon nanotube-based nonwoven composites," *Sensors*, vol. 15, no. 7, pp. 17728–17747, 2015.

[22] M. R. Taha and O. M. E. Taha, "Influence of nano-material on the expansive and shrinkage soil behavior," *Journal of Nanoparticle Research*, vol. 14, no. 10, article 1190, 2012.

[23] S. Bal, "Experimental study of mechanical and electrical properties of carbon nanofiber/epoxy composites," *Materials and Design*, vol. 31, no. 5, pp. 2406–2413, 2010.

[24] A. Duszová, J. Dusza, K. Tomášek, G. Blugan, and J. Kuebler, "Microstructure and properties of carbon nanotube/zirconia composite," *Journal of the European Ceramic Society*, vol. 28, no. 5, pp. 1023–1027, 2008.

[25] J. Njuguna, K. Pielichowski, and S. Desai, "Nanofiller-reinforced polymer nanocomposites," *Polymers for Advanced Technologies*, vol. 19, no. 8, pp. 947–959, 2008.

[26] K. P. De Jong and J. W. Geus, "Carbon nanofibers: catalytic synthesis and applications," *Catalysis Reviews*, vol. 42, no. 4, pp. 481–510, 2000.

[27] S. Parveen, S. Rana, and R. Fangueiro, "A review on nanomaterial dispersion, microstructure, and mechanical properties of carbon nanotube and nanofiber reinforced cementitious composites," *Journal of Nanomaterials*, vol. 2013, Article ID 710175, 19 pages, 2013.

[28] W. Brandl, G. Marginean, V. Chirila, and W. Warschewski, "Production and characterisation of vapour grown carbon fiber/polypropylene composites," *Carbon*, vol. 42, no. 1, pp. 5–9, 2004.

[29] K. Mukhopadhyay, D. Porwal, D. Lal, K. Ram, and G. N. Mathur, "Synthesis of coiled/straight carbon nanofibers by catalytic chemical vapor deposition," *Carbon*, vol. 42, no. 15, pp. 3254–3256, 2004.

[30] S. Iijima, "Helical microtubules of graphitic carbon," *Nature*, vol. 354, no. 6348, pp. 56–58, 1991.

[31] M. Reibold, P. Paufler, A. A. Levin, W. Kochmann, N. Pätzke, and D. C. Meyer, "Materials: carbon nanotubes in an ancient *Damascus sabre*," *Nature*, vol. 444, no. 7117, p. 286, 2006.

[32] N. Grobert, "Carbon nanotubes—becoming clean," *Materials Today*, vol. 10, no. 1-2, pp. 28–35, 2007.

[33] E. J. Duplock, M. Scheffler, and P. J. D. Lindan, "Hallmark of perfect graphene," *Physical Review Letters*, vol. 92, no. 22, Article ID 225502, 2004.

[34] Y.-S. Wang, X.-S. Cheng, and L. Wang, "Regularity of VEGF expression in bone marrow mesenchymal stem cells under hypoxia," *Chinese Journal of Applied Physiology*, vol. 25, no. 4, pp. 475–477, 2009.

[35] M. S. Morsy, S. H. Alsayed, and M. Aqel, "Hybrid effect of carbon nanotube and nano-clay on physico-mechanical properties of cement mortar," *Construction and Building Materials*, vol. 25, no. 1, pp. 145–149, 2011.

[36] H. Yan, A. Gong, H. He, J. Zhou, Y. Wei, and L. Lv, "Adsorption of microcystins by carbon nanotubes," *Chemosphere*, vol. 62, no. 1, pp. 142–148, 2006.

[37] H.-M. Cheng, Q.-H. Yang, and C. Liu, "Hydrogen storage in carbon nanotubes," *Carbon*, vol. 39, no. 10, pp. 1447–1454, 2001.

[38] A. C. Dillon and M. J. Heben, "Hydrogen storage using carbon adsorbents: past, present and future," *Applied Physics A: Materials Science and Processing*, vol. 72, no. 2, pp. 133–142, 2001.

[39] V. Ilango and A. Gupta, "Carbon nanotubes—a successful hydrogen storage medium," *World Academy of Science, Engineering and Technology*, vol. 7, no. 9, pp. 678–681, 2013.

[40] D. K. Black and K. L. Lee, "Saturating laboratory samples by back pressure," *Journal of the Soil Mechanics and Foundations Division, ASCE*, vol. 99, no. 1, pp. 75–93, 1973.

[41] K. H. Head, *Manual of Soil Laboratory Testing: Effective Stress Tests*, Wiley, New York, NY, USA, 1998.

[42] A. A. Firoozi, M. R. Taha, A. A. Firoozi, and T. A. Khan, "Assessment of nano-zeolite on soil properties," *Australian Journal of Basic & Applied Sciences*, vol. 8, no. 19, p. 292, 2014.

[43] P. Wang, M. Du, H. Zhu, S. Bao, T. Yang, and M. Zou, "Structure regulation of silica nanotubes and their adsorption behaviors for heavy metal ions: PH effect, kinetics, isotherms and mechanism," *Journal of Hazardous Materials*, vol. 286, pp. 533–544, 2015.

[44] Y. Huang and L. Wang, "Experimental studies on nanomaterials for soil improvement: a review," *Environmental Earth Sciences*, vol. 75, no. 6, pp. 1–10, 2016.

[45] M. L. Silveira, N. B. Comerford, K. R. Reddy, W. T. Cooper, and H. El-Rifai, "Characterization of soil organic carbon pools by acid hydrolysis," *Geoderma*, vol. 144, no. 1-2, pp. 405–414, 2008.

10

Integrated Nanozero Valent Iron and Biosurfactant-Aided Remediation of PCB-Contaminated Soil

He Zhang, Baiyu Zhang, and Bo Liu

Northern Region Persistent Organic Pollution Control (NRPOP) Laboratory, Faculty of Engineering and Applied Science, Memorial University, St. John's, NL, Canada A1B 3X5

Correspondence should be addressed to Baiyu Zhang; bzhang@mun.ca

Academic Editor: Ezio Ranieri

Polychlorobiphenyls (PCBs) have been identified as environmental hazards for years. Due to historical issues, a considerable amount of PCBs was released deep underground in Canada. In this research, a nanoscale zero valent iron- (nZVI-) aided dechlorination followed by biosurfactant enhanced soil washing method was developed to remove PCBs from soil. During nZVI-aided dechlorination, the effects of nZVI dosage, initial pH level, and temperature were evaluated, respectively. Five levels of nZVI dosage and two levels of initial pH were experimented to evaluate the PCB dechlorination rate. Additionally, the temperature changes could positively influence the dechlorination process. In soil washing, the presence of nanoiron particles played a key role in PCB removal. The crude biosurfactant was produced using a bacterial stain isolated from the Atlantic Ocean and was applied for soil washing. The study has led to a promising technology for PCB-contaminated soil remediation.

1. Introduction

As family members of chlorinated hydrocarbons, polychlorinated biphenyls (PCBs) are a group of manmade chemicals which were first synthesized in 1881 and commercialized in North American industries from the 1930s to the late 1970s [1–3]. Although never manufactured in Canada, PCBs have been imported and widely used in hundreds of industrial and commercial applications (e.g., electric insulators, plasticizers for adhesives, lubricants, hydraulic fluids, sealants, cutting oils, and flame retardants) due to their nonflammability and electrical insulating properties as well as chemical stability at high temperature and low vapor pressure [4, 5].

These compounds did not exist in nature. After synthesis, they were found in the environment in 1966 [6]. Since then, PCBs were so widely discovered in the global environment where trace concentrations were detected even in remote areas such as the atmosphere of the Arctic and the Antarctic and the hydrosphere and biosphere [3].

Exposure to PCBs can lead to cancer and a variety of serious noncancer health effects on different systems. Hence, Canada restricted the use of PCBs in 1977 and prohibited the import of PCBs in 1980 [7, 8]. Current legislation allows PCB-containing electrical equipment manufactured before 1980 to remain in use until the end of their service life; however, strict maintenance and handling procedures and regulatory control by governments are required to prevent any release into the environment [2]. As specified in the Federal Contaminated Sites Inventory (FCSI), PCB-contaminated sites are recognized in all the provinces and territories throughout Canada. In fact, most of these sites are contaminated as a consequence of inappropriate handling, storage, and disposal [4].

Federal and provincial governments, as well as associated industries, have been obliged to endeavour research efforts and provide financial support for site identification, remediation, and long term monitoring. Since 1994, the number of PCB-contaminated sites has been reduced under provincial jurisdiction. However, the large amount of remaining untreated sites and the revived problems in the treated sites are still risking the provincial ecosystems and environment. The preliminary assessment process estimates the volume of free products could be 15–20 million litres and the majority of the PCB pollutants are deep underground [9]. Industries have been making efforts to solve individual problems and/or processes related to site remediation practices during the past years. Among the existing technologies, incineration and landfill were frequently applied. However, the remediation was usually long term and costly, and the exhaust could cause secondary pollution [10]. There is a shortage of effective technologies to treat and remove PCB contaminants from soils and sediments. This situation has hindered the efforts to effectively protect the environments of this region. Therefore, it is desired that innovative technologies that can enhance the efficiencies and effectiveness of remediation of PCB-contaminated sites be developed within Canadian context.

Nanoscale zero valent iron (nZVI) particles have been widely applied in removing chloridized hydrocarbons including PCBs due to their extraordinarily reductive property [11, 12]. Some recent research has revealed that nZVI particles are effective in the transformation of a large variety of environmental contaminants, while they are inexpensive and nontoxic [13]. nZVI may chemically reduce PCBs effectively through reductive dechlorination, allowing the pollutant to be readily biodegradable after treatment. Studies by Mueller and Nowack [14] have shown that nZVI as a reactive barrier is very effective in the reduction of chlorinated methane, chlorinated ethane, chlorinated benzenes, and other polychlorinated hydrocarbons. Varma [15] has successfully applied nZVI in soil columns with a wide range of plant phenols as additives, which allows greater access to the contaminant and creates less hazardous waste in the manufacturing process. The application of nZVI to the contaminated soil could enhance the dechlorination of PCBs; nevertheless, higher chlorinated biphenyls require much longer time than lower ones to be completely dechlorinated. Biphenyls as the final product of PCBs are still environmental and health hazards which need further treatment. A time-saving technology that can completely degrade PCBs in the soils or remove PCBs from the soils is consequently in demand.

Soil washing has been applied to effectively and rapidly remove soil contaminants. This technology provides a closed system that remains unaffected by external conditions [16], and the system permits the control of the conditions (e.g., additive concentration) under which the soil particles are treated [17]. Soil washing is cost-effective and often combined with other remediation technologies. Solvents are critical for soil washing and selected on the basis of their ability to solubilize specific contaminants and on their environmental and health effects [18]. However, although soil washing can provide a high efficiency when extracting contaminants from the soil, there are still some limitations when dealing with PCBs. One of the constraints is that PCBs have low water solubility—0.0027–0.42 ng/L [19]; they are soluble in organic or hydrocarbon solvents, oils, and fats. Moreover, PCBs tend to stay in the soils instead of flushing with solvents or water. Since high-chlorinated biphenyls are less water-soluble than low-chlorinated ones and PCBs often preferentially adhere to the clay or silt fraction of the soils [20], removal of the high-chlorinated biphenyls in clayey or silty soils will become extremely difficult. It is thus very hard to find an appropriate washing solvent for PCB removal from soil.

Biosurfactants are surface-active compounds from biological sources, usually extracellular, produced by bacteria, yeast, or fungi [21]. Compared with chemical surfactants, biosurfactants have been applied in contaminated soil remediation due to the advantages of low toxicity, high specificity, biodegradability and biocompatibility, and functionality under extreme conditions [22–24]. Applying biosurfactants as the solvents in soil washing systems to treat PCB-contaminated soil has the following benefits: (1) it would effectively enhance solubilization of PCBs in the washing solution, leading to increased removal efficiency; and (2) it could stimulate microbial activity that enhances biodegradation of PCBs which are soil bound [24]. However, although the application of biosurfactants with soil washing can significantly increase the solubility of PCBs that increase the extraction efficiency [23], PCBs that dissolved in the washing solution need to be further treated before being released into the environment. As persistent organic pollutants, PCBs are hard to degrade, leading to costly and complex posttreatment processes before discharge [24]. In addition, larger volumes of washing solution may be needed when additives like biosurfactants are used. A high biosurfactant concentration in the washing solution can cause foaming problems and inhibit the ability to remove PCBs from the soil [25]. Increasing attention has been received on the combination of different technologies in recent years. These technologies can be applied in sequence to enhance the cost effectiveness [26]. Effective dechlorination approaches which can be integrated with soil washing and facilitate PCB biodegradation are thus desired.

This study is essential for the applications to the removal of PCBs from soil. It aims to combine nanotechnology and an existing soil washing system with biosurfactants as the solvent to better clean up the PCB-contaminated sites. Since higher chlorinated biphenyls have lower aqueous solubilities than lower chlorinated ones, biosurfactant-aided soil washing could have higher removal efficiencies on lower chlorinated biphenyls than that on higher ones [27]. Therefore, the sequence of the combined technologies would be better started with nZVI-aided dechlorination and followed by biosurfactant-aided soil washing. Through the experimental study of various factors (one factor at a time) affecting PCB dechlorination (nZVI dosage, pH, and temperature) and soil washing effectiveness (nZVI and concentrations of biosurfactant solution), the research output is expected to generate environmentally friendly and economically/technically feasible solutions for helping solve the challenging site contamination problem in Canada.

2. Materials and Methods

2.1. Materials. The materials were as follows:

(1) Soil: soil used in this research was fine sands purchased from a local company City Sand & Gravel Ltd., St. John's, NL.

(2) PCBs: commercial PCB products are no longer manufactured and traded in Canada. The contaminants used in this study were in the form of transformer oil obtained from local industry. The overall PCB

concentration in the transformer oil was measured to be 120 ppm by a commercial lab.

(3) nZVI particles: nanofer star, one kind of commercialized air-stable nanoiron powders, was purchased from NANO IRON, sro, Czech Republic.

(4) Biosurfactants: a *Bacillus* sp. bacterial strain isolated from the Atlantic Ocean [28] was cultured to generate biosurfactants in the NRPOP Lab. After culturing and extraction, the crude biosurfactants were separated from the media and characterized through testing the critical micelle concentration (CMC). These crude biosurfactants were then ready for use.

(5) Other materials and chemicals: anhydrous sodium sulfate (ACS reagent); hexane (CHROMASOLV® Plus, for HPLC, ≥ 95%); acetone (CHROMASOLV Plus, for HPLC, ≥99.9%); Supelclean™ Sulfoxide SPE Tube (PE frit, bed wt. 3 g, volume 6 mL); biphenyl-d_{10} (99 atom% D); EPA 525, 525.1 PCB Mix (500 μg/mL each component in hexane, analytical standard); barium chloride dihydrate ($BaCl_2 \cdot 2H_2O$, ACS reagent, ≥99.0%); magnesium sulfate heptahydrate ($MgSO_4 \cdot 7H_2O$, ReagentPlus®, ≥99.0%); sulfuric acid concentrate (0.1 M H_2SO_4 in water (0.2 N)); chloroform (CHROMASOLV Plus, for HPLC, ≥99.9%); methanol (CHROMASOLV, for HPLC, ≥99.9%); ammonium sulfate (($NH_4)_2SO_4$, ReagentPlus, ≥99.0%); sodium chloride (NaCl, BioXtra, ≥99.5% (AT)); iron(II) sulfate heptahydrate ($FeSO_4 \cdot 7H_2O$, BioReagent, ≥99%); monopotassium phosphate (KH_2PO_4, ≥99%); dipotassium hydrogenphosphate (K_2HPO_4, ≥99%); sucrose (BioXtra, ≥99.5%); select yeast extract; zinc sulfate heptahydrate ($ZnSO_4 \cdot 7H_2O$, BioReagent); manganese(II) sulfate tetrahydrate ($MnSO_4 \cdot 4H_2O$, BioReagent); Boric acid (H_3BO_3, BioReagent, ≥99.5%); Copper(II) sulfate pentahydrate ($CuSO_4 \cdot 5H_2O$, BioReagent, ≥98%); sodium molybdate dihydrate ($Na_2MoO_4 \cdot 2H_2O$, ACS reagent, ≥99%); cobalt(II) chloride hexahydrate ($CoCl_2 \cdot 6H_2O$, BioReagent); EDTA (ethylenediaminetetraacetic acid, ACS reagent ≥ 99%); nickel(II) chloride hexahydrate ($NiCl_2 \cdot 6H_2O$, BioReagent); and potassium iodide (KI, BioXtra, ≥99.0%), all of them were purchased from Sigma-Aldrich Canada Co., ON, Canada.

2.2. Methods

2.2.1. nZVI-Aided PCB Dechlorination

(1) PCB-Contaminated Soil Preparation. The synthesized soil that contaminated by transformer oil was applied to simulate PCB-contaminated oil spills such as uncontrolled waste disposals or leakage of storage tanks. The soil was dried at room temperature for one week and passed through a 2 mm stainless steel sieve to remove any coarse sand and gravel particles as well as to improve the homogeneity before use. Then soil characterization was conducted. After characterization, PCB-contaminated soil was prepared in two

20 L-stainless steel trays. Each tray was filled with 10 kg of soil and 2 L of transformer oil. The soil and oil were mixed thoroughly until it reached a homogenous phase. The trays were then covered with tin foil and stored for one week. After that, the oil in the tray was drained off until there was no fluid in soil, and the soil was ready for nZVI treatment.

(2) Air-Stable nZVI Powder Activation. Before any experiment, the surface character and crystal structure of these commercial nZVI particles were examined by scanning electron microscopy (SEM) and X-ray Diffraction (XRD), respectively, in the Core Research Equipment & Instrument Training Lab (CREAIT) at Memorial University. For the activation, the air-stable nanopowder of nZVI was mixed with deionized water at a ratio of 1:4. The mixture was then activated by a Branson Sonifier™ S-450D digital ultrasonic homogenizer for 2 mins at 50% amplitude. The treated mixture was sealed and stored at room temperature for two days before dechlorination experiments.

(3) Dechlorination. Activated nZVI slurry was transferred into a 500 mL wide neck amber glass bottle with 200 g PCB-contaminated soil; and 30 mL deionized water was added as well. The solid and liquid phases were thoroughly mixed and each bottle was covered with a solid-top cap. The homogenous mixture was then stored at room temperature and let the reaction between nZVI particles and PCBs last for 75 days. The effects of the nZVI dosage (5, 7.5, 10, 12.5, and 15 g per kg PCB-contaminated soil), initial pH (2, 5), and temperatures (0°C, 35°C, and 100°C) were investigated, respectively.

2.2.2. Biosurfactant-Aided Soil Washing

(1) Batch-Scale Washing System Design and Setup. The experimental setup used to perform soil washing experiments consists of a washing fluid reservoir, a soil column, a peristaltic pump, and an effluent collection system (Figure 1). The peristaltic pump contains variable speed drives that can run from 0.4 to 85.0 mL/min. The soil column is made of glass to avoid any interference from phthalate esters when contacting with plastic materials, and with a cylindrically diameter of 19 mm and 15 cm in length. The column was packed with 25 g of nZVI-treated soil and the outlet end of the column was fitted with glass beads and glass wool to prevent soil loss during washing. The system assembly is shown in Figure 1.

(2) Biosurfactant Production and Washing Fluid Preparation. The bacteria used to generate biosurfactant were isolated from the Atlantic Ocean recently. Till now, no commercial biosurfactant products associated with this strain were available. Thus, biosurfactants need to be produced before conducting washing experiments. For the media and cultivation conditions, a medium modified from Peng et al. [29] was used, which contains the following composition (g/L): sucrose (10), KH_2PO_4 (3.4), K_2HPO_4 (4.4), ($NH_4)_2SO_4$ (10.0), $FeSO_4 \cdot 7H_2O$ (2.8×10^{-4}), NaCl (2.2), $MgSO_4 \cdot 7H_2O$ (1.02), yeast extract (0.5), and 0.5 mL of trace element solution including (g/L) $ZnSO_4 \cdot 7H_2O$ (2.32), $MnSO_4 \cdot 4H_2O$ (1.78),

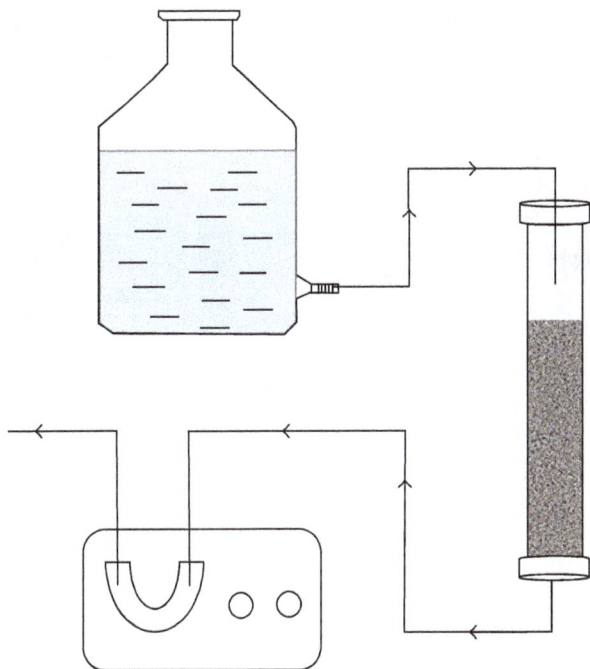

FIGURE 1: Sketch of soil washing system assembly.

H_3BO_3 (0.56), $CuSO_4 \cdot 5H_2O$ (1.0), $Na_2MoO_4 \cdot 2H_2O$ (0.39), $CoCl_2 \cdot 6H_2O$ (0.42), EDTA (1.0), $NiCl_2 \cdot 6H_2O$ (0.004), and KI (0.66). Cultivations were performed in 1 L Erlenmeyer flasks containing 750 mL medium at 30°C and stirred in a rotary shaker for 7 days. Enriched culture medium after 7 days was centrifuged at 10,000 rpm for 10 min, and the supernatant layer was extracted using chloroform-methanol (1:2) on a magnetic stirrer for 8 hours. The solvent layer was separated from the aqueous phase and the solvent was removed by rotary evaporation at 40°C and 60 rpm under reduced pressure. The CMC of the resulting crude biosurfactants was determined through measuring the surface tension in accordance with ASTM D1331-14 method. The surface tension of a crude biosurfactant solution was measured by using Du Noüy Tensiometer. The CMC value of the crude biosurfactants was estimated by the surface tension curve over a wide concentration range. They were determined by noting the concentrations at which the surface tension reaches the minimum.

(3) Soil Washing. Parallel experiments were conducted to investigate the effect of nanoiron particles on soil washing treatment. Both of the PCB-contaminated soil samples treated with 7.5 g/kg and without nZVI particles were loaded into the washing columns, respectively, to test whether the presence of nanoparticles would have any effect on soil washing. Twenty-five grams of the soil was washed with deionized water in a down flow mode for 1.5 hours at a steady flow rate controlled by the peristaltic pump. To investigate the effect of crude biosurfactant, the soil samples were washed with the crude biosurfactant solution (0.25%, 0.5%, and 3% v/v) in a down flow mode for 3 hours at a steady flow rate controlled by the peristaltic pump. The washing effluents

were sampled at 0, 10, 20, 30, 60, 90, 120, and 180 minutes of washing to investigate the change of PCB concentration with time. The change of PCB concentration in soil was determined by measuring the soil sample before and after washing.

2.2.3. Sample Analysis. The PCBs in each soil sample were first extracted into the solvent phase according to EPA method 3550B ultrasonic extraction. A modification of the method was conducted to achieve a better testing performance. Two grams of soil sample was transferred to a 30 mL beaker. Two grams of anhydrous sodium sulfate was added to the sample and the solution was well mixed. Two surrogates, 500 μL 10 ppm biphenyl-d_{10} and 200 μL 10 ppm EPA 525, 525.1 PCB Mix, were spiked to the sample. A hexane solvent of 9.3 mL was immediately added to the matrix in order to bring the final volume to 10.0 mL. This was followed by disrupting the sample with a Branson Sonifier ultrasonic probe for 2 minutes at 50% amplitude. After ultrasonic extraction, 1 mL extract was filtered by glass wool and ready for solid phase extraction (SPE) cleanup. Supelclean Sulfoxide SPE cartridges purchased from Sigma-Aldrich were used for transformer oil cleanup. The SPE normal procedure of conditioning, loading, washing, and elution was followed. The conditioning was accomplished by eluting 10 mL of acetone to remove residual moisture from the Supelclean Sulfoxide cartridges. This was followed by adding 20 mL of n-hexane to equilibrate the cartridges. The pretreated 1 mL sample was loaded onto the cartridge and washed with 5.5 mL of n-hexane. Elution was done with 13 mL of n-hexane. The eluate was concentrated to 1 mL by gentle air blow. The cleanup extracts were transferred into GC vials ready for analysis.

The PCB concentration in liquid phase was analyzed using modified Liquid-Liquid Microextraction (LLME) [30] followed by the GC-MS analysis. For modified LLME, 25 μL of 10 ppm biphenyl-d_{10} and 10 μL of 10 ppm EPA 525, 525.1 PCB Mix were spiked as surrogates to each 10 mL water sample (25 ng/L biphenyl-d_{10} and 10 ng/L EPA 525, 525.1 PCB Mix aqueous solution), which was treated by vortex mixing for 10 sec. This was followed by adding 500 μL of hexane and the vortex mixing for 1 min. The water sample was then centrifuged at 4,000 rpm for 5 min. Ten μL of extract was transferred to Microvials for GC analysis. Instrumental analysis was performed using an Agilent 7890A/5975C gas chromatograph–mass spectrometer (GC-MS) equipped with an Agilent 7693 autosampler. GC conditions were set up based on EPA method 8082A. A few adjustments were made to ensure that no PCB congener was retained in the column.

Total ion current (TIC) chromatogram was acquired to examine the changes of PCBs in soil samples. The analysis of each congener and its surrogate was carried out in selected-ion monitoring (SIM) chromatogram. The ratio of sample congener response to standard congener response was defined as the relative concentration, which was used in the results and discussion. All the samples were treated and analyzed in duplicate.

TABLE 1: Soil properties.

Properties	Results
Soil pH	7.53
Bulk density	$1.78 \, g/cm^3$
Particle density	$2.71 \, g/cm^3$
Pore space	34.3%
Moisture content	0.069%
Cation exchange capacity	95.22 cmol/kg
Hydraulic conductivity	0.024 cm/s

TABLE 2: Soil particle size distribution determined by sieve analysis.

Particle	Diameter (mm)	Size distribution (%)
Gravel	>2.0	4.5
Sand	0.05–2.0	92.5
Silt	0.002–0.05	2.5
Clay	<0.002	0.5

3. Results and Discussion

3.1. nZVI-Aided PCB Dechlorination

3.1.1. Soil Characterization. Before the nZVI-aided PCB dechlorination experiments, basic soil properties including particle size distribution, soil pH, bulk density, particle density, pore space, cation exchange capacity, hydraulic conductivity, and moisture content of the purchased plain soil were measured. The results are shown in Tables 1 and 2. The plain soil used in this research was mainly composed of sand, which was suitable for soil washing. The bulk density, particle density, pore space, hydraulic conductivity, and moisture content are physical properties which can be greatly influenced by soil composition and particle size distribution. The pH of the soil was slight alkalinity, which could result in a higher cation exchange capacity (CEC) value. In an environmental context, CEC stands for the ability of soil to adsorb contaminants. The pH and CEC are two important chemical properties which could affect the soil remediation process and thus need to be examined before remediation.

Metal substances of the plain soil sample were characterized by ICP-MS. Table 3 displays the analytical results. It is noticed that a high concentration of iron was present, which was of 33.6 g per kg soil. The addition of nZVI for PCB dechlorination thus would not much influence the composition of soil.

3.1.2. Analysis of PCB Concentrations in the Original Spiked Soil. The concentrations of PCBs in the spiked soil sample were evaluated before conducting the dechlorination and soil washing experiments. Four PCB congeners were selected as analytes due to their high abundances in the transformer oil, namely Penta-17.8, Penta-18.7, Penta-20.0, and Hexa-20.8. The former parts of the names represent the numbers of chlorine atoms in the congener compounds, while the latter ones are their corresponding retention times (minutes) in the

TABLE 3: Metal substances in the soil sample determined by ICP-MS.

Metals	Concentration in soil (mg/kg)
Arsenic	5.306
Barium	643.918
Cadmium	0.146
Chromium	16.815
Copper	13.727
Iron	33,562.114
Lead	17.681
Mercury	<LDL
Nickel	9.142
Selenium	<LDL
Thallium	0.444
Uranium	1.675
Vanadium	46.084
Zinc	71.958

Note: LDL = lower detection limit.

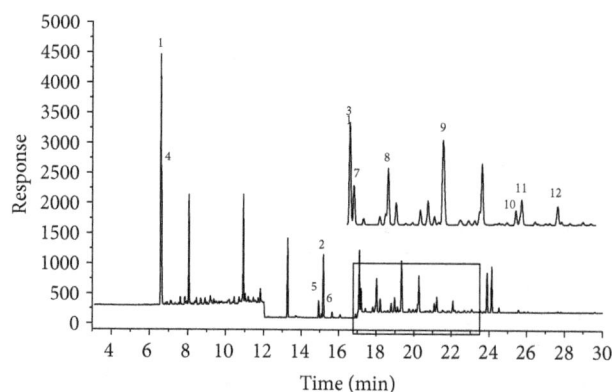

FIGURE 2: GC-MS SIM spectra of PCBs in contaminated soil sample spiked with bipenyl-d_{10} and EPA 525,525.1 PCBs standard. (1) Biphenyl-d_{10}. (2) 2,2′,4,4′-Tetrachlorobiphenyl. (3) 2,2′,3′,4,6-Pentachlorobiphenyl. (4) Biphenyl. (5) Tetra-15.6. (6) Tetra-16.3. (7) Penta-17.8. (8) Penta-18.7. (9) Penta-20.0. (10) Hexa-20.8. (11) Hexa-21.8. (12) Hexa-22.8.

TABLE 4: The initial relative concentrations of PCBs in the spiked soil.

Analytes	Surrogate	Response ratio
Penta-17.8	2,2′,3′,4,6-Pentachlorobiphenyl	0.384
Penta-18.7	2,2′,3′,4,6-Pentachlorobiphenyl	0.551
Penta-20.0	2,2′,3′,4,6-Pentachlorobiphenyl	0.736
Hexa-20.8	2,2′,3′,4,6-Pentachlorobiphenyl	0.262

MS spectra (Figure 2). The average response ratios of PCBs to their corresponding surrogates are listed in Table 4.

3.1.3. nZVI Characterization and Activation. The commercial nZVI particles were characterized by SEM and XRD prior to their applications in PCB dechlorination in soil. Figure 3 shows the SEM image of nZVI particles. It can be seen that

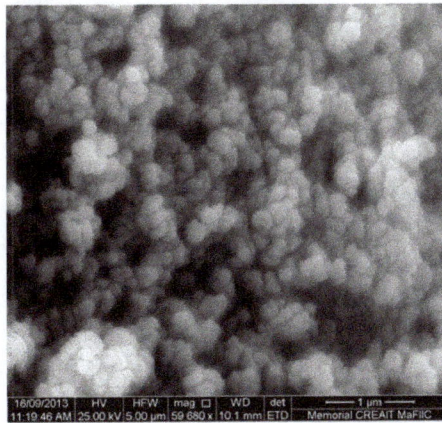

FIGURE 3: SEM Image of the commercial nZVI particles.

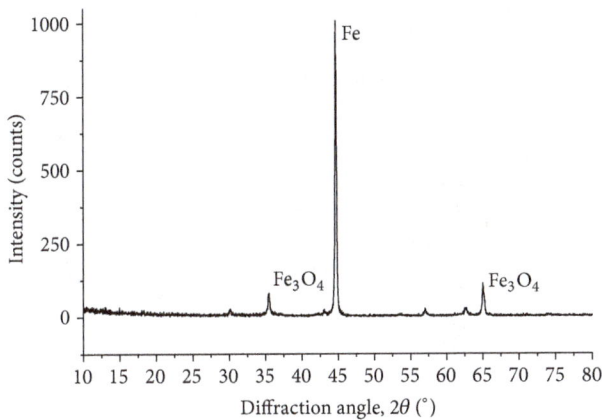

FIGURE 4: XRD of the nZVI particle.

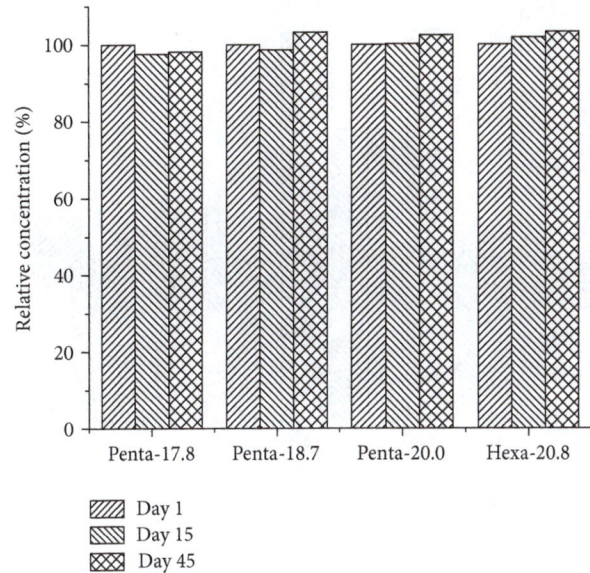

FIGURE 5: Natural attenuation of PCBs in contaminated soil.

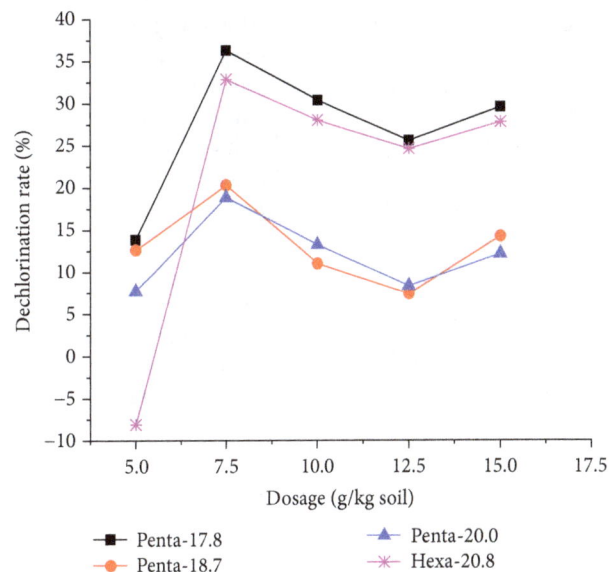

FIGURE 6: Effect of nZVI dosage on PCB dechlorination in the contaminated soil.

the majority of the particles were nearly spherical in shape and uniform in size. The particle size was in the range of 20–100 nm with an average particle size of 50 nm.

Figure 4 displays the XRD pattern of the nZVI particles and it proved that there were crystal iron particles existed in the commercial product. The 2θ values of the peaks were compared with the standard data for iron and its oxides such as magnetite and α-Fe. Apparent peak at the 2θ of $44.9°$ indicates the presence of α-Fe, while other apparent peaks show the presence of iron oxides. The redox potential of nZVI slurry was decreased from 360 mV to −300 mV after the activation.

3.1.4. Natural Attenuation of PCBs. The changes of PCB concentrations in the contaminated soil were tracked on the 1st, 15th, and 45th day, respectively, during the natural attenuation process. As depicted in Figure 5, the concentrations of all the four congeners did not change significantly within the 45-day period. It illustrated that the dechlorination rates of PCBs during the natural attenuation process were extremely slow. It also proved that the PCBs were not able to be degraded without any additional treatment.

3.1.5. Effect of nZVI Dosage. The performance of PCB dechlorination using different nZVI dosages is shown in Figure 6.

The trends of the PCB dechlorination rate versus nZVI dosage were similar based on the results of all the four congeners. The overall PCB dechlorination rate was first increased as nZVI dosage increased from 5 to 7.5 g/kg, indicating that the increase of nZVI dosage can accelerate the dechlorination of PCBs. The overall dechlorination rate of PCBs was then decreased when the nZVI dosage increased higher than 7.5 g/kg. The maximum dechlorination rates of Penta-17.8, Penta-18.7, Penta-20.0, and Hexa-20.8 during 75 days period were 36.3%, 20.3%, 18.9%, and 32.9%, respectively. The results indicated that when choosing 7.5 g/kg as the nZVI dosage, the highest dechlorination rates were achieved in all four congeners. Adding more nZVI particles had shown a negative

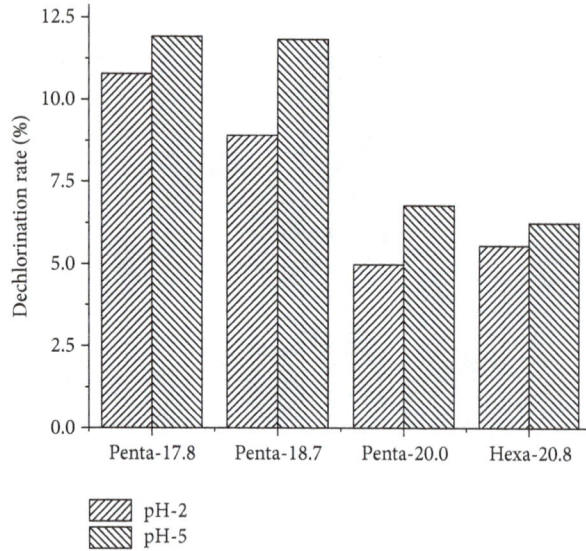

FIGURE 7: Effect of pH on PCB dechlorination in the contaminated soil.

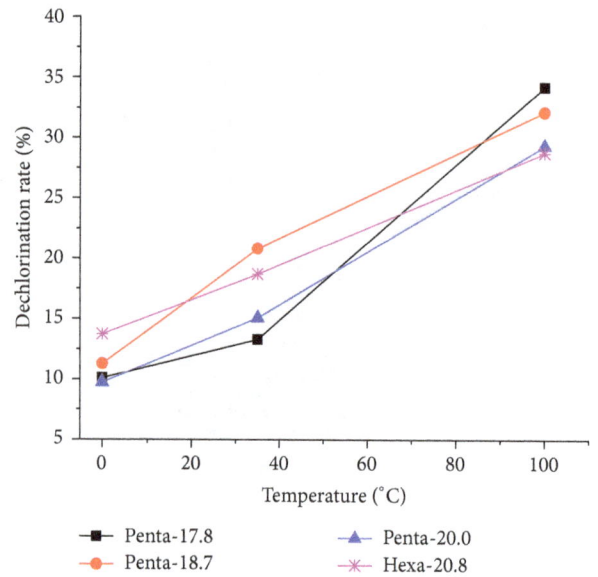

FIGURE 8: Effect of temperature on nZVI-aided PCB dechlorination.

influence on PCB dechlorination. This was possibly due to the particle aggregation formed during mixing [31]. Besides the nZVI aggregation, the biotransformation from higher chlorinated biphenyls to lower ones may also affect the PCB dechlorination rate under multiple nZVI dosages. Based on the experimental results, the nZVI dosage of 7.5 g/kg with the best PCB dechlorination performance was selected for the following treatments.

3.1.6. Effect of pH Level. Two levels of pH were selected to evaluate the effect of pH on PCB dechlorination. The result is shown in Figure 7. After 75 days monitoring, the average dechlorination rates of Penta-17.8, Penta-18.7, Penta-20.0, and Hexa-20.8 at pH of 2 were 10.8%, 8.9%, 5.0%, and 5.6%, respectively; while their average dechlorination rates at pH of 5 were 11.9%, 11.8%, 6.8%, and 6.2%, respectively. The dechlorination rates of each PCB congener were higher at pH of 5 than those at pH of 2. Previous studies have shown that an acid environment with more protons could accelerate the PCB dechlorination [32]. The results of this study led to a different conclusion. It might be because, in this case, the protons were sufficient at pH of 5 so that pH was not a dominating factor on PCB dechlorination anymore. In addition, the addition of H_2SO_4 would have more interference with the mass transfer of PCBs from the soil to the iron (Fe) surface [32]. The pH of 5 was thus selected to be the initial pH condition in the following experiments.

3.1.7. Effect of Temperature. The effect of temperature on PCB dechlorination after 75 days was investigated with results shown in Figure 8. The PCB dechlorination was greatly enhanced when the temperature increased from 0 to 100°C. As the temperature increased, the PCB dechlorination of Penta-17.8 improved the most, with a rate change from 10.1% to 34.2%. The dechlorination rates of Penta-18.7, Penta-20.0, and Hexa-20.8 were enhanced from 11.3% to 32.2%, from

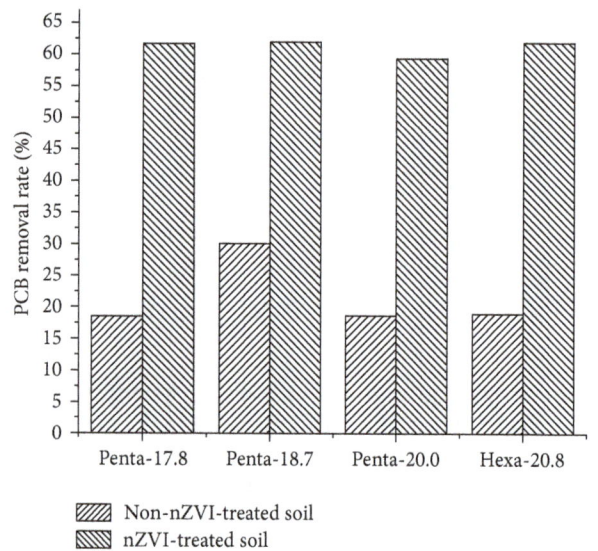

FIGURE 9: Effect of the nZVI particles on PCB removal during soil washing.

9.8% to 29.4%, and from 13.7% to 28.8%, respectively. These results showed that a temperature increase would enhance the mobility of PCBs from the soil to the iron surfaces and thus accelerate the dechlorination reaction [32].

3.2. Biosurfactant-Aided Soil Washing

3.2.1. Effect of nZVI Particles on Soil Washing. The effect of the nZVI particles on PCB removal during the soil washing treatment was investigated. Figure 9 showed the results. Although the insolubility of PCBs makes their distribution negligible in water phase, the PCBs in the transformer oil could be flushed out of the column due to the high flow rate during direct soil washing without using any biosurfactants.

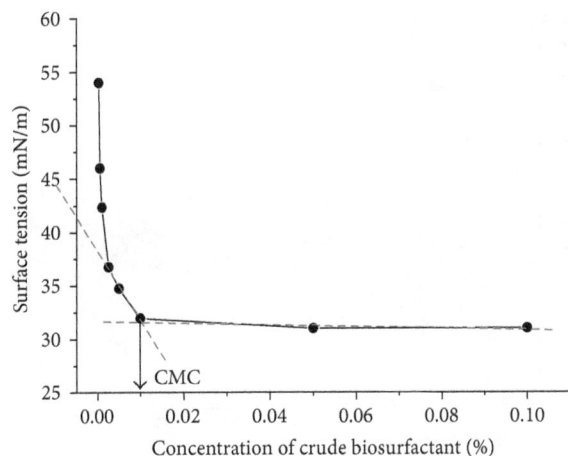

FIGURE 10: CMC of the crude biosurfactant.

As shown in Figure 9, after 1.5 hours of operation, about 18% to 30% of PCBs in the congeners were removed by direct washing of the non-nZVI-treated soil.

After the nZVI-aided dechlorination, a red color was observed in the treated soil, implying the formation of ferric hydroxides or ferric oxides. It indicated that the nZVI particles were transferred to their oxidative forms after the reaction. During washing of the nZVI-treated soil, the PCB concentration of each congener was significantly decreased. In Figure 9, the removal rates of PCBs after washing were between 60% and 62% in the nZVI-treated soil. It was illustrated that the treatment by the nZVI particles greatly enhanced the soil washing efficiency. Besides, the presence of nanoscale ferric oxides in the system plays a key role in PCB removal [33]. The contaminated soil trapped a certain amount of transformer oil, and the oil droplets were blocked by the pore throat of soil due to the high interfacial tension between oil and soil [34]. With the presence of nanoscale ferric oxides, the interfacial tension would be reduced and the mobility of oil droplets would be increased [33]. As a result, more oil droplets were desorbed from the soil, resulting in an increased effectiveness of soil washing. This experiment confirmed that the combination of the nZVI-aided dechlorination and soil washing is reasonable and feasible.

3.2.2. CMC of the Crude Biosurfactant. The surface tension of a series of biosurfactant solutions with different biosurfactant concentrations was tracked. The trend of surface tension versus biosurfactant concentration was shown in Figure 10. The value of surface tension was decreased sharply till the biosurfactant concentration reached 0.01%. When the biosurfactant concentration was higher than 0.01%, the surface tension changes became relatively stable. Therefore, the CMC of the crude biosurfactant was determined to be 0.01%.

3.2.3. Effect of Biosurfactant Concentration on Soil Washing. The nZVI-treated soil sample was washed by crude biosurfacant solutions. The concentration of crude biosurfactant in the washing fluid was set as 3%, 0.5%, and 0.25%. The initial flow rate of the column washing fluid was set within

the range of 18–20 mL/min. The results of relative PCB concentrations (the ratio of sample congener response to standard congener response) in column effluent were shown in Figures 11(a)–11(c). The elution of PCBs started at 10 min. The PCB concentrations in effluents were sharply increased and reached their peaks at 15–45 min. Steep declines were followed by the peaks and the gentle deduction appeared in the final stage.

The overall PCB removal rates after washing of the nZVI-treated soil were examined. As shown in Figure 12, the higher the concentration of the crude biosurfactant solution used, the higher the removal rate achieved. The maximum removal rate was found when using 3% crude biosurfactant and 90% of the total four PCB congeners were removed from the soil. The final removal rates using 0.5% and 0.25% crude biosurfactant solutions were 80% and 75%, respectively. The PCB removal rates using all the three crude biosurfactant solution were higher than 75%, indicating the promising effectiveness of biosurfactant-aided soil washing.

The decreasing of washing flow rates occurred especially with higher biosurfactant doses. When supplying the washing fluid with the 3% crude biosurfactant, a backwashing step was required within 30–60 min washing time. It was observed that the bounce of PCBs at 60 min in Figure 11(a) was because of backwash. The most possible explanation is that the crude biosurfactant contained insoluble particulate matter could block the pathway of washing flow, thus reducing both flow rate and emulsification rate. It could further lead to a longer treatment period. Results indicated that the crude biosurfactant solution with a concentration of 0.5% could remove the majority of the four PCB congeners with the shortest treatment time (within 60 minutes). Therefore, 0.5% was selected as an appropriate biosurfactant concentration for further applications.

The SIM spectrum shows the removal of almost all the PCBs in the soil sample after washing. As shown in Figure 13, the peaks of PCBs had almost disappeared after washing with 0.5% crude biosurfactant solution; only the peaks of surrogates were left. Besides, the contents of the transformer oil that generated the baseline wander were also removed. As a consequence, the crude biosurfactant solution was able to remove almost all the organic components including PCBs in transformer oil unselectively.

4. Conclusions

This research has focused on the development of a two-step treatment consisting of nZVI-aided dechlorination followed by biosurfactant-based soil washing technology to remove PCBs from soil. In nZVI-aided dechlorination, the effects of nZVI dosage, initial pH, and temperature on PCB transformation were evaluated one at a time, respectively. The selected dosage of nZVI was 7.5 g/kg soil. Adding more nZVI particles could have negative influence on PCB dechlorination, since the aggregates could be easily formed as the nZVI dosage increases. An environment with pH lower than 5 did not much influence the removal rates of PCBs, indicating the presence of sufficient protons in the system. The results showed that the lower pH would actually inhibit

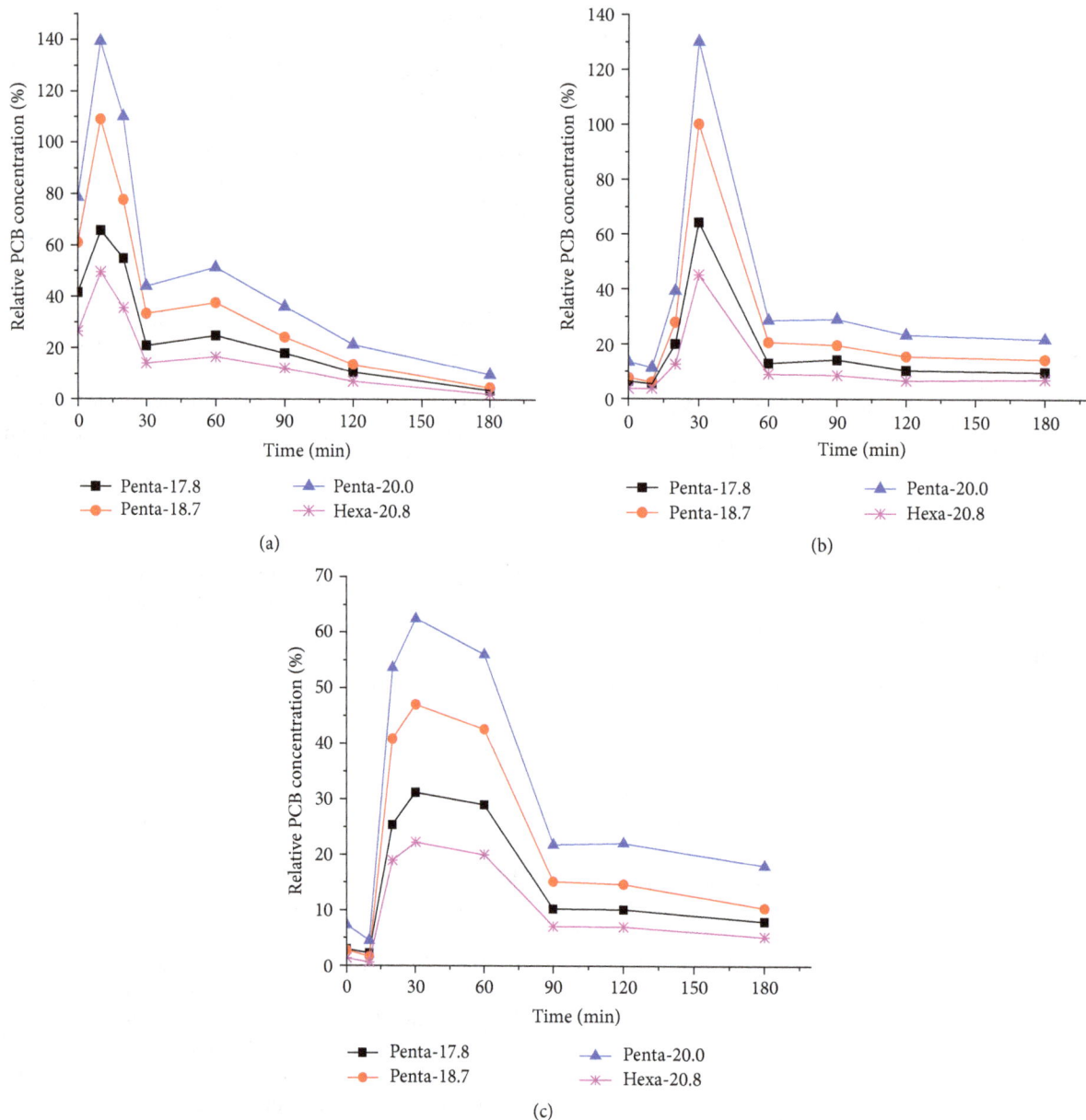

FIGURE 11: Relative concentrations of PCBs in washing effluent with (a) 3%, (b) 0.5%, and (c) 0.25% crude biosurfactant solution.

the dechlorination by the presence of H_2SO_4, which has an effect on the reduction of mass transfer. An improvement of dechlorination was observed as the temperature increased, since higher temperature would accelerate the dechlorination reaction.

In the soil washing system, the presence of nZVI particles plays a key role in PCB removal. They can greatly enhance the soil washing efficiency because the interfacial tension between the oil phase and the soil phase would be reduced and the mobility of oil droplets would be increased. Soil washing of nZVI remediated soil can be enhanced by biosurfactant. Higher biosurfactant concentration could increase the solubilization of PCBs from soil phase to liquid phase. The overall PCB removal rates using all the three crude biosurfactant concentrations (3%, 0.5%, and 0.25%) were 90%, 80%,

and 75%, respectively, indicating the promising effectiveness of this biosurfactant. Compared with the 3% biosurfactant solution, the crude biosurfactant concentrations of 0.5% and 0.25% were more cost-effective. The 0.5% crude biosurfactant solution could remove the majority of PCBs within a shorter time than the solution with a concentration of 0.25%. Therefore, 0.5% was recommended as an appropriate biosurfactant concentration for future application. This study shows a great potential in developing a promising treatment technology for PCB-contaminated soil remediation. Pilot-scale applications will be carried out to demonstrate the technology transfer.

Competing Interests

The authors declare that they have no competing interests.

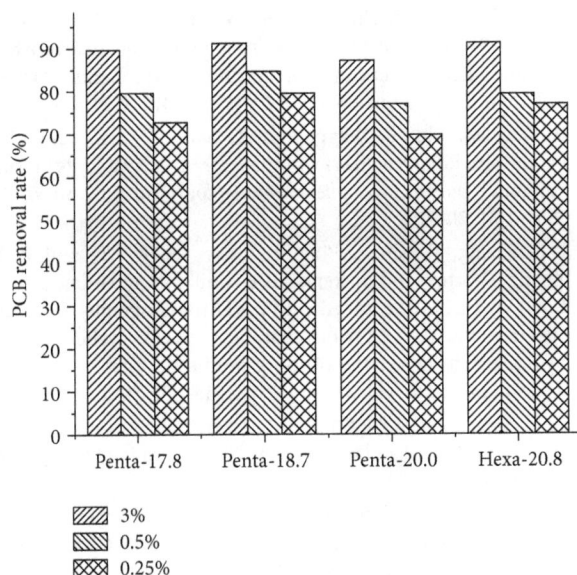

FIGURE 12: The washing efficiencies of PCBs in the nZVI-treated contaminated soil by different concentrations of crude biosurfactant solution.

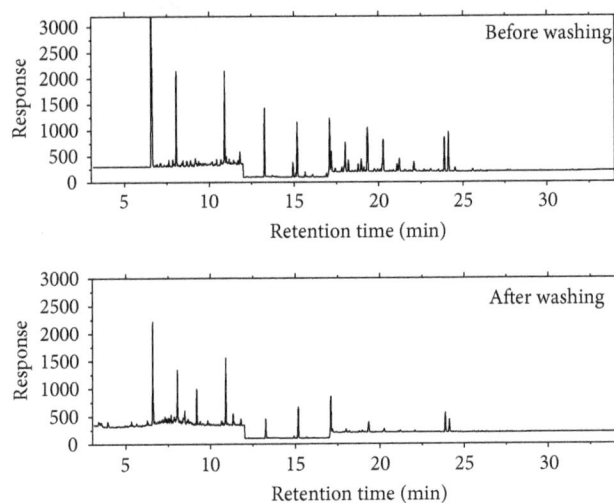

FIGURE 13: GC-MS SIM spectra of PCBs in the contaminated soil before and after washing by 0.5% crude biosurfactant solution.

Acknowledgments

The authors would like to express gratitude to The Natural Sciences and Engineering Research Council of Canada (NSERC) and the Canada Foundation for Innovation (CFI) for their support.

References

[1] D. Pal, J. B. Weber, and M. R. Overcash, "Fate of polychlorinated biphenyls (PCBs) in soil-plant systems," in Residue Reviews, pp. 45–98, Springer, New York, NY, USA, 1980.

[2] Canadian Council of Resource and Environment Ministers (CCREM), The PCB Story, (CCREM), Toronto, Canada, 1986.

[3] S. Tanabe, "PCB problems in the future: foresight from current knowledge," Environmental Pollution, vol. 50, no. 1, pp. 5–28, 1988.

[4] Canadian Council of Ministers of the Environment (CCME), "Canadian soil quality guidelines for the protection of environmental and human health: polychlorinated biphenyls (total)," in Canadian Environmental Quality Guidelines, Canadian Council of Ministers of the Environment, Winnipeg, Canada, 1999.

[5] US Environmental Protection Agency, PCBs in the United States industrial use and environmental distribution, p. 24, 1976, http://nepis.epa.gov/Exe/ZyPDF.cgi/2000I275.PDF?Dockey=2000I275.PDF.

[6] S. Jensen, "Report of a new chemical hazard," New Scientist, vol. 32, no. 612, p. 445, 1966.

[7] W. M. J. Strachan, "Polychlorinated biphenyls (PCBs): fate and effects in the Canadian environment," EPS Report 4/HA/2, Environment Canada, 1988.

[8] L. A. Barrie, D. Gregor, B. Hargrave et al., "Arctic contaminants: sources, occurrence and pathways," Science of The Total Environment, vol. 122, no. 1-2, pp. 1–74, 1992.

[9] AMEC Earth and Environmental Ltd, Commission 88.1 Environmental Site Investigation B371 CHPP Tanks, Prepared for Defence Construction Canada, Goose Bay, Canada, 2008.

[10] Federal Contaminated Sites Portal, Federal Contaminated Sites Action Plan (FCSAP), February 2014, http://www.federal-contaminatedsites.gc.ca/default.asp?lang=en.

[11] A. Mikszewski, Emerging Technologies for the in Situe Remediation of PCB-contaminated Soils and Sediments: Bioremediation and Nanoscale Zerovalent Iron, US Environmental Protection Agency, 2004.

[12] S. M. Cook, Assessing the Use and Application of Zero-Valent Iron Nanoparticle Technology for Remediation at Contaminated Sites, Jackson State University, 2009.

[13] W.-X. Zhang, "Nanoscale iron particles for environmental remediation: an overview," Journal of Nanoparticle Research, vol. 5, no. 3-4, pp. 323–332, 2003.

[14] N. C. Mueller and B. Nowack, "Nanoparticles for remediation: solving big problems with little particles," Elements, vol. 6, no. 6, pp. 395–400, 2010.

[15] R. Varma, "Greener synthesis of noble metal nanostructures and nanocomposites," in Presented at the U.S. EPA Science Forum: Innovative Technologies—Key to Environmental and Economic Progress, Washington, DC, USA, 2008.

[16] W. Chu and K. H. Chan, "The mechanism of the surfactant-aided soil washing system for hydrophobic and partial hydrophobic organics," Science of The Total Environment, vol. 307, no. 1–3, pp. 83–92, 2003.

[17] K. Urum, T. Pekdemir, and M. Gopur, "Optimum conditions for washing of crude oil-contaminated soil with biosurfactant solutions," Process Safety and Environmental Protection, vol. 81, no. 3, pp. 203–209, 2003.

[18] D. Feng, L. Lorenzen, C. Aldrich, and P. W. Maré, "Ex situ diesel contaminated soil washing with mechanical methods," Minerals Engineering, vol. 14, no. 9, pp. 1093–1100, 2001.

[19] H. I. Gomes, C. Dias-Ferreira, L. M. Ottosen, and A. B. Ribeiro, "Electrodialytic remediation of polychlorinated biphenyls contaminated soil with iron nanoparticles and two different surfactants," Journal of Colloid and Interface Science, vol. 433, pp. 189–195, 2014.

[20] T. Lyons, D. W. Grosse, and R. A. Parker, EPA engineering issue: technology alternatives for the remediation of PCB

contaminated soils and sediments (No. EPA/600/S-13/079), Washington, DC, USA, 2013.

[21] B. Zhang, G. H. Huang, and B. Chen, "Enhanced bioremediation of petroleum contaminated soils through cold-adapted bacteria," *Petroleum Science and Technology*, vol. 26, no. 7-8, pp. 955–971, 2008.

[22] X. Qin, B. Chen, G. Huang, and B. Zhang, "A relation-analysis-based approach for assessing risks of petroleum-contaminated sites in Western Canada," *New Developments in Sustainable Petroleum Engineering*, vol. 1, no. 2, pp. 183–200, 2009.

[23] P. F. Amaral, M. A. Z. Coelho, I. M. Marrucho, and J. A. Coutinho, "Biosurfactants from yeasts: characteristics, production and application," in *Biosurfactants*, pp. 236–249, Springer, New York, NY, USA, 2010.

[24] H. Xia and Z. Yan, "Effects of biosurfactant on the remediation of contaminated soils," in *Proceedings of the 4th International Conference on Bioinformatics and Biomedical Engineering (iCBBE '10)*, pp. 1–4, IEEE, Chengdu, China, June 2010.

[25] Unified Facilities Guide Specifications (USGS), Soil Washing through Separation/Solubilization, UFGS-02 54 23, 2010, https://www.wbdg.org/ccb/DOD/UFGS/UFGS%2002%2054%2023.pdf.

[26] H. I. Gomes, C. Dias-Ferreira, and A. B. Ribeiro, "Overview of in situ and ex situ remediation technologies for PCB-contaminated soils and sediments and obstacles for full-scale application," *Science of the Total Environment*, vol. 445-446, pp. 237–260, 2013.

[27] W. Y. Shiu and D. Mackay, "A critical review of aqueous solubilities, vapor pressures, Henry's law constants, and octanol–water partition coefficients of the polychlorinated biphenyls," *Journal of Physical and Chemical Reference Data*, vol. 15, no. 2, pp. 911–929, 1986.

[28] Q. Cai, B. Zhang, B. Chen, Z. Zhu, W. Lin, and T. Cao, "Screening of biosurfactant producers from petroleum hydrocarbon contaminated sources in cold marine environments," *Marine Pollution Bulletin*, vol. 86, no. 1-2, pp. 402–410, 2014.

[29] F. Peng, Z. Liu, L. Wang, and Z. Shao, "An oil-degrading bacterium: *Rhodococcus erythropolis* strain 3C-9 and its biosurfactants," *Journal of Applied Microbiology*, vol. 102, no. 6, pp. 1603–1611, 2007.

[30] J. S. Zheng, B. Liu, J. Ping, B. Chen, H. J. Wu, and B. Y. Zhang, "Vortex- and shaker-assisted liquid–liquid microextraction (VSA-LLME) coupled with gas chromatography and mass spectrometry (GC-MS) for analysis of 16 polycyclic aromatic hydrocarbons (PAHs) in offshore produced water," *Water, Air, and Soil Pollution*, vol. 226, no. 9, pp. 318–331, 2015.

[31] N. C. Müller and B. Nowack, "Nano zero valent iron—the solution for water and soil remediation," Report of the Observatory NANO, 2010.

[32] P. Varanasi, A. Fullana, and S. Sidhu, "Remediation of PCB contaminated soils using iron nano-particles," *Chemosphere*, vol. 66, no. 6, pp. 1031–1038, 2007.

[33] L. Hendraningrat and O. Torsæter, "Metal oxide-based nanoparticles: revealing their potential to enhance oil recovery in different wettability systems," *Applied Nanoscience*, vol. 5, no. 2, pp. 181–199, 2015.

[34] A. Roustaei, S. Saffarzadeh, and M. Mohammadi, "An evaluation of modified silica nanoparticles' efficiency in enhancing oil recovery of light and intermediate oil reservoirs," *Egyptian Journal of Petroleum*, vol. 22, no. 3, pp. 427–433, 2013.

11

Long-Term Dynamics of Urban Soil Pollution with Heavy Metals in Moscow

N. E. Kosheleva and E. M. Nikiforova

Faculty of Geography, Moscow State University, Leninskie Gory, Moscow 119991, Russia

Correspondence should be addressed to N. E. Kosheleva; natalk@mail.ru

Academic Editor: Francesco Sdao

Results of 21-year-long (1989–2010) observations of the concentrations and the spatial distribution patterns of nine heavy metals (HMs) in topsoils of the Eastern district of Moscow are presented. The quantitative parameters of soil pollution include the annual increase rates of HM concentrations in several land-use zones. The maps of geochemical anomalies were compiled using the data collected in 1989, 2005, and 2010. The growth of the total volume of industrial and vehicles' emissions between 1989 and 2005 caused significant deposition of Pb, Zn, Cu, and Cd. The additional input of Cd to the soils is attributed to the application of sewage sludge as fertilizers. The relative increment of concentrations was the highest for Pb, Co, Cu, Ni, and Cr. In 2005–2010, the relative annual increment rate was the highest for Cr, Cd, Co, and Ni, and it increased by an order of magnitude as compared to the previous period. By contrast, Pb and Cu concentrations decreased owing to the soil reclamation, the exclusion of leaded gasoline as a fuel for vehicles and closing some hazardous enterprises. Joint analysis of snow and soil geochemical maps allows identification of the zones of actual, permanent, and relict pollution.

1. Introduction

Development of industrial cities and rapid growth of urban population aggravates ecological problems caused by air and water contamination and permanent accumulation of pollutants in soils. Heavy metals (HMs) are of special concern since they belong to very hazardous substances. The content and distribution of HMs in Moscow soils have been analyzed by many authors [1–7]. At the same time, such urgent issues of urban ecology as a long-term dynamics of soil contamination with HMs in different land-use zones have not yet been considered. The aim of this research is to study spatial-temporal trends in soil contamination with HMs in one of the most polluted districts of Moscow.

The specific purposes of the study are as follows:

(i) to investigate the main features of long-term dynamics of HM content in the soil cover in relation to land-use type and fluctuations in urban activities;

(ii) to map technogenic geochemical anomalies in urban soils for different time periods;

(iii) to determine the character and intensity of HM accumulation in urban soils by combined analysis of snow and soil geochemical maps;

(iv) to evaluate the environmental risk of the contamination on the basis of integral indices.

2. Materials and Methods

The Eastern district has been chosen as a study object; it is located on the Meshchera outwash plain. The southern, most polluted, part of the Moscow Eastern district was investigated. This territory belongs to the southern taiga Meshchera landscapes with a temperate continental climate. The snow cover appears in late October–early January and reaches the maximum height in February-March. A flat plain with altitudes of 150–160 m a.s.l. is composed of glaciofluvial sands and loams with low content of HMs [8]. Soil cover has been severely disturbed; natural sod-podzolic, podzolic-bog, and bog soils are preserved only locally in the suburbs and green areas. The major part of the area is occupied by

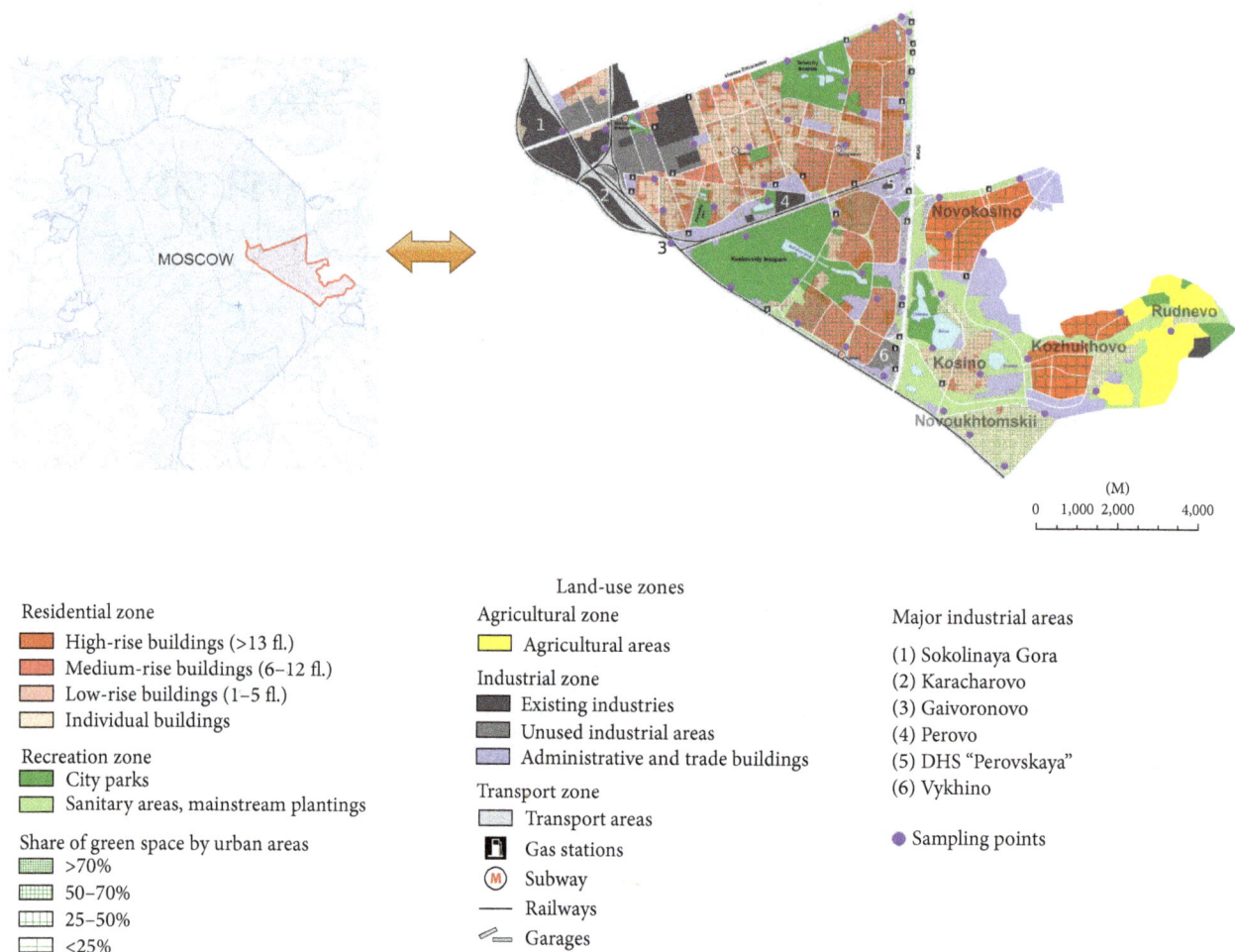

FIGURE 1: The map of land-use zones for the territory of the Eastern district of Moscow [12].

specific urban soils, that is, urbanozems and technozems [9], developed on filled redeposited substrates and the cultivated layer; in some parks and recreation zones, soils develop on the natural parent rocks.

The most widespread soils are urbanozems that are deeply transformed anthropogenic soils, the profiles of which change to a depth of more than 50 cm. Their main difference from natural soils is the presence of a diagnostic "urbic" horizon. This surface horizon is a filled mixed layer with an admixture (>5%) of anthropogenic inclusions (construction and industrial waste, domestic garbage); as a rule, it contains organic matter. The physical and chemical properties of this horizon change owing to the input of eolian dust [5].

The territory has multispecialized industries, including instruments and equipment producing, chemical and oil-processing factories, heating power stations, incineration plant, and a dense network of roads. Highways are the Moscow Ring Road (MKAD) and Shosse Entuziastov. Motor transport is the main source of pollution in the district [10]. The largest industrial zones are Karacharovo and Sokolinaya Gora in the north-east of the study area; Perovo in the center; Vykhino and Gaivoronovo in the south. Owing to

the fall in production and decrease in the volume of wastes and emissions, the input of stationary air pollution sources decreased from 26-27% of the total mass of technogenic load, in the early 1990s to 10% at the end of the decade [11]. Nevertheless Moscow still ranks 10th–13th among Russian cities by the total pollution level from stationary sources; industry remains the most important factor responsible for the contamination of the urban environment with specific emissions.

The geochemical studies and soil sampling were performed during the summers of 1989, 2005, and 2010 at the same points (Figure 1). Samples were collected at approximately uniform grid with a sampling spacing of 700–900 m. This scheme is widely used in the world for geochemical mapping of cities [13]. Altogether 153 soil mixed samples were taken from the most humus enriched and contaminated upper (0 to 15 cm) horizons of the urban soils. Mixed samples were composed of 4 individual ones. Soddy-podzolic soils of the Meshchera Plain (45–50 km east of the city, where its influence was not displayed) were studied as background analogues. These light-textured soils develop under meadow grass-forb and forest spruce-birch communities. Ten samples

TABLE 1: Levels of landscape pollution with HMs and dust [14].

Levels of pollution	Dust deposition P_n, kg/km² per day	HM immission Z_d	Total indices of	
			Snow pollution Z_c	Soil pollution Z_c
Low	<200	<1000	<32	<16
Medium	200–300	1000–2000	32–64	16–32
High	300–500	2000–4000	64–128	32–64
Very high	500–800	4000–8000	128–256	64–128
Maximum	>800	>8000	>256	>128

were taken from the surface horizons of the reference soils. The bulk content of Zn, Cd, Pb, Cu, Cr, Co, Ni, Mn, and Cs was determined by atomic absorption spectrometry using an Hitachi 180–70 spectrometer. To determine all the elements soil samples were transferred into the state of the solution using the mixture of acids HCl, HClO$_4$, and HNO$_3$, with triple evaporation and complexation of HF with boric acid. The residue was then dissolved in HCl.

The type of land use is of primary importance for the formation of technogenic geochemical anomalies; therefore land-use zoning of the territory was performed using satellite high-resolution images [12]. The following areas were specified (Figure 1): highways and industrial zones, medium-rise residential blocks within MKAD, recreation zone (parks, forest parks, and leisure areas), new high-rise buildings (Novokosino, Kozhukhovo), old residential blocks (private low-rise buildings in Novoukhtomskii, Kosino, and Rudnevo settlements), and agricultural area (arable lands of the former Mossovet sovkhoz). The zones differ in duration and intensity of anthropogenic stress. Geochemical maps were compiled using spline interpolation (Geostatistical Analyst, ArcGIS 10).

The accumulation or dispersion of dust and HMs in urban soils was evaluated using enrichment or depletion factors estimated via comparison with the data on the HM concentrations Cf in the reference soddy-podzolic soils: EF = C/Cf and DF = Cf/C, where C is the element concentration in urban soils. The results are displayed on a special plot named *geochemical spectrum*. The immission of the solid fraction of HMs to soil D equals the dust load P_n in the snow multiplied by HM concentration in it; $D = P_n \cdot C$. Exceedance of HM fallout above background values Df is characterized by a factor $K_d = D/Df$. Total multielemental pollution of snow and soil cover is defined by integral indices of immission $Z_d = \sum_{i=1}^{n} K_d - (n-1)$ and total pollution $Z_c = \sum_{i=1}^{n} EF - (n-1)$, respectively, where n is number of elements with K_d or EF > 1. They summarize the excess of the HM contents in urban soils over background levels and thus display the degree of their anthropogenic geochemical transformation [14, 25, 26]. Integral indices have 5 levels which determine the degree of pollution (Table 1). Maps of these parameters reveal spatial patterns of pollution in these depositing media.

3. Results and Discussion

3.1. Accumulation of HMs in Urban Soils. The intensity of HM accumulation in the soils of the Eastern district (Figure 2) was evaluated in relation to the background soils

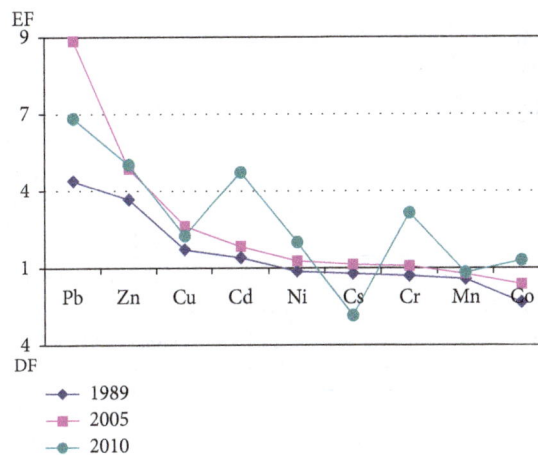

FIGURE 2: Geochemical spectra of urban soils in the Eastern district of Moscow showing enrichment and depletion factors (EF, DF) of metals in relation to background soils.

of Meshchera plain which have trace-level concentrations of all HMs under study, except for Cd and Cs. Low values of metal concentrations in background soils are explained by their contents in parent fluvioglacial sands [8] and the small intensity of fallout from the atmosphere [2, 9].

In 1989 the group of four elements accumulated in urban soils with the following sequence according to their enrichment factors EF (inferior indices): Pb$_{4.4}$Zn$_{3.7}$Cu$_{1.7}$Cd$_{1.4}$. Other metals dissipated with intensity which was defined by their depletion factors DF (top indices): Co$^{2.3}$Mn$^{1.4}$Cr$^{1.3}$Cs$^{1.2}$Ni$^{1.1}$. The content of HMs in urban soils was very variable; coefficients of variation CV of particular HMs changed from 41 to 89%, reaching 102% for Cd because of differences between the individual land-use zones.

In 2005 metals pollutants showed the increasing content while preserving the above sequence: Pb$_{9.9}$Zn$_{4.9}$Cu$_{2.6}$Cd$_{1.8}$. During 1989–2005 the average concentrations of two main pollutants, Pb and Zn, increased by a factor of 2.2 and 1.3, respectively. The content of Cr, Cs, and Ni came close to their background levels; only Co and Mn had concentrations below the background values (DF 1.6 and 1.2). Soil concentrations of all HMs under study were slightly less variable than in 1989 (CV 32–66% for Co, Ni, Mn, Cs, Zn, Cr, and Cu and 91-92% for Cd and Pb).

In 2010, 21 years after the beginning of investigations, further growth of soil pollution with HMs was observed, and

TABLE 2: Mean content of HMs (mg/kg) in topsoils of natural background and various land-use zones in the Eastern district of Moscow in 1989, 2005, and 2010.

Year (number of samples)	Pb	Zn	Cu	Cd	Co	Cr	Ni	Mn	Cs
Natural background of the Meshchera lowland									
1989 (10)	9.04	37.5	27.9	0.34	6.60	32.0	15.2	585	5.40
Highways (A) and industrial zones (P)									
1989 (12)	60.5	224	80.7	0.65	3.85	34.0	17.6	442	5.58
2005 (14)	174	305	130	0.81	5.82	51.7	26.9	518	8.08
2010 (13), A	59.5	183	67.5	1.23	8.48	89.9	32.2	412	1.90
2010 (5), P	114	286	88.0	1.37	9.3	92.2	35.4	513	1.88
Residential zone with low-rise buildings (L)									
1989 (5)	72.4	138	39.4	0.64	2.67	32.1	16.2	463	4.53
2005 (5)	159	180	65.7	0.81	3.84	47.9	22.6	549	6.34
2010 (5)	68.8	163	60.6	1.33	8.12	95.6	33.6	544	1.64
Residential zone with medium-rise buildings (M)									
1989 (16)	23.4	130	40.1	0.35	2.48	20.5	12.1	340	4.65
2005 (17)	39.9	165	58.8	0.48	3.76	29.2	17.8	414	6.47
2010 (10)	66.7	223	62.3	1.43	8.6	106	30.0	493	2.09
Residential zone with high-rise buildings (H)									
1989 (5)	16.4	76.5	24.6	0.14	1.61	16.3	9.14	260	2.43
2005 (5)	20.2	88.3	28.1	0.15	1.82	18.2	10.6	309	2.91
2010 (7)	44.1	162	54.0	1.45	8.57	104	26.1	389	1.99
Recreational zone (R)									
1989 (6)	21.1	76.7	24.3	0.17	2.44	9.24	7.58	512	4.04
2005 (6)	32.9	82.7	28.9	0.20	2.87	11.6	9.20	538	4.44
2010 (9)	49.4	147	50.4	2.05	8.8	101	29.0	482	1.87
Agricultural landscapes (AG)									
1989 (5)	45.6	96.3	57.7	1.07	3.20	34.0	18.0	535	3.74
2005 (5)	84.7	122	71.2	1.44	4.09	37.9	21.5	640	5.19
2010 (3)	34.0	173	67.0	3.70	7.7	137	27.3	492	1.47

the group of metals concentrated in topsoils was expanded up to seven elements. According to EF values, they form the following sequence: $Pb_{6.8}Zn_{5.0}Cd_{4.7}Cr_{3.1}Cu_{2.3}Ni_{2.0}Co_{1.3}$. Mn and Cs have the contents below background (DF 1.2 and 2.8, resp.). Since Cs is concentrated in hydroxides of Mn, its removal (washout) from the topsoils is accompanied by a significant decrease in the content of Cs [27]. The concentration of Pb—the main pollutant—decreased by a factor of 1.4 in comparison to 2005 that could be the result of soil reclamation in combination with the exclusion of leaded gasoline as a fuel for vehicles and the closing some hazardous enterprises which produced machines and electronic equipment. The intensive growth of Cd content in urban soils is caused by its presence in the motor vehicle emissions; in agricultural and recreation zone it is related to the application of large amounts of mineral and organic fertilizers and sewage sludge which contain Cd as an admixture. Concentrations of other metals increased slightly or remained at the same level. The contents of Co, Cs, and Pb in soils became less variable, while that of Zn and Cr increased up to 68–70% and of Cd, up to 136%.

3.2. HM Content in Soils with Different Type of Land Use. Distinctions in the levels of HM concentration in soils of different land-use zones were revealed through the analysis of values in Table 2. Differences in mean HM content were evaluated by t-test. They were significant at P 90–95% for the majority of the elements and land-use zones since mixed samples which were composed of 4 individual ones were used. In 1989 Pb and Zn were accumulated in soils of all land-use zones, particularly within old low-rise residential areas (EF 8.0 and 3.7, resp.), near highways (6.7 and 6.0), and agricultural areas (5.0 and 2.6). The third priority pollutant—Cu—was mostly accumulated in the industrial and transport zones (2.9). Soils of the recreational zone and new high-rise residential areas showed low contents of HMs. Among them the greatest dispersion of Cr (DF 3.5 and 2.0) and Co (2.7 and 4.1, resp.) in comparison with background soils was observed.

In 2005 the overall picture of HMs distribution in the soils of individual zones was the same, but the concentrations of main pollutants—Pb and Zn—significantly increased (Table 2). The highest concentrations were found in the soils of highways and industrial zone (EF 19.2 and 8.1, resp.), old residential blocks (17.6 and 4.8), and agricultural areas (9.4 and 3.3). Low concentrations of the majority of HMs (Co, Cr, Cd, Ni, Mn, and Cs) in comparison with background values

TABLE 3: Mean surplus/decrease rate of HM contents in topsoils of the Eastern district of Moscow.

Period	Increment rate	Pb	Zn	Cu	Cd	Co	Cr	Ni	Mn	Cs
1989–2005	mg/kg per year	3.1	2.8	1.6	0.01	0.08	0.64	0.36	4.40	0.11
	%	0.08	0.03	0.04	0.03	0.04	0.03	0.03	0.02	0.03
1989–2005	mg/kg per year	−5.4	1.2	−2.1	0.20	0.90	13.0	2.2	3.8	−0.86
	%	−0.36	0.03	−0.15	0.89	0.71	0.97	0.45	0.04	−1.1

were observed in the soils of recreational zone and new high-rise residential quarters (DF 3.6–1.1).

In 2010 geochemical contrasts in the soils of individual land-use zones became less pronounced. The concentrations of Cr, Co, and Mn increased in comparison with 2005 practically in all land-use zones, and those of Pb, Cu, Zn decreased in transport and industrial zones, old residential and agricultural areas (except for Zn). The content of Pb in these zones decreased by a factor of 2.3–2.5 that is apparently related to the soil reclamation, as well as to improvement of fuel and changes of vehicle structure. The content of Cd sharply increased, especially in the soils of recreational and agricultural zones. This could be explained by the application of high doses of sewage sludge and phosphoric fertilizers.

3.3. Long-Term Dynamics of HM Anomalies in Urban Soils.
The time-spatial patterns of HMs distribution in soils are presented on the maps of Z_c index compiled for several years of the 21-year-long period of observations (Figure 3). In 1989 the territory was characterized by low level of total pollution ($Z_c < 16$) as a result of rather weak industrial and transport influence on urban soils. On this background there were several small spots of low-contrast technogenic anomalies of HMs with medium level of pollution located in northwest, southeast, and central parts of the district. In the centers of these anomalies the values of Z_c indicator did not exceed 32–35.

In 2005 the existing anomalies of HMs in soil cover of the district considerably increased their size and contrast; the total pollution of soils increased by a factor of 1.7 for the 16-year period. The most contrast technogenic anomaly of HMs with Z_c about 48 in its center was formed in the northwest under the influence of emissions from the large industrial zone "Sokolinaya Gora" and Shosse Entuziastov Highway. Several anomalies along the Moscow Ring Road (MKAD at Figure 1) which were poorly expressed in 1989 became larger with maximum Z_c 32–40. Southeast anomaly of HMs in soils also became more pronounced, up to Z_c 48. Thus, the pollution level in anomalies has changed from the medium to high.

In 2010 a further increase in the areas and intensity of HM anomalies was observed. The northwest anomaly with the total index of soil pollution Z_c 48–80 expanded beyond the industrial zone "Sokolinaya Gora" and partly covered residential areas to the south of Shosse Entuziastov Highway. In the center of the district large anomaly with $Z_c > 100$ was formed under the influence of the industrial zone "Perovo" and district heating station (DHS) "Perovskaya." In the southeast part of the district technogenic anomaly of HMs

was extended to the east while the Z_c indicator increased up to 50–70. It is the result of emissions of recently constructed incineration plant near the settlement of Rudnevo.

Growing contamination of urban soils with HMs is caused not only by an increase in emissions from industry and transport in particular, but also by a change in absorption capacity of the soils. Anthropogenic transformation of the physical and chemical properties of the soils leads to the formation of complex physical and chemical barriers. They arise as a result of precipitation of carbonate construction dust enriched by fine particles, application of soil peat compost mixtures with a high content of humic substances for planting of greenery as well as flooding and soil sealing which alters its gas and redox regimes. The HM accumulation on these man-made barriers is not necessarily accompanied by a reduction in their mobility; analysis of mobile forms of HMs showed that their precipitation from the atmosphere exceeds the speed of their washout from topsoils [28].

3.4. Trends of HM Accumulation in Topsoils and Snow Cover.
Tendencies of HM accumulation in the urban soils were evaluated for two periods: 1989–2005 and 2005–2010 by calculation of average annual growth of HM content in the topsoils (Table 3) which differ considerably depending on the type of land use (Figure 4).

From 1989 to 2005 the growth of HM content in soils of the Eastern district was relatively small. The maximum absolute increase was inherent of Mn (4.4 mg/kg per year) > Pb (3.1) > Zn (2.8) > Cu (1.6) > Cr (0.64). Thus, the relative rate was the greatest for Pb, Co, Cu, Ni, and Cr (0.07–0.03% per year) in the soils of industrial zones and near highways, old low-rise residential quarters, and agricultural areas (Figure 4).

In 2005–2010, the rates of HM accumulation in the soils increased and became more differentiated among the land-use zones (Figure 4). Depending on the average annual values HMs form the following sequence: Cr (13.1 mg/kg per year) > Mn (3.8) > Ni (2.2) > Zn (1.2) > Co (0.9). Cr, Cd, Co, and Ni had the maximum relative rate of growth (up to 0.5–1.0%), in comparison with the previous period it increased by tens times (Table 2). The content of Pb, Cs, and Cu significantly decreased showing negative values of the annual rate mainly owing to soil reclamation.

Uneven distribution of HM accumulation rates within the territory of the district is well seen on Figure 4. Cr and Cd show the overall growth, especially in less polluted zones (recreational, agricultural, and new residential areas), while Pb and Zn dissipate in the soils of industrial zones and highways, old low-rise residential and agricultural areas.

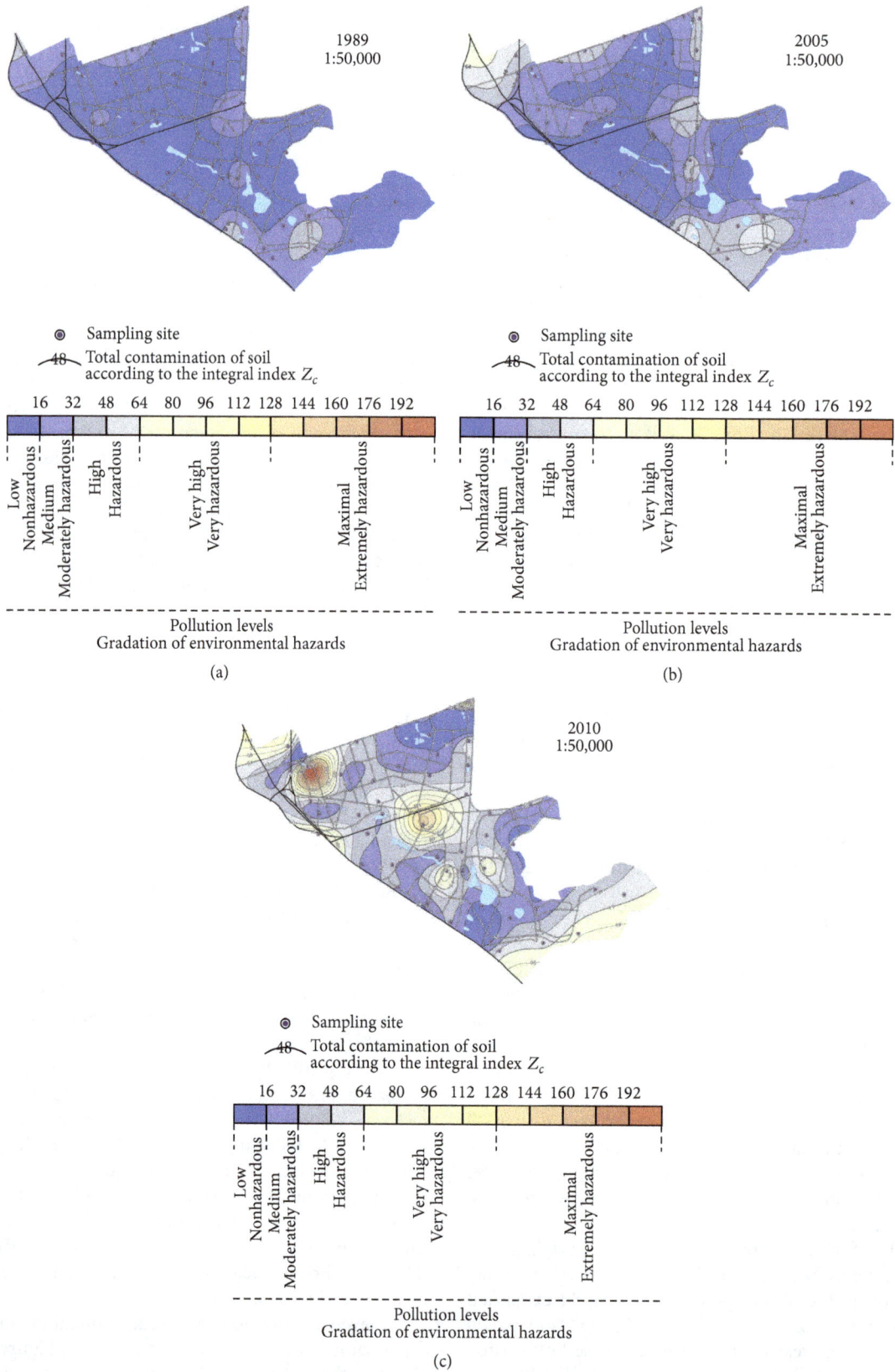

FIGURE 3: Total technogenic anomalies of HMs (according to Z_c index) in topsoils of the Eastern district of Moscow in 1989, 2005, and 2010.

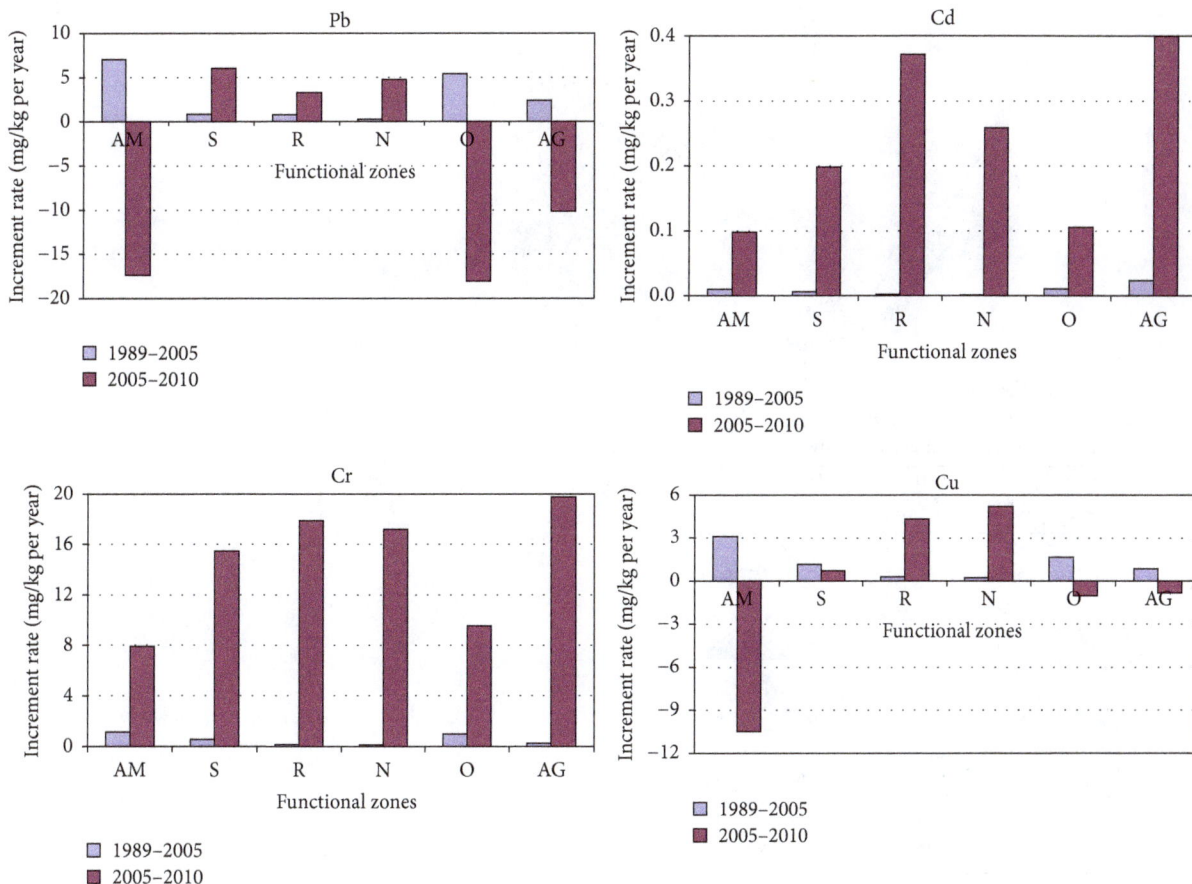

FIGURE 4: Average increase or reduction of the content of Pb, Cd, Zn, and Cr in the topsoils of different land-use zones in the Eastern district of Moscow for two periods of observations (1989–2005 and 2005–2010). Symbols of land-use zones are given in Table 2.

TABLE 4: Long-term dynamics of total HM pollution of soils in the Eastern district of Moscow depending on the type of their use.

| Observation year | Average values of total index of pollution Z_c in soils of the land-use zones | | | | | | | |
	Highways	Industrial	Medium-rise residential	High-rise residential	Recreational	Low-rise residential	Agricultural	Total for the district
1989	15.6		6.3	2.9	3.5	13.2	10.4	9.02
2005	33.4		10.3	3.7	5.0	25.5	17.2	17.3
2010	17.9	28.3	20.4	15.4	17.9	18.6	22.9	19.4

The dynamics of soil pollution was compared with the distribution of HMs in the snow. The snow cover reflects actual state of the environment while the soil represents more stable long-term pollution. Joint analysis of snow and soil geochemical maps allows identification of the areas of actual, permanent, and relict pollution. The relict techno-genic anomalies occur only in soil cover, modern ones, only in snow, and permanent ones in snow and soils simultaneously [25]. The snow pollution with HMs in the Eastern district of Moscow was defined using the integral index of immission Z_d [29]. It characterizes both the composition and amount of dust which accumulates in snow. Overlay of the maps of Z_c and Z_d integral indices resulted in the identification of permanent anomalies within the major part of the district (Figure 5). In the center of the area a relict anomaly was

formed as a result of elimination and reprofiling of the stationary sources of emissions. On the north, along the Shosse Entuziastov Highway a modern anomaly of HMs was discovered in the vicinity of Terletskiy park.

3.5. Assessment of Environmental State of Urban Soils. The analysis of average zonal values of the total indicator of pollution Z_c (Table 4) showed that in 1989 the most part of the territory under study was characterized by a low level of HM pollution. The average value of Z_c was only 9.0 that indicates harmless ecological condition of soils. Z_c values exceeded this level only in 10 of 49 sampling sites (i.e., 20.4%), mainly in the industrial zones and near highways.

In 2005 much less sampling points showed Z_c values < 16 (31 of 52, i.e., 59.6%). Moreover, 14 points (26.9%) shifted

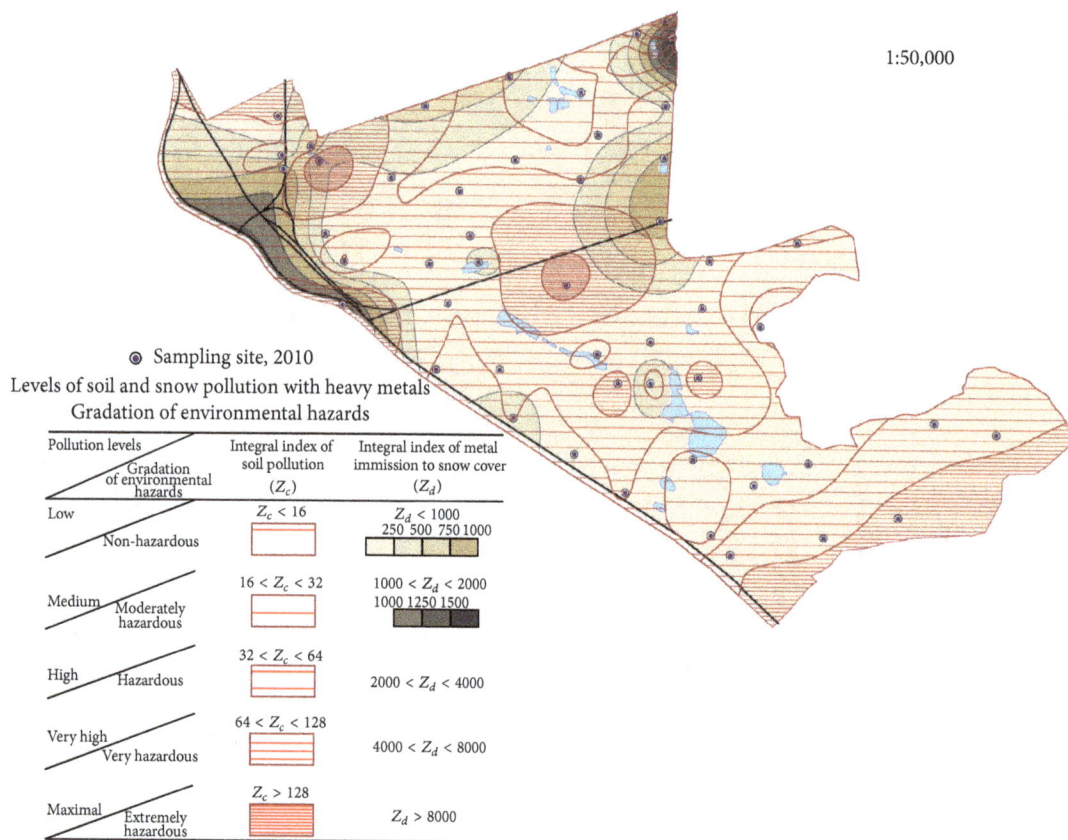

FIGURE 5: Location of areas of actual, permanent, and relict pollution in the Eastern district of Moscow according to anomalies in soils and snow cover.

TABLE 5: Clarks in the lithosphere [15] and the average contents of Pb, Zn, Cr, and Ni in soils (mg/kg) of the various cities of the world (compiled by the authors according to [16–24] and to own data for the Eastern district of Moscow, 2010).

HM	Lithosphere clark	Stockholm	Madrid	Seville	Berlin	Hamburg	London	Hong Kong	Palermo	Naples	Belgrade	Da Nang	Moscow, Eastern district
Pb	16	101	161	161	119	218	294	93.4	253	262	55.5	20.0	61.8
Zn	83	171	210	107	243	516	183	168	151	251	118	142	189
Cr	83	34.0	74.7	42.8	35.0	95.4	—	<40.0	39.0	74.0	32.1	92.2	100
Ni	58	12.8	14.1	23.5	10.7	62.5	—	<20.0	19.1	—	68.0	22.6	30.6

to medium and 7 points (13.5%) to high level of pollution. Average Z_c value for the district was 17.3 that corresponds to medium level of HM pollution of soils. There were only three land-use zones with low level of soil contamination, residential zones with medium- and high-rise buildings, and recreational zone (Table 4). In 2005 soils of the residential low-rise zone and agricultural landscapes had medium level of HM pollution, and those of highways and industrial zones had high one.

In 2010 the average Z_c value for the district increased up to 19.4. In contrast to the decrease of soil pollution level for highways, industrial and residential zones with low-rise buildings, a significant increase of HM pollution occurred in recreational, agricultural, and residential high-rise zones.

Thus it is only in the latter zone where the level of soil contamination corresponds to harmless category; other zones are moderately polluted (Z_c 17.9–28.3).

To assess the ecological rank of Moscow in terms of the level of HM accumulation in soils among the cities of the world, we compared the average contents of four most common HMs (Pb Zn, Cr, and Ni) in the soils of large and capital cities of several countries with their values in the soils of the Eastern district of Moscow. Soils of the large cities of the world have rather high average HM concentrations (Table 5) comparable to Moscow quantities, especially Pb and Zn. Most intensive accumulation of Pb is typical for the soils of London, Naples, Palermo, and Hamburg, while that of Zn is typical for the soils of Hamburg, Naples, Berlin, and

Madrid. The contents of Cr and Ni are below their clarks in the lithosphere; the minimum values are observed in the soils of Madrid, Berlin, Palermo, and Stockholm. So, according to 2010 data, soils of the Eastern district of Moscow do not show particularly high levels of HM accumulation. The average content of Pb in the topsoils of the district is close to its values in the soils of Belgrade and that of Zn is close to its values in the soils of Stockholm and London. The same levels of the content of Cr, as in soils of the Eastern district, are recorded in Palermo and Hamburg, and that of Ni is recorded in Seville and Palermo.

4. Conclusions

Soils of the Eastern district of Moscow show the gradual increase of HM pollution levels. Among them Pb and Zn are the leaders with enrichment factors EF 3.7–9.9. In 2010 Cd joined them (EF 4.7); the intensive growth of its content is caused by its presence in the motor vehicle emissions as well as by application of mineral and organic fertilizers and sewage sludge containing Cd as an admixture. Over the last 5 years the contents of Pb and Cu in soils of the district decreased owing to the soil reclamation and adding uncontaminated material, the exclusion of leaded gasoline as a fuel for vehicles, and also conversion and closing a number of hazardous industrial enterprises which emit these metals.

In 1989–2005 the annual increase of HM concentrations in soils of the Eastern district was rather low. Pb and Cu demonstrated the maximum rates of increase (0.08 and 0.04% per year, resp.). In 2005–2010 rates of the annual increase of HM concentrations in soils were higher, with more pronounced differentiation in particular land-use zones. Cr, Cd, Co, and Ni had the maximum rates of increase (1.0–0.45% per year). The contents of Pb and Cu in the soils of industrial zones and highways, agricultural, and old residential areas went down. Nevertheless, these zones still have the highest contents and the widest range of elements pollutants.

At present, a medium total pollution level is common for the soils of the Eastern district, with the Z_c index reaching 28.3 in the industrial zone. The priority contaminants are Pb, Zn, and Cd. The maximum level of Pb and Zn concentrations (114 and 286 mg/kg, resp.) in topsoils has been revealed in the industrial zone, while that of Cd (3.7 mg/kg) has been revealed in the recreational zone.

To represent the stages of development of the technogenic multielemental anomalies in urban soils maps of a total index of pollution Z_c were compiled for particular years of the 21-year period of observations. Their analysis revealed a significant increase of the contrast and the area of HM anomalies in recent years, which could be attributed to the growth of the total amount of industrial and motor transport emissions and change in absorption capacity of the urban soils.

Acknowledgments

Field and chemical works were supported by the Russian Geographic Society and the Russian Foundation for Basic Research (Project no. 13-05-41191), and analysis and interpretation of geochemical data were accomplished with the financial support of the Russian Scientific Foundation (Project no. 14-27-00083). The authors thank Dr. D. V. Vlasov and Mrs. G. L. Shinkareva who took part in the field investigations. Dr. T. S. Khaybrakhmanov is much appreciated for the geochemical maps he compiled.

References

[1] E. M. Nikiforova and G. G. Lazukova, "Moscow. Perovo District. Plain landscapes," in *Ecogeochemistry of Urban Landscapes*, N. S. Kasimov, Ed., pp. 57–90, Moscow University Publishing House, Moscow, Russia, 1995 (Russian).

[2] S. B. Samaev, L. S. Sokolov, and A. S. Panteleev, "Soil pollution caused by motor traffic in Moscow," in *Automotive Complex and Ecological Safety*, pp. 266–270, Prima-Press-M, Moscow, Russia, 1999 (Russian).

[3] Yu. N. Nikolaev, T. V. Shestakova, V. V. Nefed'ev et al., *Assessment of Geochemical Contamination in the Losinyi Ostrov National Park*, Prima-Press-M, Moscow, Russia, 2000 (Russian).

[4] Kh. G. Yakubov, *Environmental Monitoring of Plantations in Moscow*, Stagirit-N, Moscow, Russia, 2005 (Russian).

[5] E. M. Nikiforova and N. E. Kosheleva, "Dynamics of contamination of urban soils with lead in the Eastern district of Moscow," *Eurasian Soil Science*, vol. 40, no. 8, pp. 880–892, 2007.

[6] E. M. Nikiforova and N. E. Kosheleva, "Fractional composition of lead compounds in soils of Moscow and Moscow region," *Eurasian Soil Science*, vol. 42, no. 8, pp. 874–884, 2009.

[7] O. V. Plyaskina and D. V. Ladonin, "Heavy metal pollution of urban soils," *Eurasian Soil Science*, vol. 42, no. 7, pp. 816–823, 2009.

[8] I. A. Avessalomova, "Landscapes of Meshchera lowplain," in *Landscape-Geochemical Principles of Background Monitoring of Natural Environment*, M. A. Glazovskaya and N. S. Kasimov, Eds., pp. 79–90, Nauka, Moscow, Russia, 1989 (Russian).

[9] T. V. Prokofyeva, I. A. Martynenko, and F. A. Ivannikov, "Classification of Moscow soils and parent materials and its possible inclusion in the classification system of Russian soils," *Eurasian Soil Science*, vol. 44, no. 5, pp. 561–571, 2011.

[10] N. I. Ivanov, Ed., *Engineering Ecology and Environmental Management*, Logos, Moscow, Russia, 2002 (Russian).

[11] V. R. Bityukova and R. Argenbrayt, "Regional peculiarities of modern ecological situation in Moscow: premises, contemporary tendencies, prospect," *Eurasian Geography and Economics*, no. 4, pp. 14–31, 2002.

[12] E. M. Nikiforova, N. E. Kosheleva, I. A. Labutina, and T. S. Khaybrakhmanov, "Geoinformation landscape-geochemical mapping of city territories (the case study of Eastern District of Moscow)," *Journal of Civil Engineering and Science*, vol. 3, no. 3, pp. 142–151, 2014.

[13] C. C. Johnston, A. Demetriades, J. Locutra, and R. T. Ottesen, *Mapping the Chemical Environment of Urban Areas*, John Wiley & Sons, Chichester, UK, 2011.

[14] N. S. Kasimov, V. R. Bityukova, A. V. Kislov et al., "Problems of urban geochemistry," *Exploration and Conservation of Mineral Resources*, no. 7, pp. 8–13, 2012 (Russian).

[15] A. P. Vinogradov, "Average content of chemical elements in the main types of igneous rocks," *Geochemistry*, no. 7, pp. 555–572, 1962 (Russian).

[16] P. Tume, J. Bech, B. Sepulveda, L. Tume, and J. Bech, "Concentrations of heavy metals in urban soils of Talcahuano (Chile): a preliminary study," *Environmental Monitoring and Assessment*, vol. 140, no. 1–3, pp. 91–98, 2008.

[17] M. Linde, H. Bengtsson, and I. Oborn, "Concentrations and pools of heavy metals in urban soils in Stockholm, Sweden," *Water, Air, & Soil Pollution: Focus*, vol. 1, pp. 83–101, 2001.

[18] H. T. T. Thuy, H. J. Tobschall, and P. V. An, "Distribution of heavy metals in urban soils—a case study of Danang-Hoian Area (Vietnam)," *Environmental Geology*, vol. 39, no. 6, pp. 603–610, 2000.

[19] X. Li, C.-S. Poon, and P. S. Liu, "Heavy metal contamination of urban soils and street dusts in Hong Kong," *Applied Geochemistry*, vol. 16, no. 11-12, pp. 1361–1368, 2001.

[20] M. Birke and U. Rauch, "Urban geochemistry: investigations in the Berlin metropolitan area," *Environmental Geochemistry and Health*, vol. 22, no. 3, pp. 233–248, 2000.

[21] M. Imperato, P. Adamo, D. Naimo, M. Arienzo, D. Stanzione, and P. Violante, "Spatial distribution of heavy metals in urban soils of Naples city (Italy)," *Environmental Pollution*, vol. 124, no. 2, pp. 247–256, 2003.

[22] L. Madrid, E. Diaz-Barrientos, E. Ruiz-Cortés et al., "Variability in concentrations of potentially toxic elements in urban parks from six European cities," *Journal of Environmental Monitoring*, vol. 8, no. 11, pp. 1158–1165, 2006.

[23] D. Crnković, M. Ristić, and D. Antonović, "Distribution of heavy metals and arsenic in soils of Belgrade (Serbia and Montenegro)," *Soil & Sediment Contamination*, vol. 15, no. 6, pp. 581–589, 2006.

[24] F. Ajmone-Marsan and M. Biasioli, "Trace elements in soils of urban areas," *Water, Air, & Soil Pollution*, vol. 213, no. 1, pp. 121–143, 2010.

[25] Yu. E. Saet, B. A. Revich, E. P. Yanin et al., *Geochemistry of the Environment*, Nedra, Moscow, Russia, 1990 (Russian).

[26] N. S. Kasimov, N. E. Kosheleva, O. I. Sorokina, S. N. Bazha, P. D. Gunin, and S. Enkh-Amgalan, "Ecological-geochemical state of soils in Ulaanbaatar (Mongolia)," *Eurasian Soil Science*, vol. 44, no. 7, pp. 709–721, 2011.

[27] A. I. Perel'man, *Geochemistry of Elements in the Hypergenesis Zone*, Nedra, Moscow, Russia, 1972 (Russian).

[28] N. E. Kosheleva, N. S. Kasimov, and D. V. Vlasov, "Factors of the accumulation of heavy metals and metalloids at geochemical barriers in urban soils," *Eurasian Soil Science*, vol. 48, no. 5, pp. 476–492, 2015.

[29] N. S. Kasimov, N. E. Kosheleva, D. V. Vlasov, and E. V. Terskaya, "Geochemistry of snow cover in the Eastern District of Moscow," *Moscow University Vestnik. Series 5. Geography*, no. 4, pp. 14–25, 2012 (Russian).

12

Pesticides Usage in the Soil Quality Degradation Potential in Wanasari Subdistrict, Brebes, Indonesia

Tri Joko,[1,2] **Sutrisno Anggoro,**[1,3] **Henna Rya Sunoko,**[1,4] **and Savitri Rachmawati**[2]

[1]*Doctoral Program of Environmental Science, School of Postgraduate Studies, Diponegoro University, Semarang City, Indonesia*
[2]*Public Health Faculty, Diponegoro University, Semarang City, Indonesia*
[3]*Faculty of Fisheries and Marine Science, Diponegoro University, Semarang City, Indonesia*
[4]*Faculty of Medicine, Diponegoro University, Semarang City, Indonesia*

Correspondence should be addressed to Tri Joko; trijokoundip@gmail.com

Academic Editor: Marco Trevisan

Uncontrolled application of pesticides can contaminate soil and may kill other nontarget organisms. This study aims to determine the usage pattern of pesticides by farmers in Wanasari Subdistrict and study the soil quality degradation potential. This study was a quantitative and qualitative research. Sources of data were collected from observation, questionnaire, and in-depth interview methods. The respondents were shallot farmers who planted shallot during 2013–2016 ($n = 60$). In-depth interview was done with three respondents from the local agricultural extension center (BPP). This study found that there were some different types of insecticides and fungicides that were used in every planting season. The farmers applied pesticides in large amount once every three or four days. They mixed minimally three insecticides and fungicides types about 30–40 ml for each type. Organophosphate residues that were found in soil samples were methidathion residue about 0.014 mg/kg, malathion residue ranging around 0.1370–0.3630 mg/kg, and chlorpyrifos residue in the range of 0.0110–0.0630 mg/kg. The excessive application of pesticides showed the land degradation potential. Soil quality laboratory testing is recommended to ensure the agricultural land condition. Routine assessment of soil quality and pesticide usage control is recommended to keep sustainable ecosystem.

1. Introduction

Uncontrolled application of pesticides can contaminate soil and may kill other nontarget organisms. Pesticides can damage soil biomass and microorganism such as bacteria, fungi, and earthworms. Microbial biomass is a labile component of soil organic matter and has an important role in soil nutrient element cycle [1]. S. A. Reinecke and A. J. Reinecke (2007) had studied earthworm biomass and cholinesterase activity affected by pesticides [2]. The authors concluded that earthworms were affected detrimentally by the pesticides due to chronic (chlorpyrifos) and intermittent exposure (azinphos methyl). Other research also showed that malathion exposure gave the significant reduction in body weight and decreased sperm viability of *Eisenia fetida* adults species. The organism also had an adverse impact on growth and reproduction by the chlorpyrifos exposure, and the cypermethrin exposure also gave the significant reduction in cocoon production [3–6]. Pesticides which are applied to the soil could have an impact on nontarget organisms and damage the local metabolism that is required by soil fertility and pesticide degradation itself [7–11]. The large pesticides application and its impact can be identified in Indonesian shallot farming.

Some of the data showed that shallot productivity by Indonesian farmers became nonoptimal because of several factors. The factors that affected the productivity of shallot farming are improper cultivation technique, uncontrollable environmental factor, pest attack, and plant disease [12, 13]. In order to control the pest and plant disease, farmers apply pesticides excessively. Wanasari Subdistrict is one of the agricultural centers in Brebes which produces shallot (*Allium cepa* L). Wanasari located at least 2 km from the downtown. The usage of pesticides increases every year, including in this region. The emergence of new pesticides enables farmers to try any type of them. The habit of trying every type of new pesticides has happened in Wanasari Subdistrict. On the other hand, socialization and advocacy regarding the procedure for safety usage of pesticides have been carried out by the local agricultural institute, in Indonesia known as Balai Penyuluh Pertanian (BPP) or called agricultural extension center. However, the socialization is not optimal and the farmers still apply pesticides to the land with an excessive amount. The enhancement of pesticides that were used by farmers happens not only in a rural area of Indonesia but also in urban area.

The enhancement of agricultural production associated with more intensive use of pesticides including insecticides. It is assumed that in 2050 the use of pesticides will be 2.7 times greater than in 2000. This will cause humans and the environment to be in a dangerous condition [14, 15]. There were many references about the impact of pesticides on soil and water environment. The excessive application of pesticides also causes pest resistance, and this condition also occurred in Wanasari Subdistrict. Resistance is a natural phenomenon. Even as the introduction of new classes of insecticides such as cyclodienes, carbamates, formamidines, organophosphates, and pyrethroids, new resistant cases that appeared were reported in the period of 2–20 years. Pest resistance resulted in the fact that farmers must increase the dose of pesticides that are used; even the addition could be 2-3 times the amount added before.

This study aims to determine the usage pattern of pesticides by farmers in Wanasari Subdistrict. In this study, potential degradation for agricultural soil quality was also reported. Data and information obtained from this research then can be used as the basis for further research to determine the relationship between the pesticide applications to soil parameter degradation quantitatively.

2. Materials and Methods

2.1. Time and Study Area. This research was conducted from August 2016 until January 2017 located at Wanasari Subdistrict, Brebes District, Central Java Province, Indonesia. The location of the study was selected based on the shallot productivity criteria level and pesticide usage behavior.

2.2. Interview and Questionnaire. This study was a quantitative and qualitative research. Sources of data were collected from observation, questionnaire, and in-depth interview. The respondents for quantitative study were shallot farmers who planted shallot during the last three years (n = 60). The interviews were conducted with three official officers from the local agricultural extension center, known as Balai Penyuluh Pertanian (BPP). Data were presented descriptively by outlining the existing findings of observations and respondents interview. Interviews were conducted to determine the pattern of shallot planting in the study area, the pattern of pesticide usage, which includes the types of pesticides, how to mix the pesticides for spraying, and time of pesticides use based on the planting period. Instruments used in this study are questionnaires, notebook, and recorder. The questionnaires were inputted to Microsoft Office Excel for descriptive analysis. The data analysis was presented in tables and graphs. In-depth interview was carried out to determine the role of BPP in the use of pesticides and also to validate the findings of the questionnaires.

2.3. Soil Sampling and Organophosphate Residue Testing. Soil sampling methods were done randomly. Soil samples were taken throughout the sampling locations at the same time. The sampling was conducted during the dry season. Soil samples were taken at 25–50 cm depth, and the top soil was not included in the sample. Soil samples were collected using a tube sampler and then stored in a clean plastic bag. Samples are labeled according to the date of sampling, sample locations, and types to be tested. Then the samples were sent to the lab to be tested. The measurement method of pesticide residues is using Gas Chromatography-Mass Spectrometry (GC-MS) in the Agrochemicals Residue Laboratory, Indonesian Ministry of Agriculture. The result is expressed in units of mg/kg.

2.3.1. Tools and Materials. The tools used in this study were mechanical shaker (shaker), Florisil columns, glassware, rotary evaporator, homogenizer, vacuum pump, Buchner filter, water bath, analytical scales, and a Gas Chromatography set (GC-2014 Shimadzu) equipped with ECD and Rtx-1 columns (column length 30 m, inner diameter 0.25 mm). Materials used in the study were acetone (E. Merck), n-hexane (E. Merck), anhydrous Na_2SO_4 (E. Merck), Florisil, activated carbon, Celite 545-magnesium, filter paper, aquadest, standard solution of organophosphate pesticides, and soil samples.

2.3.2. Soil Sampling. The ground surface to be sampled is flattened and cleaned; then the soil was dug with a soil corer perpendicular to the layer. The soil was taken to 25–30 cm depth. Next, the soil around the tube was dug up with a shovel. About 1-2 cm of the topsoil was removed. Then ± 250 grams of soil was leveled and sampled. The soil samples were then placed on the sample plastic and labeled with sample codes, sampling dates, sample locations, and test types. Samples that had been prepared then were immediately sent to the laboratory. Sample delivery should be safe to avoid damage or torn plastic. The delivery time was about 2-3 days.

2.3.3. Preparation of Soil Samples. Dried soil sample (by aerated method) weighed about 25 grams and then was put into a round bottom flask and 100 ml acetone solvent was added and then closed. The round bottom flask containing the sample was shaken with a shaker for 20 minutes at a sufficient rate. After 20 min, the extraction was repeated by rewinding with the same time and velocity; then the separating funnels were placed on the iron stand with clamp and remained until separation occurred between the solvent and the sample; then separation between filtrate and residue occurred. The filtrate was evaporated with an evaporator, extracted with 25 ml of n-hexane 2 times, and then cleaned up by passing the sample on the chromatographic column filled with Florisil and anhydrous sodium sulfate. Eluate was evaporated until ±1 ml then the flask was rinsed with acetone gradually and the result was collected in test tube up to 10 ml volume and the sample solution was ready to be injected into the Gas Chromatography.

2.3.4. Organophosphate Residue Testing. Analysis of pesticide residues of organophosphate groups in soil samples was carried out by the method used by the Agrochemical Residue Laboratory of Balingtan (Balai Penelitian Lingkungan Pertanian). The solutions obtained from the extraction of soil samples determine the residual content of the organophosphate by the results of Gas Chromatography (GC-2014 Shimadzu).

3. Results and Discussion

3.1. Planting Time Pattern in the Study Area. Wanasari Subdistrict is one of the shallot agricultural centers in Brebes. Wanasari Subdistrict drained Pemali River which is used in the irrigation. Pemali River flows throughout the year, but the reduction of water discharge may occur in dry season. This hydrology conditions influence the shallot planting time and productivity based on its area. Shallot planting times vary depending on the regions. Wanasari Subdistrict stretches from north to south covering an area of 7,226 ha consisting of 2,123 ha of agricultural land using technical irrigation area, 632 ha using semitechnical irrigation, and rainfed area about 1,346 ha. In general, Wanasari Subdistrict has four shallots planting periods per quarter throughout the year, the first planting is in January–March, the second planting is in April to June, the third growing season is in July–September, and the fourth growing season is in October–December. In addition, farmers also plant paddy, red peppers, green beans, bitter melon, peanuts, eggplants, tomatoes, cucumbers, and beans. Shallot was grown throughout the year, while rice and chili are planted during two periods. Paddy field for planting shallot has the widest area compared to other crops. Data from the Agricultural Extension Center of Wanasari Subdistrict showed in 2014 shallot harvest in Wanasari reached 6120 hectares; then the paddy areas are about 3800 hectares.

The irrigated areas by Pemali River were Dukuhwringin, Dumeling, Glonggong, Jagalempeni, Kertabesuki, Lengkong, Pesantunan, Siasem, Sidamulya, and Wanasari village. The small part areas of the irrigation river are Pebatan and Sawojajar village. Areas that do not irrigate by the Pemali River are Keboledan, Klampok, Kupu, Sigentong, Sisalam, Siwungkuk, Tanjungsari, and Tegalgandu, but there are discharge channels as a source of irrigation. Pemali River flow across most of villages in Wanasari Subdistrict has benefit for agricultural area. Farmers can plant shallot along the year because of the abundance of water. The rainy season which occurs in the range from October to March also brings benefits for the area as a part of irrigation system in the location. With the availability of sufficient water, it can guarantee that shallot can be continued to be planted in this area.

3.2. Pesticides Usage Pattern among Farmers in the Study Area. Observations and interviews with farmers showed the classified patterns of pesticides excessive application. When spraying, farmers mixed the pesticide at least 2-3 types and there could even be 5–7 types. The doses used were

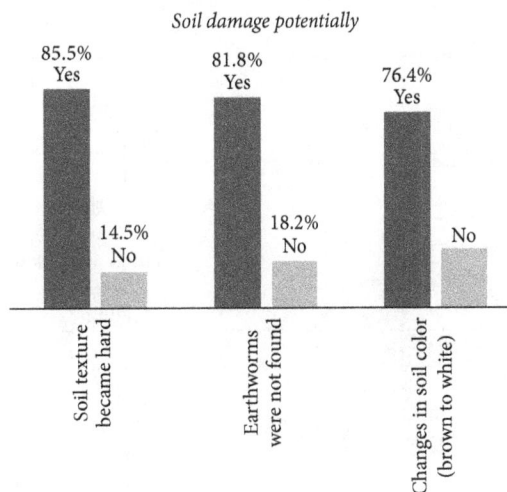

FIGURE 1: Degradation potential of soil quality.

30–40 ml per type. Certainly the application of pesticides was tailored to the pest attack and shallot plant disease. The excessive pesticide applications become uncontrolled in a pest explosion event, when the pests attack less than usual the usage of insecticides reduced. When spraying, farmers used the spray tank that has 15 or 17 liters' capacity. For the fields of 1000 m^2, it takes at least two tanks for spraying once. If farmers used pesticides of at least 3 mixtures of 30 ml per species, then a tank containing 90 ml of pesticides was sprayed. For the 1000 m^2 land at least 180 ml pesticides were sprayed. When this amount is calculated up to the time of the year it could be demonstrated that there were many pesticides absorbed by the ground.

The pattern of insecticides usage in shallot agriculture varies. The amount of insecticides used has different mixed pattern in each cropping period, or with the fungicides usage. However, farmers often mixed between insecticides and fungicides. In the first growing season in January to March farmers generally use four types of mixtures (variation between insecticides and fungicides). The types often used in this season were Arjuna, Tumagon 100 EC, Daitin, and Vondozeb. The second planting season is in April, May, and June; an increasing number of the mixtures are used, which can be up to five mixtures. In the third growing season from July to September the farmers do not use fungicides. When the second and third growing season are in dry season and pests attack increased, the use of insecticides is done in large numbers. Shallot had little of plant disease and it was not found during the dry season. This has led to the reduction of fungicide usage. In the fourth growing season in October–December, the opposite things occurred. These months already entered the rainy season so that pests diminish; shallot plant disease increases. These conditions led to the fact that the usage of insecticides was not much required, and the usage of fungicides increased (Table 1).

3.3. Types of Pesticides Mostly Used by Farmers. Insecticides and fungicides are used by many farmers in the shallot crop. Basically insecticides are used to control pests of plants, and

TABLE 1: Agriculture planting time in Wanasari Subdistrict.

Planting time			
I	II	III	IV
Insecticides used Arjuna Tumagon Fungicides used Daitin Vondozeb	Insecticides used Dursban Arjuna Marshal Fungicides used Folicur Vondozeb	Insecticides used Arjuna Trigard Fungicides used Not required	Insecticides used Dursban Fungicides used Amistar top Dithane

Jan	Feb	Mar	Apr	May	Jun	Jul	August	Sept	Oct	Nov	Dec
Pest Beet armyworm *Spodoptera* exigua L *Liriomyza* *huidobrensis* (serpentine leafminer) Disease Purple blotch (*Alternaria porri*) Fusarium vascular wilt (*Fusarium oxysporum*) Downy mildew Anthracnose Downy mildew (*Peronospora destructor*)				Pest Beet armyworm *Spodoptera* exigua L *Liriomyza* *huidobrensis* Grasshopper pest Purple blotch (*Alternaria porri*) Fusarium vascular wilt (*Fusarium oxysporum*) Downy mildew (*Peronospora destructor*)			Pest Beet armyworm *Spodoptera* exigua L *Liriomyza* *huidobrensis* Taro caterpillar *Spodoptera litura* *Thrips parvispinus*Disease Hardly found			Pest Hardly found Disease Purple blotch (*Alternaria porri*) Fusarium vascular wilt (*Fusarium oxysporum*) Downy mildew Anthracnose Downy mildew (*Peronospora destructor*)	

FIGURE 2: Hard soil texture and brown-white color soil, taken from agriculture area in Wanasari. Source: researcher documentation.

fungicides are used to save crops from disease. The study showed that Arjuna was the type of favorite pesticide most used by farmers (96%). Arjuna is a kind of insecticide that contains chlorfenapyr active ingredient from pyrrole group. The insecticide was classified as moderately hazardous by WHO. Furthermore, there was Tumagon insecticide used by 90% of respondents. Trigard insecticide that contains cyromazine active ingredient is widely used by 83% of farmers, and 81% of farmers use Antracol insecticides. Other types of insecticides that are widely used are Xtreme (80%), Dursban (75%), Sumo (52%), and Marshal (48%). Active ingredients contained in widely used insecticide are pyrrole, carbamates, tiadiazin, and organophosphates (Table 2, Figure 1).

The chlorfenapyr active ingredient is used to decontaminate armyworm (*Spodoptera litura*) and pest thrips (*Thrips parvispinus*). Chlorpyrifos and *beta-cyfluthrin* are also used to eradicate armyworm. Propineb active ingredient is used to treat purple spots disease on shallot. The usage of insecticides increased in three to six months (March–June) and decreased in July to September; even the usage of these insecticides was getting slight in October to December. The results of in-depth interviews with farmers in Wanasari Subdistrict showed that there were many insecticides' usage in the dry season months because pests attacking occurs in dry season. Meanwhile, when the usage of insecticide was declining, the usage of fungicide increased. Fungicides were widely used, especially in the rainy season. Increased rainfall causes crop diseases emergence, especially the shallot crop.

Table 2 showed that the average dose of insecticide with chlorfenapyr active ingredient has the highest amount about

TABLE 2: List of most insecticides and fungicides previously approved for use on shallot cultivation in Wanasari Subdistrict.

Number	Types of pesticides	Formulation name	Active ingredients	Group	Class by WHO	Pesticides usage ($n = 60$) (%)	Mean ± SD*
(1)	Insecticides	Arjuna	Chlorfenapyr	Pirol	II	96	34.47 ± 0.93
(2)	Insecticides	Tumagon	Buprofezin	Tiadiazin	III	85	33.51 ± 1.70
(3)	Insecticides	Dursban	Chlorpyrifos	Organophosphate	II	60	34.64 ± 2.91
(4)	Insecticides	Marshal	Carbosulfan	Carbamate	II	52	34.42 ± 2.27
(5)	Insecticides	Sumo	beta-Cyfluthrin	Pyrethroid	Ib	45	34.81 ± 3.06
(6)	Insecticides	Trigard	Cyromazine	Urea	III	23	33.57 ± 4.00
(7)	Fungicides	Antracol	Propineb	Carbamate, organozinc	U	88	34.85 ± 1.51
(8)	Fungicides	Delsene	Mancozeb	Dithiocarbamate, organomanganese	U	82	34.06 ± 1.75
(9)	Fungicides	Vondozeb	Mancozeb	Dithiocarbamate, organomanganese	U	63	34.05 ± 2.78
(10)	Fungicides	Folicur	Tebuconazole	Triazole	II	53	34.19 ± 2.24
(11)	Fungicides	Amistar top	Azoxystrobin	Pyrimidine	U	47	34.29 ± 3.06
(12)	Fungicides	Dithane	Mancozeb	Dithiocarbamate, organomanganese	U	27	33.69 ± 3.91

*Average dose used per 15-liter tank (ml); classes information: Ib is highly hazardous, II is moderately hazardous, III is slightly hazardous, and U is unlikely to present acute hazard in normal use. U = unlikely to present acute hazard in normal use.

34.47 ml compared to other insecticides and it was used by 96% of respondents. Insecticides with buprofezin active ingredient are used by 85% of respondents, with an average of 33.51 ml dose used. Chlorpyrifos active ingredient is used by 60% of respondents with average dose 34.64 ml used. Carbosulfan insecticide with beta-cyfluthrin and cyromazine active ingredients is used by 52% of respondents (34.42 ml), 45% of respondents (34.81 ml), and 23% of respondents (33.57 ml), respectively.

Table 2 showed that 88% of respondents use Antracol fungicide with propineb active ingredient. The average dose used in this type of fungicide was 34.85 ml. Fungicides with mancozeb active ingredient have different formulation names including Delsene, Vondozeb, and Dithane. A total of 82% respondents used Delsene fungicide with an average dose of 34.06 ml, 63% of respondents used Vondozeb fungicide with an average dose of 34.05, and 27% of respondents used Dithane with an average dose of 33.69 ml. In addition, Folicur fungicide with tebuconazole active ingredient was used by 53% of respondents with an average dose of 34.19 ml, and Amistar top fungicide with azoxystrobin active ingredient was used by 47% of respondents with an average dose of 34.29 ml.

3.4. *Pesticides Residue in Soil.* We took soil samples and tested the pesticide residues to determine whether there are pesticides that are still accumulating in the soil or not after spraying activity. Soil samples were taken at several locations of shallot agricultural area in Wanasari Subdistrict. Pesticide residue of the samples was tested by agricultural research centers of the Indonesian Ministry of Agriculture (Table 3).

In this study, chlorpyrifos residue active ingredient was found in samples from all locations about 0.01 to 0.06 mg/kg.

Insecticide which contains chlorpyrifos active ingredient is still used widely by shallot farmers. One type of pesticides which contains chlorpyrifos active ingredient is called Dursban. Adsorption of pesticides to the soil increases the persistence of chlorpyrifos in the environment by reducing its availability against dissipative capability and its degradative. While the adsorption effect on the environment toxicity depends on the route of exposure [16]. Chlorpyrifos has low water solubility (1.4 mg/l), high soil absorption coefficient (average Koc = 8498 ml/g), and medium vapor pressure (2.7×10^{-3} Pa at 25°C) [17, 18].

In this study, methidathion residue active ingredient only is found in one sampling location from Sisalam village about 0.014 mg/kg. Methidathion is a nonsystemic insecticide. Methidathion degradation in the environment is influenced by temperature and pH. The half degradation of methidathion is reported during half day up to 41 days depending on the temperature and pH. The higher the ambient temperature is, the easier the methidathion will be degraded. Methidathion easily degraded at alkaline pH. In acidic conditions hydrolytic cleavage occurs mainly on the C-S bond, and under alkaline conditions cleavage occurs on the P-S bond. Methidathion has low water solubility of about 240–250 mg/l at 20°C. These conditions increase the potential of methidathion compound to move off-site into water surface depending on the conditions and environmental factors. Methidathion compound adsorption coefficient on clay is about 2.8% with Koc = 310 [19].

In this study, the level of malathion residue active ingredient was found in the range of 0.13 to 0.36 mg/kg and it has higher amount compared with chlorpyrifos and methidathion residues. The persistence of malathion in the environment is affected by sunlight and ultraviolet light exposure.

TABLE 3: Pesticides residue found in soil samples from agricultural land in some of Wanasari's village (organophosphate group).

Active ingredients	Concentration of residue in some villages (mg/kg)				LOD (mg/kg)
	Tanjungsari	Sisalam	Dukuhwringin	Wanasari	
Diazinon	<LOD	<LOD	<LOD	<LOD	0.0101
Fenitrothion	<LOD	<LOD	<LOD	<LOD	0.0100
Methidathion	<LOD	0.0140	<LOD	<LOD	0.0104
Malathion	0.1370	0.1450	0.3630	0.1490	0.0104
Chlorpyrifos	0.0630	0.0150	0.0140	0.0110	0.0101
Parathion	<LOD	<LOD	<LOD	<LOD	0.0100
Profenofos	<LOD	<LOD	<LOD	<LOD	0.0101

Information: LOD: *Limit of Detection*.

Photolysis reaction occurs to reduce the toxicity of malathion. Malathion residue in the soil may also occur due to improper handling of current pesticide formulations during the mixing process, including the spill or practices usage and poor storage. Malathion is released into the atmosphere (in the form of gas) and can be transported back to the surface of the soil and water through wet deposition. Malathion rapidly degraded in the soil; the results of previous studies showed that half degradation began within hours to about 1 week. Malathion can also evaporate from the soil and in addition malathion moves very easily in the soil. The process of malathion leaching from the soil into the water is not possible due to the rapid degradation of this compound in the environment. Malathion is soluble in water and can move quickly into the ground despite its low persistence (1–25 days in the ground) [20–23].

Other active ingredients such as diazinon, fenitrothion, parathion, and profenofos are not detected by Gas Chromatography (GC-MS) caused by the differences of each compound persistence. In addition, previous research showed that physical properties of organophosphate chemistry are also expected to affect the concentration of some organophosphates which are not detected by GC-MS. The presence of pesticide residues in the soil could adversely affect soil fertility level. One indicator is to determine soil fertility using earthworm presence. Previous studies had investigated the effect of pesticides on earthworms and showed a 50% reduction of earthworm activity when incorporated into the soil contaminated by pesticides and 90% reduction occurs when pesticides are placed on the soil surface. The presence of earthworms is very important for the agroecosystem sustainability, but earthworms can be degraded by the intensive use of pesticides [24, 25].

3.5. The Decrease of Soil Quality (Qualitative Approach). Land degradation might occur after the application of large quantities of pesticides continuously. The results of interviews with shallot farmers in Wanasari Subdistrict indicated potential damage to the soil quality in their farms. As 85.5% of respondents said that the soil became hard, 81.8% stated that earthworm biomass was not found, and 76.4% of respondents admitted that the ground color changed to brown-white (Figures 1 and 2).

4. Conclusions

The study showed that different insecticides and fungicides were used in each growing season. This study found that farmers apply a lot of pesticides. Pesticides are applied once every three or four days. The mixtures for insecticides and fungicides used at least three types of variations with doses of 30–40 ml for each type. Excessive application of pesticides shows the potential for land degradation. Organophosphate residues that have been found in shallot farm are methidathion, malathion, and chlorpyrifos active ingredients at Tanjungsari village, Sisalam, Dukuhwringin, and Wanasari. The methidathion residue is about 0.014 mg/kg, malathion residue revolves around 0.1370–0.3630 mg/kg, and chlorpyrifos residue ranged from 0.0110 to 0.0630 mg/kg. Researchers suggest testing the quality of soil in the laboratory to ensure the condition of the soil in the current agricultural environment. Routine assessment for soil quality and pesticides usage control can be considered to maintain a sustainable ecosystem.

References

[1] F. Azam, S. Farooq, and A. Lodhi, "Microbial biomass in agricultural soils—determination, synthesis, dynamics and role in plant nutrition," *Pakistan Journal of Biological Sciences*, vol. 6, no. 7, pp. 629–639, 2003.

[2] S. A. Reinecke and A. J. Reinecke, "The impact of organophosphate pesticides in orchards on earthworms in the Western Cape, South Africa," *Ecotoxicology and Environmental Safety,* vol. 66, no. 2, pp. 244–251, 2007.

[3] S. Yasmin and D. D'Souza, "Effects of pesticides on the growth and reproduction of earthworm: a review," *Applied and Environmental Soil Science*, vol. 2010, Article ID 678360, 9 pages, 2010.

[4] O. Espinoza-Navarro and E. Bustos-Obregón, "Effect of malathion on the male reproductive organs of earthworms, Eisenia foetida," *Asian Journal of Andrology*, vol. 7, no. 1, pp. 97–101, 2005.

[5] S.-P. Zhou, C.-Q. Duan, H. Fu, Y.-H. Chen, X.-H. Wang, and Z.-F. Yu, "Toxicity assessment for chlorpyrifos-contaminated soil with three different earthworm test methods," *Journal of Environmental Sciences*, vol. 19, no. 7, pp. 854–858, 2007.

[6] S. Zhou, C. Duan, X. Wang, W. H. G. Michelle, Z. Yu, and H. Fu, "Assessing cypermethrin-contaminated soil with three different earthworm test methods," *Journal of Environmental Sciences*, vol. 20, no. 11, pp. 1381–1385, 2008.

[7] E. T. Topp and G. Vallaeys Soulas, "Pesticides: Microbial degradation and effects on microorganism," in *Modem Soil Microbiology*, J. A. van Elsas, Ed., Modem Soil Microbiology, pp. 547–573, Marcel Dekker Inc, New York, 1997.

[8] D. S. Jenkinson and D. S. Powlson, "The effects of biocidal treatments on metabolism in soil-I. Fumigation with chloroform," *Soil Biology and Biochemistry*, vol. 8, no. 3, pp. 167–177, 1976.

[9] S. P. Kale and K. Raghu, "Relationship between microbial numbers and other microbial indices in soil," *Bulletin of Environmental Contamination and Toxicology*, vol. 43, no. 6, pp. 941–945, 1989.

[10] P. C. Kearney and S. Kellogg, "Microbial adaptation to pesticides," *Pure and Applied Chemistry*, vol. 57, no. 2, pp. 389–403, 1985.

[11] M. M. Andrea, T. B. Peres, L. C. Luchini, M. A. Marcondes, A. Pettinelli Jr., and L. E. Nakagawa, "Impact of long term applications of cotton pesticides on soil biological properties, dissipation of [^{14}C]-methyl parathion and persistence of multipesticide residues," in *Proceedings of the International Atomic Energy Agency*, Vienna, Austria, 2001.

[12] B. Waryanto, M. A. Chozin, Dadang, and E. I. K. Putri, "Environmental efficiency analysis of shallot farming: a stochastic frontier translog regression approach," *Journal of Biology, Agriculture and Healthcare*, vol. 4, no. 19, pp. 2224–3208, 2014.

[13] G. W. Sasmito, *Simulation Application Diagnosis Expert System for Pests and Diseases Plant Shallots and Chili Using Chaining foreward and Rule-Based Approach*, Postgraduate Program, Diponegoro University, Semarang, Indonesia, 2010.

[14] S. S. Sexton, Z. Lei, and D. Zilberman, "The economics of pesticides and pest control," *International Review of Environmental and Resource Economics*, vol. 1, no. 3, pp. 271–326, 2007.

[15] V. V. Oberemok, K. V. Laikova, Y. I. Gninenko, A. S. Zaitsev, P. M. Nyadar, and T. A. Adeyemi, "A short history of insecticides," *Journal of Plant Protection Research*, vol. 55, no. 3, pp. 221–226, 2015.

[16] S. Y. Gebremariam, M. W. Beutel, D. R. Yonge, M. Flury, and J. B. Harsh, "Adsorption and desorption of chlorpyrifos to soils and sediments," in *Reviews of Environmental Contamination and Toxicology*, vol. 215, pp. 123–175, Springer, New York, NY, USA, 2012.

[17] K. D. Racke and J. R. Coats, "Comparative degradation of organophosphorus insecticides in soil: specificity of enhanced microbial degradation," *Journal of Agricultural and Food Chemistry*, vol. 36, no. 1, pp. 193–199, 1988.

[18] J. H. Cink, "Degradation of chlorpyrifos in soil: effect of concentration, soil moisture, and temperature," in *Retrospective Theses and Dissertations*, Iowa State University, Ames, Iowa, USA, 1995.

[19] CDPR, "Methidathion risk characterization document," Department of Pesticide Regulation California Environmental Protection Agency, 2007, http://www.cdpr.ca.gov/docs/emon/pubs/tac/tacpdfs/methidathion/envfate_mthd.pdf.

[20] V. I. Tsipriyan and N. I. Martsenyuk, "Toxicological evaluationof photolytic degradation products of pesticides," *Cig. Sanit*, vol. 8, pp. 77–80, 1984.

[21] R. H. Neal, P. M. McCool, and T. Younglove, *Assessment of Malathion and Malaoxon Concentration and Persistence in Water, Sand, Soil and Plant Matrices Under Controlled Exposure Conditions*, Environmental Hazards Assessment Program State Of California Environmental Protection Agency Department of Pesticide Regulation, 1993.

[22] EPA, "Malathion," 2000, https://archive.epa.gov/pesticides/reregistration/web/pdf/malathion-red-revised.pdf.

[23] K. L. Newhart, *Environmental Fate of Malathion*, California Environmental Protection Agency Department of Pesticide Regulation Environmental Monitoring Branch, 2006.

[24] B. Y. H. Nugroho, S. Y. Wulandari, and A. Ridlo, "Analisis Residu Pestisida Organofosfat Di Perairan Mlonggo Kabupaten Jepara," *Jurnal Oseanografi*, vol. 4, no. 3, pp. 541–544, 2014.

[25] C. H. Hogger and H. U. Ammon, "Testing the toxicity of pesticides to earthworms in laboratory and field tests," *Bulletin OILB/SROP*, vol. 17, pp. 157–178, 1994.

13

Long and Midterm Effect of Conservation Agriculture on Soil Properties in Dry Areas of Morocco

Malika Laghrour,[1,2] Rachid Moussadek,[2] Rachid Mrabet,[2] Rachid Dahan,[2] Mohammed El-Mourid,[3] Abdelmajid Zouahri,[2] and Mohamed Mekkaoui[1]

[1] Groupe Physicochemistry of Materials, Nanomaterials and Environment, Faculty of Sciences, Mohammed V University, Ibn Battuta Avenue, P.O. Box 1014, Rabat, Morocco
[2] National Institute of Agricultural Research (INRA), BP 6356, Rabat, Morocco
[3] International Center for Agricultural Research in the Dry Areas (ICARDA), Rabat Instituts, North-Africa Platform, P.O. Box 6299, 10112 Rabat, Morocco

Correspondence should be addressed to Malika Laghrour; laghrour.malika@gmail.com
and Rachid Moussadek; rachidmoussadek@yahoo.fr

Academic Editor: Artemi Cerda

In Morocco, conservation agriculture, particularly no tillage systems, has become an alternative strategy to mitigate land degradation caused by conventional tillage in semiarid to arid regions. This paper is based on behaviour to tillage treatments of two Vertisols in Morocco. After 11 years of testing, soil organic matter content results showed a significant difference ($P < 0.05$) only at soil surface (0–10 cm) in favour of no tillage and a variation of 30% at this depth. The results obtained after 32 years of testing showed a significant soil profile difference ($P < 0.05$), up to 40 cm under no tillage compared to conventional tillage, and a variation of 54% at 5–10 cm. For total nitrogen, there was no significant effect between no tillage and conventional tillage at the soil surface after 11 years unlike the result obtained after 32 years. There are no significant differences in bulk density between tillage treatments at soil surface for both sites. The measurement of soil structural stability showed a significant effect ($P < 0.05$) for all three tests and for both sites. This means that no tillage helped Vertisols to resist different climatic constraints, preserving environmental soil quality.

1. Introduction

Soil organic matter (SOM) is a fundamental attribute of soil quality [1–4]. The presence of sufficient quantities of SOM in the soil changes the nutrients and physical properties of soil considerably [5–10]. It improves soil fertility [11], reinforces the cohesion between mineral particles, and contributes to improved water infiltration and good structure. SOM plays an important role in soil protection against drought, erosion, and runoff and ensures sustainable food production [12]. However, a fear of a excessive loss of SOM has been expressed in recent decades in the world [13]. This loss of SOM is of natural and particularly anthropogenic origin, due to the use of inappropriate soil management techniques. Regretfully, the farmers believe in these inappropriate techniques, such as

deep tillage, and are unaware of the fact that, in the long term, these agricultural practices are contrary to sustainable agriculture. Some recent studies have shown that intensive plowing of the soil exposes SOM to rapid mineralisation, reduces soil fertility, destroys the aggregates, and deteriorates aeration and moisture conditions. Additionally, this plowing leads to increased soil erosion and crop residues can reduce soil loss [14–19].

For this purpose, the adoption of a strategy to protect natural resources should be an internationally strong preoccupation. Bessam and Mrabet [20], Mrabet [21], Moussadek et al. [22], and Sheehy et al. [23] have revealed, in several studies, the importance of adopting conservation agriculture, including no tillage (NT) as an alternative remedy for addressing this alarming situation. In terms of agricultural technology, NT is considered one of the greatest revolutions of this century. This practice is based on direct seeding, where the residues from antecedent cultures are left on the soil surface to ensure the accumulation of carbon and its sequestration in different types of climate in the world [24–26].

In Morocco, as in many parts of the world, adaptation to the NT system has become an alternative strategy, to mitigate land degradation in semiarid to arid regions through carbon accumulation on the soil surface. The organic matter under most Moroccan regions is less than 2%, and after ten years, the average loss of soil organic matter due to tillage was estimated to be 30% [25]. Generally, there are many benefits of the NT vis-a-vis climate hazards in semiarid and arid regions (drought, erosion, etc.). NT's usefulness and its importance as a conservative technique for soil quality have prompted Moroccan researchers at the National Institute for Agricultural Research (INRA) to adopt NT immediately. This began in the Jemaat Shaim region of Morocco in 1983 and then at the experimental station of Sidi El Aidi Chaouia, and finally NT was implemented in the experimental station of Merchouch in Zaer-Rabat at 2004 [27]. The objective of this work is to study the long- and medium-term effects of tillage modes, that is, NT and conventional tillage (CT), on pertinent soil quality:

(i) Soil organic matter content profile.

(ii) Structural stability and bulk density of the soil.

(iii) Total nitrogen and carbon/nitrogen ratio (C/N).

2. Materials and Methods

2.1. Study Areas. Two experimental sites of Vertisol were chosen for this study (Figure 1).

The first site (site I) is located at the experimental station of Merchouch INRA. This experimental field is attached to the Regional Center of Agricultural Research in Rabat (RCARR). It is located in the Central Plateau of the rural commune of Merchouch, circle of Rommani, Province of Khemisset, about 68 km to the southeast of Rabat and 16 km to the North-West of Rommani. It is crossed by the secondary road N° 218 that divides it into two parts: northern and southern part. The geographical coordinates of this site are as follows: altitude: 339; longitude: 06°71′ west; latitude: 33°60′ north.

The second experimental site (site II) is located at 12 km North-East of the village of Jemaat Shaim (in Safi region) and 40 km from the coast of the Atlantic ocean (N 32°40′ × 10°0′, 170 m). This site is characterized by an annual precipitation ranging from 147.2 to 527.8 mm with a mean of 30 years estimated at 300 mm. The temperature reaches −4°C in December and January and 47°C in July and August. The annual averages of rainfall and temperature in 2014-2015 for two sites are shown, respectively, in Figure 2.

This figure shows that the Jemaat Shaim site is characterized by a higher annual temperature and less rainfall than those recorded in the Merchouch site. The climate in the first site is semiarid, while the climate in the second site is

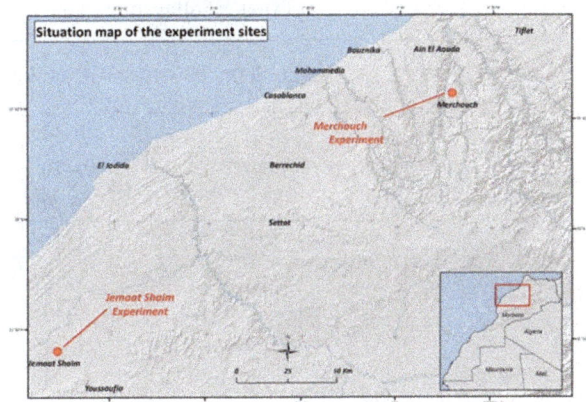

FIGURE 1: Maps showing the two sites.

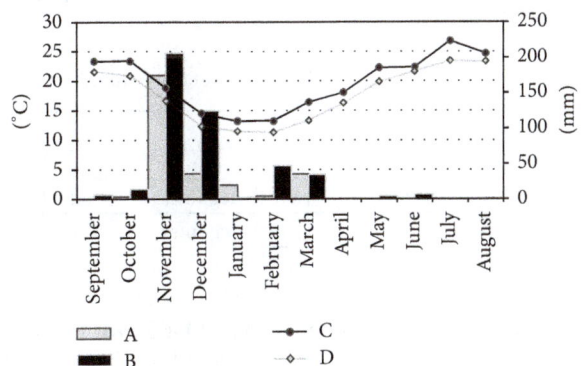

FIGURE 2: Mean temperatures and precipitations during 2014-2015 for both study sites in Morocco. (A) mean precipitations (mm) for 2nd site; (B) mean precipitations (mm) for 1st site; (C) mean temperatures (°C) for 2nd site; (D) mean temperatures (°C) for 1st site.

classified as semiarid to arid. According to the particle size results shown in Table 1, in the first site, soil has more than 50% of clay in 0–80 cm depth, and in the second site, soil clay quantity decreased less than 50% in depth horizon.

2.2. Methodology. The plots of the two experimental sites were differentiated with two tillage treatments: CT and NT. The first treatment consists in a conventional tillage (plowing up to 30 cm deep) followed by shallow tillage (10–15 cm). The objective was to prepare a fine seedbed and to bury plant residues. In contrast to the second treatment, this one comprises a single operation. It was realised with a special NT drill that consists of an opening of 2-3 cm from the ground to put the seeds at 5 cm depth. Both sites are based on the agricultural rotation cereals/legumes. This rotation was started in the second site since 2009.

2.3. Soil Measurements. In order to study the impact of NT and CT on the main components of soil quality, different methods of soil sampling were followed according to our objectives and parameters that used in study measurement.

For the first experiment, thirty-six samples have been collected from the soil surface, including the following

TABLE 1: Soil texture characteristics at the soil profile for the both sites.

Field	Duration	Soil depth (cm)	Texture (%)			Texture
			Clay	Slit	Sand	
Site I Merchouch	11 years	0–10	54.5	32.3	11.2	Clay
		0–20	50.0	37.3	12.7	Clay
		20–80	52.5	35.1	12.4	Clay
Site II Jemaat Shaim	32 years	0–2.5	56.5	32.23	11.38	Clay
		2.5–5	46.2	30.95	22.9	Clay
		5–10	45	22.9	32.1	Clay
		10–20	42.1	21.01	36.91	Clay
		20–30	39.7	23.32	37.02	Clay/loam clay
		30–40	31.4	33.24	35.05	Loam clay
		40–60	42.1	21.01	36.82	Clay
		60–80	42	20.21	36.84	Clay

TABLE 2: Summary of soil samples collected from two sites.

Field	Sampling time	Rotation	Crops	Number of samples		
				Total samples	For chemical analysis	For physical analysis
Site I	May 26, 2015	Cereals/legumes	Soft wheat	36	24	12
Site II	June 23, 2015	Cereals/legumes	Chickpea	104	80	24

depths: (0–10 cm) to (10–20; 20–40; 40–60 cm), based on three replicates for each parameter just for SOM analysis. For this site, the samples collection was in May 26, 2015. The plots were based on cereal (soft wheat). In the oldest field experiment, sampling concerned eight depths, 0–2.5; 2.5–5; 5–10; 10–20; 20–30; 30–40; 40–60; and 60–80 cm, for five replicates. There were twenty-four samples for the soil physical analysis (structural stability and BD) with 3 replications, for each analysis considering only two depths (0–10 and 10–20 cm). Samples collection was performed on June 23, 2015, on two plots of chickpea. The summary of soil samples collected from two sites is presented in Table 2.

2.4. Analytical Methods

2.4.1. Organic Matter.
The collected samples were dried aerobically and sieved through 0.2 mm in order to determinate soil organic carbon content. These contents were estimated using the modified method of Walkley and Black [28] which is based on the principle that the potassium dichromate oxidizes the contained carbon in the soil. Potassium dichromate changes its colour depending on the amount of reduced products; this change in colour can be compared to the amount of organic carbon in the soil. We deduce the organic matter content by multiplying the carbon content by 1.724.

2.4.2. Total Nitrogen.
TN was determined by the modified Kjeldahl method [29]. The mineralisation of soil organic nitrogen as ammonium sulfate was performed in the presence of concentrated sulfuric acid. By stripping to water vapor, there is liberation of ammonia in an alkaline medium and its dosage is performed by acidimetry.

2.4.3. Dry Bulk Density (BD).
The measure of BD was performed only at the soil surface (0–10 cm), using a cylinder (core sample). The samples were weighed before and after passing through the oven at 105°C for 48 hours. This measure is in $g \cdot cm^{-3}$ and is evaluated according to the method of Grossman and Reinsch [30].

2.4.4. Aggregate Stability.
The soil aggregate stability estimates the capacity of a soil to retain its structure when it is subjected to different constraints. It is determined by the method proposed by Le Bissonnais [31], which combines three tests describing the behaviour of the soil under different climatic and water conditions. Le Bissonnais methods consist in subjecting soil sampling to 3 treatments:

(i) Fast wetting by immersion.

(ii) Wetting slow capillary.

(iii) Mechanical disaggregation agitation after rewetting.

Moreover, the soil organic matter is the main factor of cohesion between soil aggregates of these three tests.

2.5. Statistical Analysis.
Statistical analysis of the data have been realised for the purpose to study the effect of tillage treatments (NT and CT) on soil physical and chemical properties. For this, our results measurements were performed by comparing their means according to the t-test (Student's t-test). The software used for statistical processing is the SPSS STATISTICS 21.

TABLE 3: Tillage effect on SOM content (mean ± standard deviation).

Field	Soil depth (cm)	Organic matter content (g·Kg⁻¹)		CT to NT % SOM change
		NT	CT	
Site I	0–10	22.23 ± 0.20a*	17.13 ± 0.25b*	30
	10–20	19.11 ± 0.03a	17.38 ± 0.12a	10
	20–40	17.23 ± 0.10a	16.43 ± 0.05a	5
	40–60	15.80 ± 0.04a	15.36 ± 0.03a	3
Site II	0–2.5	19.12 ± 0.47a	13.02 ± 0.21b	47
	2.5–5	15.38 ± 0.27a	10.74 ± 0.14b	44
	5–10	15.62 ± 0.06a	10.06 ± 0.17b	54
	0–10	16.44 ± 0.22a	10.97 ± 0.17b	50
	10–20	12.68 ± 0.06a	9.42 ± 0.10b	35
	20–30	12.48 ± 0.09a	9.58 ± 0.10b	30
	30–40	13.22 ± 0.17a	9.80 ± 0.10b	35
	40–60	11.18 ± 0.06a	9.440 ± 0.17a	19
	60–80	10.20 ± 0.10a	12.94 ± 0.22a	−21

*In the same row, the values followed by different letters are significantly different within each study site according to Student's test ($P < 0.05$).

3. Results and Discussion

3.1. Impact of Tillage Systems on Soil Organic Matter Content. The importance of SOM is its positive influence on many physical and chemical properties of soil. This is why the impact of NT, in which residues of the antecedent culture are not removed and decomposed at soil surface, on SOM content has been studied by different authors for different climatic conditions and different types of soils. But no studies have been made on SOM of Vertisols after 11 and 32 years under NT.

For the first site (Table 3), it was found that the SOM content is higher under NT than CT with a variation of 30%. The statistical analysis shows a significant difference only in 0–10 cm between tillage systems (P value = 0.01 < 0.05) and no significant effect in the soil profile (from 10 to 60 cm of depth) under NT. This result is similar to that obtained by Moussadek et al. [22], and Laghrour et al. [32], for the same study field after seven and ten years of trial, respectively. This is also confirmed by Angers et al. [33] and Blanco-Canqui and Lal [34], who have also shown a high SOM content at 0–10 cm layer under NT compared to CT. Moreover, Hassink [35] and Shi et al. [36] found that, with the depth, the values were similar between the two tillage treatments. Wander et al. [37], Guzmane et al. [38], and Sparrow et al. [39] reported that, under NT, soil organic carbon contents are strongly stratified at soil surface and decrease rapidly with depth.

Based on student test, our results for the second site (Table 3) show significant differences in the soil profile, at depths up to 40 cm (from soil surface to 40 cm of depth) between the two tillage treatments. This result is contrary to Angers and Eriksen-Hamel [40] and Syswerda and coll. [41] who showed that the difference between NT and CT is not observed over 30 cm in depth. However, an important accumulation was recorded under NT for the 5–10 cm layer compared to those recorded for the other two layers of the soil surface (0–2.5 cm and 2.5–5 cm), where the variation of SOM between tillage treatments was 54%, 47%, and 44%,

respectively. However, Liu et al. [42] observed after 17 years under NT a greater content of organic carbon in 0–5 cm layer than 5–10 cm layer. The results of these authors were statistically significant only for 0–10 cm of depth and there was no significant difference for the other layers of depth from 10 to 60 cm. They added that, in deep layers, the organic carbon content was higher in CT versus NT contrary to our results obtained after 11 years of trial. The different variation of SOM obtained in the profile can be explained by the type of soil and this is confirmed by Paton [43] and Kovda et al. [44], who showed that the Vertisols are characterized by the presence of slickensides and deep cracks. In another study, the significant effect of these results up to 40 cm deep under NT is due to the accumulation of organic carbon at soil surface [45, 46] and its long-term distribution in the deeper layers. These results are similar to those obtained by Dimassi et al. [47].

3.2. Impact of Tillage Systems on Total Nitrogen and C/N Ratio. Eighty-five to 95 percent of TN consists of organic nitrogen. This latter becomes available for plants by mineralisation related to the activity of microorganisms. The SOM is the main soil nitrogen reserves.

For the first site, the TN is higher at soil surface under NT compared to CT (Table 4). But this result shows no significant difference between tillage treatments. Similar results were found by Angers et al. [33] and Blanco-Canqui and Lal [34]. This is probably explained by the lack of SOM accumulated on the soil surface after 11 years of experiment.

After 32 years, the great values of the TN are measured under NT with significant differences between the two tillage treatments from the surface (0–2.5, 2.5–5, and 5–10 cm) up to 10–20 cm deep layer (Table 4). The values obtained for the rest of profile have no significant effect. This significant NT effect seems influenced by the accumulation of organic matter content and adoption period of NT system under different conditions. This is clearly visible and understood in several

TABLE 4: Tillage treatments effect on total nitrogen (mean ± standard deviation) and on C/N ratio.

Field	Soil depth (cm)	Total nitrogen (g·Kg⁻¹)		C/N	
		NT	CT	NT	CT
Site I	0–10	1.20 ± 0.02a*	1.20 ± 0.01a*	10.75	8.28
Site II	0–2.5	1.22 ± 0.01a	0.92 ± 0.01b	9.02	8.31
	2.5–5	1.08 ± 0.01a	0.94 ± 0.01b	8.23	6.63
	5–10	1.10 ± 0.01a	0.94 ± 0.01b	8.28	6.25
	10–20	1.20 ± 0.01a	0.94 ± 0.01b	6.15	5.87
	20–30	1.04 ± 0.01a	0.98 ± 0.00a	6.98	6.67
	30–40	0.94 ± 0.01a	0.90 ± 0.01a	8.24	6.31
	40–60	0.88 ± 0.01a	0.98 ± 0.01a	7.39	5.61
	60–80	0.88 ± 0.01a	0.88 ± 0.01a	6.80	8.64

*In the same row, the means followed by the same letter were not significantly different at $P < 0.05$ between tillage treatments.

research works. After 5 years under NT tests, Testa et al. [48] indicated a significant difference at soil surface (0–7.5 cm). Significant differences were also found after 9 years of trial at 0–12.5 cm [49] and after 11 years at 0–17.5 cm and 20 cm by Burle et al. [50] and Mrabet et al. [51], respectively.

From these results obtained by several authors, it appears that the TN content is becoming increasingly important with soil depth and depends mainly on the adoption of the NT duration.

The C/N ratio is considered as an important soil fertility indicator which can be used to reflect the interaction between SOM and soil TN [52]. The results for the first site (Table 4) showed that soil C/N ratio in 0–10 cm layer was higher under NT than CT. For the second site (Table 4), there is a greater increase in this ratio under NT than CT at the soil profile up to 60 cm in depth. This result of C/N ratio at soil surface in favour of NT can be explained by the residues remaining from the prior crop. In addition, the soil C/N ratio declined with depth profile under both tillage treatments. This may be probably due to the clay content, which varies with depth. Some authors reported that high clay content is often associated with organic matter decomposition increases when the C/N ratio decreases [53–56].

Similar results were found in favour of NT after 22 years [52], while Mazzoncini et al. [57] did not find a difference of C/N ratio between the two tillage systems despite the variations in soil organic carbon and soil total nitrogen.

3.3. *Impact of Tillage Systems on Structural Stability.* Le Bissonnais [31] defined the structural stability or the stability of the aggregates by the capacity of a soil to maintain its arrangement between the solid and empty particles when exposed to various stresses as they may be of different types and of different intensities. The impacts of soil tillage tools, rain or wetting, are examples of these constraints. Good structural stability reduces crusting and consequently soil losses by runoff and erosion. The effects of tillage treatments on the aggregate stability measured by mean weight diameter are shown in Figure 3.

For the first site, the measurement of the structural stability of soil shows significant effect tillage for all three tests. This means that, after 11 years of testing, Vertisol can resist

different climatic constraints, specifically fast wetting. The significance of this test means that the soil can resist brutal rain while the slow-wetting test means that the Vertisol can resist moderate rainfall [22, 58]. These results are consistent with those obtained by Laghrour et al. [32], for the same experiment study after 10 years. These authors did not find a meaningful result for the mechanical disaggregation test, contrary to our results that showed, after 11 years of experiment, a significant difference between tillage treatments. This means that, after 11 years, the Vertisol of Merchouch can withstand this test, the purpose of which is to test the soil cohesion wet regardless of bursting. Sheehy et al. [23] showed that aggregate stability results for Vertic Cambisol (10 years of experiment) and Eutric Regosol (11 years of experiment) were significantly different between NT and CT at soil surface.

However, for the second site, NT results increased aggregate stability, compared to CT, of between 89 and 30 percent for the depths 0–10 cm and 10–20 cm, respectively. This great stability recorded at the soil surface shows that, after 32 years under NT, Vertisols are more stable in NT than in CT against the erosion and runoff. In addition, statistical analysis clearly indicates that all tests for 0–10 and 10–20 cm of depth are significantly different ($P < 0.05$) between the treatments in favour of NT. These results are contrary to those obtained by Kibet et al. [59], who showed that, after long-term experiment (established in 1981), they found a significant effect for a Sharpsburg silty clay loam soil only at 0–10 cm under NT compared to other tillage treatments and any significant effect was found at 10–20 cm of depth between tillage treatments.

Moreover, the significant effect observed for the fast-wetting test, which is a destructive test, shows that soil can resist breakdown aggregates [22, 58] and sudden rainfall. However, the results obtained for the slow-wetting test, characterized by a low water gradient and larger air exhaust possibilities, indicate that the NT was able to limit the speed of this wetting. This means that the Vertisol can better withstand moderate rain. A significant effect was also observed for the mechanical disaggregation test after 32 years.

Amézketa [60] showed that the aggregate stability was affected by several factors, with SOM being the most important. Indeed, SOM acts as a binder between the particles and alters the hydrophobicity of the aggregates. Still, numerous

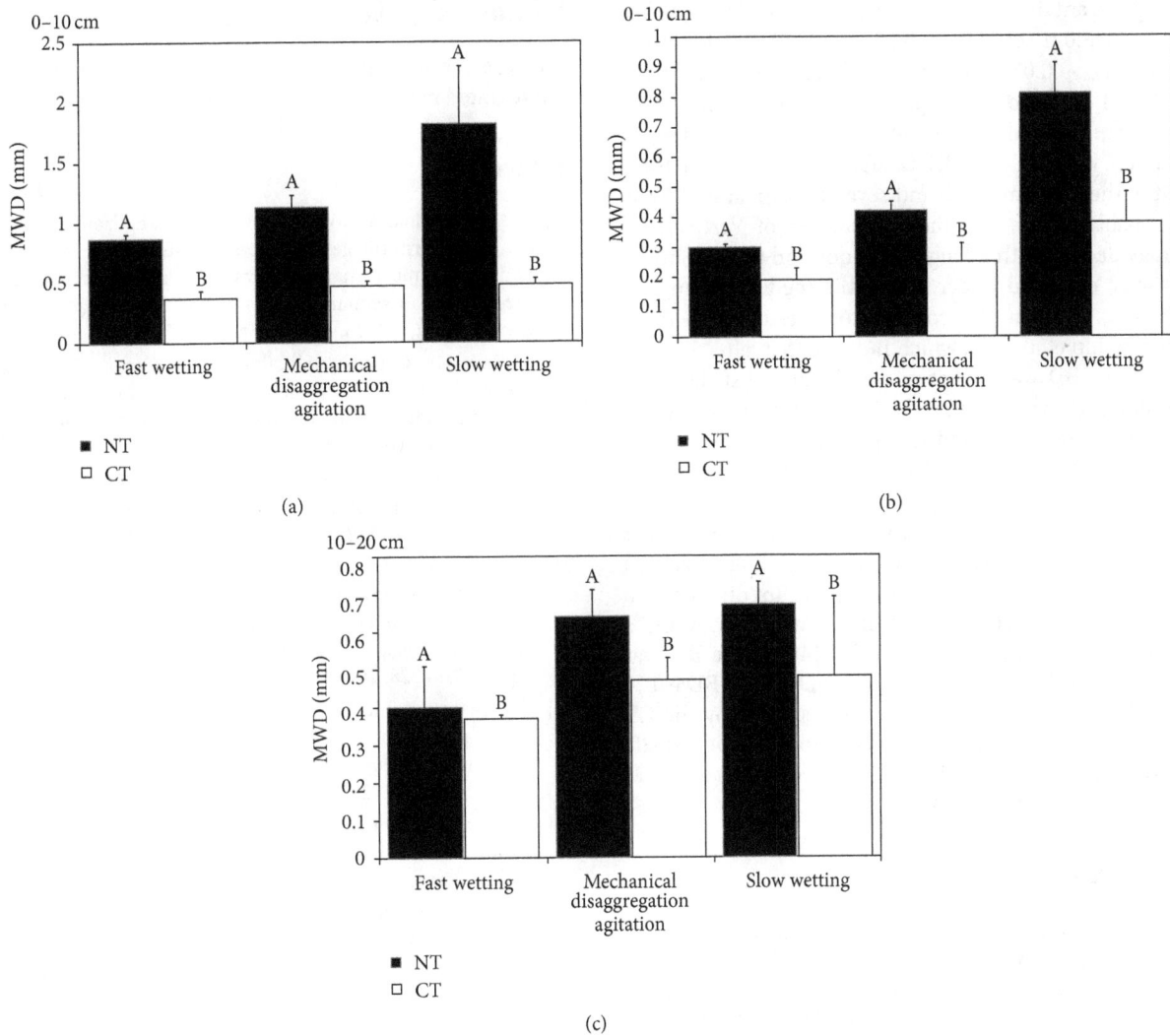

(a)

(b)

(c)

FIGURE 3: Effect of tillage treatments on structural stability at soil surface for both sits located in Morocco. (a) Mean weight diameter (MWD) for each test of structural stability for the 1st site at 0–10 cm of depth; (b and c) mean weight diameter for 2nd site. (b) At 0–10 cm of depth; (c) at 10–20 cm of depth. Bars represent standard deviation, the difference letters above the column show significant difference between no tillage (NT) and conventional tillage (CT) practices.

studies have related increased structural stability to the accumulation of SOM [61–66]. It is to be noted that, despite the significant effects of the structural stability recorded under NT versus CT, the results indicate low values of MWD (0.3 to 0.81 for NT and 0.19 to 0.48 for CT). This finding is likely due to the Vertisol.

After these results, we can say that NT helped to construct a good structure with time, which is highly desirable for sustaining agricultural productivity (it can be beneficial to plant growth) and for preserving environmental quality (it can reduce soil erosion and nutrient losses in runoff) [67].

3.4. Impact of Tillage Systems on Bulk Density.
The BD is one of the most important parameters in studies of soil structure. It allows for the diagnosis of the soil compaction as this density is related to the porosity of the soil. The result (Table 5) indicates no significant effect after 11 years of NT.

TABLE 5: Tillage treatments effect on bulk density (mean ± standard deviation).

Field	Soil depth (cm)	BD (g·cm^{-3})	
		NT	CT
Site I	0–10	1.29 ± 0.04a	1.15 ± 0.14a
Site II	0–10	1.41 ± 0.07a	1.37 ± 0.22a
	10–20	1.45 ± 0.05a	1.35 ± 0.14a

The same result was found after 10 years of trial [32], contrary to Moussadek et al. [22] showing a significant effect of this measure BD (CT) < BD (NT) under NT in 2011. This explains that the soil under NT usually becomes more porous and less compacted and thus more water permeable.

The results of the BD for the second site are presented also in Table 5. The result of this analysis indicates that there are no significant differences between the two tillage systems at soil surface 0–10 cm ($P = 0.78 \gg 0.05$) and 10–20 cm ($P = 0.311 > 0.05$). At 0–10 cm, there is a value of BD under NT (1.41) almost equal to that seen in CT (1.37). For the 10–20 cm layer, the difference remains. In addition, in comparison with CT, the BD is higher under NT at three percent in the 0–10 cm depth and seven percent at 10–20 cm. This probably means that the compaction of Vertisol soil types may decline with a long-term adoption of NT. Similar results were observed under NT in the medium term of a Vertisol [32]. We can also compare our results with those obtained by Liu et al. [42], which indicate that, after 17 years of testing, the BD was higher under NT at soil surface and at 0–30 cm in depth. The BD of a Cambisol is higher under NT versus CT by 5, 3.5, and 13.9 percent at depths of 0–5, 5–10, and 10–20, respectively. The authors found a significant difference ($P < 0.05$) between treatments for 0–10 cm and 10–20 cm. de Moraes et al. [68] compared the BD of Oxisols under NT to that obtained under CT after 11 (NT11) and 24 years (NT24). These authors found a significant difference in comparing the NT system to that of CT and did not find a significant effect between NT11 and NT24, but the BD was low compared with the NT24 and NT11 (BD NT11 > BD NT24 > BD CT), while low BD in NT24 was recorded by the CT contributed with a significant difference in the depth 10–20. At this depth, the results obtained can be summarized as follows: BDNT11 > BDCT > BDNT24.

4. Conclusion

Among the cultural practices, NT is the subject of sustained attention in recent years. This practice of NT is recently evolved and many observations showed that it has a strong impact on the nature and evolution of SOM that influences a set of chemical properties (TN and C/N) and physical soil (BD and structural stability). This paper has allowed for studying the effect of NT and CT on the physical and chemical properties of the semiarid to arid regions. The results for the second site of Jemaat Shaim show that, after 32 years of testing, the NT promotes the accumulation of the SOM in the soil surface and in depth up to 40 cm, in contrast to CT. The accumulation at soil surface is confirmed after 11 years of testing in the first site that records significant levels of the SOM at the soil surface. On the one hand, the high content of SOM clearly explains the availability of nutrients as total nitrogen and the stability of the aggregates under NT compared with those who have been intensively tilled. On the other hand, the results obtained under the two tillage treatments can be explained by the duration of this new agricultural practice. This allowed for saying that CT is more susceptible to degradation by erosion and runoff, but under NT soil it has stable aggregates as shown in the test of structural stability.

Competing Interests

The authors declare that there is no conflict of interests regarding the publication of this paper.

Acknowledgments

This research is supported by a joint INRA-ICARDA project in integrated natural resources management (INRM project).

References

[1] V. A. Laudicina, A. Novara, V. Barbera, M. Egli, and L. Badalucco, "Long-term tillage and cropping system effects on chemical and biochemical characteristics of soil organic matter in a mediterranean semiarid environment," *Land Degradation and Development*, vol. 26, no. 1, pp. 45–53, 2015.

[2] A. Novara, L. Gristina, M. B. Bodì, and A. Cerdà, "The impact of fire on redistribution of soil organic matter on a Mediterranean hillslope under maquia vegetation type," *Land Degradation and Development*, vol. 22, no. 6, pp. 530–536, 2011.

[3] S. Mukhopadhyay, R. E. Masto, A. Cerdà, and L. C. Ram, "Rhizosphere soil indicators for carbon sequestration in a reclaimed coal mine spoil," *Catena*, vol. 141, pp. 100–108, 2016.

[4] L. Parras-Alcántara, B. Lozano-García, E. C. Brevik, and A. Cerdá, "Soil organic carbon stocks assessment in Mediterranean natural areas: a comparison of entire soil profiles and soil control sections," *Journal of Environmental Management*, vol. 155, pp. 219–228, 2015.

[5] T. B. Bruun, B. Elberling, A. de Neergaard, and J. Magid, "Organic carbon dynamics in different soil types after conversion of forest to agriculture," *Land Degradation and Development*, vol. 26, no. 3, pp. 272–283, 2015.

[6] X. Chen, Z. Duan, and M. Tan, "Restoration affect soil organic carbon and nutrients in different particle-size fractions," *Land Degradation & Development*, vol. 27, no. 3, pp. 561–572, 2016.

[7] J. C. de Moraes Sá, L. Séguy, F. Tivet et al., "Carbon Depletion by Plowing and its Restoration by No-Till Cropping Systems in Oxisols of Subtropical and Tropical Agro-Ecoregions in Brazil," *Land Degradation and Development*, vol. 26, no. 6, pp. 531–543, 2015.

[8] S. P. de Oliveira, N. B. de LaCerdà, S. C. Blum, M. E. O. Escobar, and T. S. de Oliveira, "Organic carbon and nitrogen stocks in soils of northeastern Brazil converted to irrigated agriculture," *Land Degradation and Development*, vol. 26, no. 1, pp. 9–21, 2015.

[9] M. D. M. Montiel-Rozas, M. Panettieri, P. Madejón, and E. Madejón, "Carbon sequestration in restored soils by applying organic amendments," *Land Degradation and Development*, vol. 27, no. 3, pp. 620–629, 2015.

[10] L. Deng and Z.-P. Shangguan, "Afforestation drives soil carbon and nitrogen changes in China," *Land Degradation and Development*, 2016.

[11] X.-B. Du, C. Chen, L.-J. Luo et al., "Long-term no-tillage direct seeding mode for water-saving and drought-resistance rice production in rice-rapeseed rotation System," *Rice Science*, vol. 21, no. 4, pp. 210–216, 2014.

[12] A. Bot and J. Benites, "The importance soil organic matter: key to drought-resistant soil sustained food production," *FAO Soils Bulletin*, vol. 80, 2005.

[13] J. Balesdent, "Un point sur l'évolution des réserves organiques des sols de Franc," *Etude et Gestion des Sols*, vol. 3, no. 4, pp. 245–260, 1996.

[14] T. Nishigaki, M. Shibata, S. Sugihara, A. D. Mvondo-Ze, S. Araki, and S. Funakawa, "Effect of mulching with vegetative residues on soil water erosion and water balance in an oxisol

cropped by cassava in east cameroon," *Land Degradation and Development*, 2016.

[15] A. Nawaz, R. Lal, R. K. Shrestha, and M. Farooq, "Mulching affects soil properties and greenhouse gas emissions under long-term no-till and plough-till systems in alfisol of central Ohio," *Land Degradation and Development*, 2016.

[16] S. B. Mwango, B. M. Msanya, P. W. Mtakwa, D. N. Kimaro, J. Deckers, and J. Poesen, "Effectiveness of mulching under *miraba* in controlling soil erosion, fertility restoration and crop yield in the Usambara Mountains, Tanzania," *Land Degradation & Development*, vol. 27, no. 4, pp. 1266–1275, 2016.

[17] M. N. Jiménez, E. Fernández-Ondoño, M. Á. Ripoll, J. Castro-Rodríguez, L. Huntsinger, and F. B. Navarro, "Stones and organic mulches improve the Quercus Ilex L. Afforestation success under mediterranean climatic conditions," *Land Degradation and Development*, vol. 27, no. 2, pp. 357–365, 2016.

[18] M. Prosdocimi, P. Tarolli, and A. Cerdà, "Mulching practices for reducing soil water erosion: a review," *Earth-Science Reviews*, vol. 161, pp. 191–203, 2016.

[19] M. Prosdocimi, A. Jordán, P. Tarolli, S. Keesstra, A. Novara, and A. Cerdà, "The immediate effectiveness of barley straw mulch in reducing soil erodibility and surface runoff generation in Mediterranean vineyards," *Science of the Total Environment*, vol. 547, pp. 323–330, 2016.

[20] F. Bessam and R. Mrabet, "Long-term changes in soil organic matter under conventional tillage and no-tillage systems in semiarid Morocco," *Soil Use and Management*, vol. 19, no. 2, pp. 139–143, 2003.

[21] R. Mrabet, "Effects of residue management and cropping systems on wheat yield stability in a semiarid mediterranean clay soil," *American Journal of Plant Science*, vol. 2, pp. 202–216, 2011.

[22] R. Moussadek, R. Mrabet, P. Zante et al., "Effets du travail du sol et de la gestion des résidus sur les propriétés du sol et sur l'érosion hydrique d'un Vertisol Méditerranéen," *Canadian Journal of Soil Science*, vol. 91, no. 4, pp. 627–635, 2011.

[23] J. Sheehy, K. Regina, L. Alakukku, and J. Six, "Impact of no-till and reduced tillage on aggregation and aggregate-associated carbon in Northern European agroecosystems," *Soil & Tillage Research*, vol. 150, pp. 107–113, 2015.

[24] A. Kassam, T. Friedrich, R. Derpsch et al., "Conservation agriculture in the dry Mediterranean climate," *Field Crops Research*, vol. 132, pp. 7–17, 2012.

[25] R. Moussadek, R. Mrabet, R. Dahan, A. Zouahri, M. El Mourid, and E. V. Ranst, "Tillage System Affects Soil Organic Carbon Storage and Quality in Central Morocco," *Applied and Environmental Soil Science*, vol. 2014, Article ID 654796, 8 pages, 2014.

[26] K. Kuotsu, A. Das, R. Lal, G. C. Munda, P. K. Ghosh, and S. V. Ngachan, "Land forming and tillage effects on soil properties and productivity of rainfed groundnut (Arachis hypogaea L.)-rapeseed (Brassica campestris L.) cropping system in northeastern India," *Soil and Tillage Research*, vol. 142, pp. 15–24, 2014.

[27] R. Mrabet, *No-tillage Systems for Sustainable Dryland Agriculture in Morocco*, INRA Publication, Rabat, Morocco, 2008.

[28] D. W. Nelson and L. E. Sommers, "Total carbon, organic carbon and organic matte," in *Methods of Soil Analysis, Part 2, Chemical and Microbiological Properties, Agronomy Monograph*, A. L. Page, R. H. Miller, and D. R. Keeney, Eds., vol. 9, pp. 539–579, American Society of Agronomy, Madison, Wis, USA, 2nd edition, 1982.

[29] W. B. McGill and C. T. Figueiredo, "Total nitrogen," in *Soil Sampling and Methods of Analysis*, M. R. Carter, Ed., pp. 201–211, Canadian Society of Soil Science/Lewis Publishers, 1993.

[30] R. B. Grossman and T. G. Reinsch, "Bulk density and linear extensibility," in *Methods of Soil Analysis: Part 4, Physical Methods*, J. H. Dane and G. C. Topp, Eds., SSSA Book Series, pp. 201–228, Soil Science Society of America, Madison, Wis, USA, 2002.

[31] Y. Le Bissonnais, "Aggregate stability and assessment of soil crustability and erodibility: I. Theory and methodology," *European Journal of Soil Science*, vol. 47, no. 4, pp. 425–437, 1996.

[32] M. Laghrour, R. Moussadek, A. A. Zouahri et al., "Impact of no tillage on physical proprieties of a clay soil in central Morocco," *Journal of Materials and Environmental Science*, vol. 6, no. 2, pp. 391–396, 2015.

[33] D. A. Angers, M. A. Bolinder, M. R. Carter et al., "Impact of tillage practices on organic carbon and nitrogen storage in cool, humid soils of eastern Canada," *Soil and Tillage Research*, vol. 41, no. 3-4, pp. 191–201, 1997.

[34] H. Blanco-Canqui and R. Lal, "No-tillage and soil-profile carbon sequestration: an on-farm assessment," *Soil Science Society of America Journal*, vol. 72, no. 3, pp. 693–701, 2008.

[35] J. Hassink, "The capacity of soils to preserve organic C and N by their association with clay and silt particles," *Plant and Soil*, vol. 191, no. 1, pp. 77–87, 1997.

[36] X. H. Shi, X. M. Yang, C. F. Drury, W. D. Reynolds, N. B. McLaughlin, and X. P. Zhang, "Impact of ridge tillage on soil organic carbon and selected physical properties of a clay loam in southwestern Ontario," *Soil and Tillage Research*, vol. 120, pp. 1–7, 2012.

[37] M. M. Wander, M. G. Bidart, and S. Aref, "Tillage impacts on depth distribution of total and particulate organic matter in three Illinois soils," *Soil Science Society of America Journal*, vol. 62, no. 6, pp. 1704–1711, 1998.

[38] J. G. Guzman, C. B. Godsey, G. M. Pierzynski, D. A. Whitney, and R. E. Lamond, "Effects of tillage and nitrogen management on soil chemical and physical properties after 23 years of continuous sorghum," *Soil and Tillage Research*, vol. 91, no. 1-2, pp. 199–206, 2006.

[39] S. D. Sparrow, C. E. Lewis, and C. W. Knight, "Soil quality response to tillage and crop residue removal under subarctic conditions," *Soil and Tillage Research*, vol. 91, no. 1-2, pp. 15–21, 2006.

[40] D. A. Angers and N. S. Eriksen-Hamel, "Full-inversion tillage and organic carbon distribution in soil profiles: a meta-analysis," *Soil Science Society of America Journal*, vol. 72, no. 5, pp. 1370–1374, 2008.

[41] S. P. Syswerda, A. T. Corbin, D. L. Mokma, A. N. Kravchenko, and G. P. Robertson, "Agricultural management and soil carbon storage in surface vs. deep layers," *Soil Science Society of America Journal*, vol. 75, no. 1, pp. 92–101, 2011.

[42] E. Liu, S. G. Teclemariam, C. Yan et al., "Long-term effects of no-tillage management practice on soil organic carbon and its fractions in the northern China," *Geoderma*, vol. 213, pp. 379–384, 2014.

[43] T. R. Paton, "Origin and terminology for gilgai in Australia," *Geoderma*, vol. 11, no. 3, pp. 221–242, 1974.

[44] I. Kovda, E. Morgun, and D. Tessier, "Etude de Vertisols à gilgai du Nord-Causcas: mécanismes de differenciation et aspects pédogeochimiques," *Etude et Gestion des Sols*, vol. 3, no. 1, pp. 41–52, 1996.

[45] R. Mrabet, "Stratification of soil aggregation and organic matter under conservation tillage systems in Africa," *Soil and Tillage Research*, vol. 66, no. 2, pp. 119–128, 2002.

[46] A. S. Nascente, Y. C. Li, and C. A. Costa Crusciol, "Cover crops and no-till effects on physical fractions of soil organic matter," *Soil & Tillage Research*, vol. 130, pp. 52–57, 2013.

[47] B. Dimassi, B. Mary, R. Wylleman et al., "Long-term effect of contrasted tillage and crop management on soil carbon dynamics during 41 years," *Agriculture, Ecosystems and Environment*, vol. 188, pp. 134–146, 2014.

[48] V. M. Testa, L. A. J. Teixeira, and J. Mielniczuk, "Cararcteristicas quimicas de um Podzolico Vermelho-escuro afetadas por sistemas de culturas," *Revista Brasileira de Ciência do Solo*, vol. 16, pp. 107–114, 1992.

[49] A. Pavinato, *Teores de Carbono e Nitrogenio do Solo e Productividade de Milho Afetados por Sistemas de Cultura de Mestrado em Ciencia do Solo*, PPG-Agronomia, UFRGS, Porto Alegre, Brazil, 1993.

[50] M. L. Burle, J. Mielniczuk, and S. Focchi, "Effect of cropping systems on soil chemical characteristics, with emphasis on soil acidification," *Plant and Soil*, vol. 190, no. 2, pp. 309–316, 1997.

[51] R. Mrabet, N. Saber, A. El-Brahli, S. Lahlou, and F. Bessam, "Total, particulate organic matter and structural stability of a Calcixeroll soil under different wheat rotations and tillage systems in a semiarid area of Morocco," *Soil and Tillage Research*, vol. 57, no. 4, pp. 225–235, 2000.

[52] H. Zhang, Y. Zhang, C. Yan, E. Liu, and B. Chen, "Soil nitrogen and its fractions between long-term conventional and no-tillage systems with straw retention in dryland farming in northern China," *Geoderma*, vol. 269, pp. 138–144, 2016.

[53] J. Diekow, J. Mielniczuk, H. Knicker, C. Bayer, D. P. Dick, and I. Kögel-Knabner, "Soil C and N stocks as affected by cropping systems and nitrogen fertilisation in a southern Brazil Acrisol managed under no-tillage for 17 years," *Soil and Tillage Research*, vol. 81, no. 1, pp. 87–95, 2005.

[54] E. Ouédraogo, A. Mando, and L. Stroosnijder, "Effects of tillage, organic resources and nitrogen fertiliser on soil carbon dynamics and crop nitrogen uptake in semi-arid West Africa," *Soil and Tillage Research*, vol. 91, no. 1-2, pp. 57–67, 2006.

[55] T. Yamashita, H. Flessa, B. John, M. Helfrich, and B. Ludwig, "Organic matter in density fractions of water-stable aggregates in silty soils: effect of land use," *Soil Biology and Biochemistry*, vol. 38, no. 11, pp. 3222–3234, 2006.

[56] Y. Lou, M. Xu, X. Chen, X. He, and K. Zhao, "Stratification of soil organic C, N and C: N ratio as affected by conservation tillage in two maize fields of China," *Catena*, vol. 95, pp. 124–130, 2012.

[57] M. Mazzoncini, D. Antichi, C. Di Bene, R. Risaliti, M. Petri, and E. Bonari, "Soil carbon and nitrogen changes after 28 years of no-tillage management under Mediterranean conditions," *European Journal of Agronomy*, vol. 77, pp. 156–165, 2016.

[58] M. Belmekki, M. El Gharous, O. EL Gharras, M. Boughlala, O. Iben Halima, and B. Bencharki, "Tillage effects on basic properties of an calcaeous soil under moroccan semi-arid climate," *International Journal of Advanced Research in Engineering & Technology*, vol. 5, no. 3, pp. 130–146, 2014.

[59] L. C. Kibet, H. Blanco-Canqui, and P. Jasa, "Long-term tillage impacts on soil organic matter components and related properties on a typic argiudoll," *Soil & Tillage Research*, vol. 155, pp. 78–84, 2016.

[60] E. Amézketa, "Soil aggregate stability: a review," *Journal of Sustainable Agriculture*, vol. 14, no. 2-3, pp. 83–151, 1999.

[61] M. R. Carter, "Characterizing the soil physical condition in reduced tillage systems for winter-wheat on a fine sandy loam using small cores," *Canadian Journal of Soil Science*, vol. 72, no. 4, pp. 395–402, 1992.

[62] R. J. Haynes and G. S. Francis, "Changes in microbial biomass C, soil carbohydrate composition and aggregate stability induced by growth of selected crop and forage species under field conditions," *European Journal of Soil Science*, vol. 44, no. 4, pp. 665–675, 1993.

[63] D. A. Angers, L. M. Edwards, J. B. Sanderson, and N. Bissonnette, "Soil organic matter quality and aggregate stability under eight potato cropping sequences in a fine sandy loam of Prince Edward Island," *Canadian Journal of Soil Science*, vol. 79, no. 3, pp. 411–417, 1999.

[64] J. Six, E. T. Elliott, and K. Paustian, "Aggregate and soil organic matter dynamics under conventional and no-tillage systems," *Soil Science Society of America Journal*, vol. 63, no. 5, pp. 1350–1358, 1999.

[65] N. Bissonnette, D. A. Angers, R. R. Simard, and J. Lafond, "Interactive effects of management practices on water-stable aggregation and organic matter of a Humic Gleysol," *Canadian Journal of Soil Science*, vol. 81, no. 5, pp. 545–551, 2001.

[66] M. R. Carter, "Soil quality for sustainable land management: organic matter and aggregation interactions that maintain soil functions," *Agronomy Journal*, vol. 94, no. 1, pp. 38–47, 2002.

[67] R. Lal, "Soil structure and sustainability," *Journal of Sustainable Agriculture*, vol. 1, no. 4, pp. 67–92, 1991.

[68] M. T. de Moraes, H. Debiasi, R. Carlesso, J. C. Franchini, V. R. da Silva, and F. B. da Luz, "Soil physical quality on tillage and cropping systems after two decades in the subtropical region of Brazil," *Soil & Tillage Research*, vol. 155, pp. 351–362, 2016.

14

Role of Inorganic and Organic Fractions in Animal Manure Compost in Lead Immobilization and Microbial Activity in Soil

Masahiko Katoh,[1,2] Wataru Kitahara,[3,4] and Takeshi Sato[2]

[1]Department of Agricultural Chemistry, School of Agriculture, Meiji University, 1-1-1 Higashi-Mita, Tama-ku, Kanagawa 214-8571, Japan
[2]Department of Civil Engineering, Faculty of Engineering, Gifu University, 1-1 Yanagido, Gifu 501-1193, Japan
[3]Department of Civil Engineering, Graduate School of Engineering, Gifu University, 1-1 Yanagido, Gifu 501-1193, Japan
[4]In Situ Solutions Co., Ltd., 2-5-2 Kandasudamachi, Chiyoda-ku, Tokyo 101-0041, Japan

Correspondence should be addressed to Masahiko Katoh; mkatoh@meiji.ac.jp

Academic Editor: Ezio Ranieri

This study aimed to identify how the ratio of inorganic-to-organic components in animal manure compost (AMC) affected both lead immobilization and microbial activity in lead-contaminated soil. When AMC containing 50% or more inorganic fraction with high phosphorous content was applied to contaminated soil, the amounts of water-soluble lead in it were suppressed by over 88% from the values in the soil without compost. The residual fraction under sequential extraction increased with the inorganic fraction in the AMC; however, in those AMCs, the levels of microbial enzyme activity were the same or less than those in the control soil. The application of AMC containing 25% inorganic fraction could alter the lead phases to be more insoluble while improving microbial enzyme activities; however, no suppression of the level of water-soluble lead existed during the first 30 days. These results indicate that compost containing an inorganic component of 50% or more with high phosphorus content is suitable for immobilizing lead; however, in the case where low precipitation is expected for a month, AMC containing 25% inorganic component could be used to both immobilize lead and restore microbial activity.

1. Introduction

Lead is one of the most common and harmful heavy-metal soil contaminants worldwide, particularly near mines and shooting ranges. Lead contamination in the soil of such sites poses a risk to human and animal health as well as plant growth. Thus, its mobility and bioavailability should be reduced by appropriate treatments. In addition, the rehabilitation of soil ecosystems that have undergone destruction by lead contamination should be accomplished through the remediation of such soil. Since contamination in these sites is extensive and asset values are extremely low, chemical immobilization capable of transforming lead into less soluble phases is a cost-effective remediational approach. Thus, various immobilization materials have been studied and developed [1–7].

Animal manure compost (AMC), which is the most abundantly found organic waste material in Japan, can immobilize heavy metals and improve plant growth and microbial activity [8–16] by supplying nutrients to plant and soil biota. Therefore, in addition to lead immobilization, the application of AMC to lead-contaminated soil can help rehabilitate soil ecosystems, which inorganic immobilization materials cannot [17]. Numerous studies have investigated the effectiveness of AMC for lead immobilization [18, 19]; however, consistent conclusions on the mechanisms behind this process have not been obtained as such immobilization that greatly depends upon the AMC type, that is, AMC's physicochemical properties. In particular, the presence of both inorganic and organic components in AMC makes it complicated to understand lead immobilization.

TABLE 1: Physicochemical properties of contaminated soil used in this study (on the basis of air-dried weight).

Sand (%)	Silt (%)	Clay (%)	pH	TC (g kg^{-1})	TN (g kg^{-1})	WSOC[a] (mg kg^{-1})	WS-Pb[b] (mg kg^{-1})	Pb (g kg^{-1})	P (g kg^{-1})	Al (g kg^{-1})	Fe (g kg^{-1})	Amorphous Fe (g kg^{-1})
									Total			
81.5	9.5	9.0	7.3	6.0	0.0	139	35	4.40	0.40	13.1	33.8	1.3

[a]Water-soluble organic carbon.
[b]Water-soluble lead.

Lead immobilization by AMC can be separated into indirect and direct mechanisms. The typical indirect mechanisms of lead immobilization are explained by the pH increase owing to the alkalinity of the inorganic component in AMC; this pH increase can promote the precipitation of lead carbonate and hydroxide minerals, resulting in a reduction of lead mobility and bioavailability [1, 20]. The direct mechanisms of lead immobilization by AMC have been considered to be the reaction of the lead with the inorganic and organic components in AMC. Phosphorus, sulfate, and iron in the inorganic component and humic substances in the organic component in AMC seem to be the sources responsible for the reaction with lead [10, 18, 21–23]. Katoh et al. [24] applied a specific method for the fractionation of the inorganic and acid-insoluble organic components from AMC to elucidate their separate contributions to lead immobilization. They indicated that the inorganic component in AMC could immobilize lead more effectively than its organic components [24]. These results imply that the inorganic component in AMC has a crucial role in lead immobilization, and AMC with a higher inorganic content is more suitable. However, to rehabilitate microbial activity, the organic component in AMC is also required since it supplies sufficient nutrients to soil microorganisms [17]. This component, however, may negatively affect lead mobility; water-soluble organic matter in AMC reacts and forms complexes with lead ions, resulting in enhancements in lead mobility. Therefore, a suitable ratio of inorganic and organic components in AMC should be clarified to reduce the mobility and bioavailability of lead and enhance the microbial activities in the soil. AMC has a wide range of the inorganic-to-organic component ratios [25]; however, to our knowledge, the optimal ratio of inorganic-to-organic components in AMC for lead immobilization and the rehabilitation of microbial activity have not been studied.

We investigated the mobility and bioavailability of the lead, lead phases, and microbial activities in soil amended with the AMCs of various inorganic-to-organic component ratios. The inorganic fraction was derived from swine manure compost and the acid-insoluble organic fraction was derived from cattle manure compost. These fractions can immobilize lead more effectively than other AMCs owing to the high content of phosphorus and mature organic matter in the inorganic and acid-insoluble organic fractions, respectively [24]. We aimed to identify how the ratio of inorganic-to-organic components in AMC affected the mobility and bioavailability of lead, lead phase, and microbial activity. On the basis of the obtained results, we discuss the optimal ratio of the inorganic and organic components in AMC to immobilize lead and rehabilitate microbial activity in soil.

TABLE 2: Chemical properties of inorganic and acid-insoluble organic fractions used in this study (on the basis of air-dried weight, mg g^{-1}) [24].

Characteristic	Inorganic fraction	Acid-insoluble organic fraction
Total calcium	152	ND[*1]
Total magnesium	44	ND
Total potassium	53	ND
Total iron	7	ND
Total phosphorus	131	ND
Total carbon	ND	452
ADF-C[*2]	ND	344
Humic acid carbon	ND	25
Fulvic acid carbon	ND	52

[*1]Not determined.
[*2]Acid detergent fiber carbon.

2. Materials and Methods

2.1. Preparation of Soil. The lead-contaminated soil used herein was collected from depths of 5–15 cm at a shooting range located at 35°28′6″N and 137°29′2″E in Gifu, Japan. The soil sample was air-dried, passed through a 2 mm sieve, and was used for chemical analysis and incubation tests. The selected physicochemical properties of the soil used are shown in Table 1. The soil had a sandy loam texture. The total lead content of the soil was 4.40 g kg^{-1} and the soil pH was 7.3.

2.2. Preparation of Animal Manure Compost. Commercial swine and cattle manure composts were used herein to obtain the inorganic and acid-insoluble organic fractions, respectively. We selected these composts since each fraction of the compost can immobilize lead more effectively than other composts [24]. The fractionation followed the method described by Katoh et al. [24]. In brief, the swine manure compost was combusted at 600°C for 2 h; the residue after combustion was used as the inorganic fraction. The cattle manure compost was subjected to extraction with 1 M HCl for 1 h to remove almost all inorganic components, and the residue after extraction was used as the acid-insoluble organic fraction. The pH of each fraction was adjusted to 7 to match that of the soil. Table 2 shows the selected chemical properties of each fraction [24]. The calcium, magnesium, potassium, iron, and phosphorus contents in the inorganic fraction were 152, 44, 53, 7, and 131 mg g^{-1}, respectively; moreover, the total, acid-detergent fiber, humic acid, and fulvic acid carbon contents in the acid-insoluble organic fraction were 452, 344, 25, and 52 mg g^{-1}, respectively [24]. After pH adjustment, the fractions were well mixed at the inorganic-to-acid-insoluble organic fraction ratios of 100/0, 75/25, 50/50, 25/75, and 0/100; the mixed samples were used as the AMCs herein (hereafter referred to AMC with inorganic-to-organic ratios of 100/0, 75/25, 50/50, 25/75, and 0/100).

2.3. Soil-Incubation Experiment. The contaminated soil was mixed with each AMC sample at 10 wt% and incubated at room temperature (25°C) for 184 days. Soil without AMC was also prepared as a control. Three replicates were prepared for each treatment. The water content in the soil was maintained at 60% of its maximum water-holding capacity during the incubation period. Soil samples were collected at 0, 7, 30, 90, and 184 days; the samples were freeze-dried and analyzed to determine the levels of water-soluble lead and water-soluble organic carbon. The soil sampled on day 184 was analyzed to assess lead phases by sequential extraction and microbial enzyme activities. Herein, the level of water-soluble lead was employed as an indicator of lead mobility and bioavailability in the soil, because its water-soluble form is easily mobile and utilized by plant and soil biota.

Another soil-incubation experiment was conducted to evaluate CO_2 emission from the soil with the AMCs with inorganic-to-organic ratios of 100/0, 50/50, and 0/100. A 100 mL polypropylene bottle including the soil with and without the AMC, and two glass beakers—one having 0.5 M NaOH and the other having 0.5 M H_2SO_4—were placed into a 5 L glass bottle and incubated at room temperature (25°C). A 5 L bottle without soil was also prepared. Three replicates were prepared for each treatment. At days 7, 14, 30, 60, 120, and 194, the glass beaker containing 0.5 M NaOH was replaced with a fresh 0.5 M NaOH-containing beaker. The collected NaOH was titrated with 0.2 M HCl and the CO_2 emission was calculated from the following equation:

$$C_s = \frac{V \times F - C_b}{S},\qquad(1)$$

where C_s is the amount of CO_2 emission (mol g-soil^{-1}), V is the amount of 0.2 M HCl titrated (mL), F is the factor of 0.2 M HCl (mol-CO_2 mL^{-1}), C_b is the amount of CO_2 emission in the blank (mol), and S is the amount of soil (10.0 g). The CO_2 emission derived from AMC was calculated by subtracting the emission by the AMC-treated soil samples from that by the control soil sample.

2.4. Analytical Methods. The soil's pH value was measured in ultra-pure water using a pH meter (MM-60M, DKK-TOA Co., Japan). Soil texture was determined using the hydrometer method [26]. The total carbon and nitrogen contents in the soil were determined using a carbon, hydrogen, and nitrogen elemental analyzer (MT-6; Yanaco New Science Inc., Japan). The water-soluble lead and organic carbon in soil were extracted using ultra-pure water (1 : 10 soil : solution ratio) and analyzed using an inductively coupled plasma-atomic emission spectrometer (ICP-AES; ULTIMA2; HORIBA Ltd., Japan) and a total organic carbon analyzer (TOC-V$_{WS}$;

Shimadzu Co., Japan). The amorphous iron was extracted using Shuman's method [27]. The total lead, phosphorus, aluminum, and iron contents in the soil were determined by acid digestion with HNO_3 and HCl using a microwave. A sequential extraction procedure was performed on the soil samples in accordance with the procedure described by Tessier et al. [28]. In brief, each fraction was extracted with a 1 M $MgCl_2$ solution (exchangeable fraction), 1 M sodium acetate solution at pH 5 (carbonate fraction), 0.04 M $NH_2OH\text{-}HCl$ in 25% (v/v) HOAc solution in a water bath at 95°C with occasional agitation (Fe/Mn oxide fraction), 0.02 M HNO_3 solution, 5 mL 30% H_2O_2 solution in a water bath at 85°C with occasional agitation (organic fraction), and 5 mL of HNO_3 and 2 mL of HCl using a microwave oven (residual fraction). All of the extracted or digested solutions were passed through a 0.45 μm filter and analyzed to determine the elemental concentration using ICP-AES.

Three different microbial enzyme activities—dehydrogenase, urease, and saccharase—were measured in the soil sample. The dehydrogenase activity was determined in accordance with the method described by Tabatabai [29] using triphenyltetrazolium chloride. The triphenylformazan produced was measured using spectrophotometry at 485 nm. The urease activity was monitored using the method described by Kandeler and Gerber [30] by measuring the NH_4^+ produced after the incubation of soil with urea. The saccharase activity was determined in accordance with the method described by Frankenberger and Johanson [31].

2.5. Statistical Analysis. Statistical analyses were performed using JMP Ver. 8.0.2 (SAS Institute Inc.). An analysis of variance was used to measure microbial enzyme activities. The differences between the mean values were determined using Tukey's honestly significant difference test at a 95% confidence level.

3. Results

3.1. Soil pH, Water-Soluble Lead, and Organic Carbon in Soil with AMC. The pH in the control soil was not significantly changed and ranged from 7.5 to 7.7 throughout the incubation period (Table 3). Moreover, the soil pH levels in all the AMCs were not significantly different from that in the control, indicating that the soil pH did not affect the enhancement or reduction in the lead mobility and bioavailability in the soil with AMC.

Figure 1(a) shows the amount of water-soluble lead in the soil with AMC. The amount of water-soluble lead in the control soil ranged from 35 to 44 mg kg^{-1} throughout the incubation period. The amount of water-soluble lead in the soil treated with AMC with an inorganic-to-organic ratio of 0/100 was higher than that in the control soil throughout the incubation period. The amount of water-soluble lead in the soil treated by AMC with an inorganic-to-organic ratio of 25/75 was equal to or at a higher level than that in the control soil before day 30, but it decreased to 36% of that in the control soil on day 184. The amounts of water-soluble lead in the soil samples treated by AMC with inorganic-to-organic

TABLE 3: pH in soils amended with simulated compost during incubation period ($n = 3$).

Compost (Inorganic/organic)	Days of incubation				
	0	7	30	90	184
100/0	7.7 ± 0.0	7.5 ± 0.0	7.6 ± 0.1	7.8 ± 0.0	7.8 ± 0.0
75/25	7.7 ± 0.0	7.5 ± 0.0	7.7 ± 0.0	7.8 ± 0.0	7.8 ± 0.1
50/50	7.6 ± 0.0	7.6 ± 0.0	7.7 ± 0.1	7.8 ± 0.0	7.6 ± 0.0
25/75	7.4 ± 0.0	7.4 ± 0.0	7.8 ± 0.0	7.8 ± 0.0	7.6 ± 0.0
0/100	7.2 ± 0.0	7.3 ± 0.0	7.6 ± 0.1	7.8 ± 0.0	7.5 ± 0.2
Control	7.5 ± 0.1	7.6 ± 0.0	7.7 ± 0.0	7.5 ± 0.1	7.5 ± 0.2

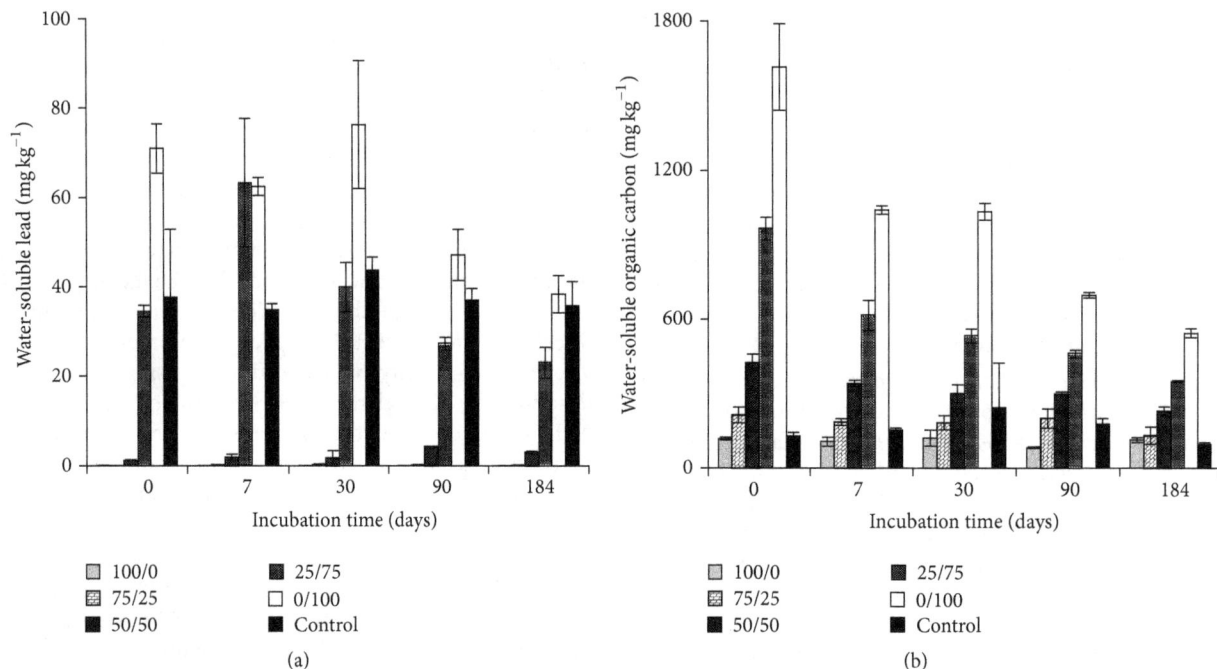

FIGURE 1: Water-soluble lead (a) and organic carbon (b) in soil amended with AMC of inorganic/organic = 100/0, 75/25, 50/50, 25/75, and 0/100 and without AMC (control). Vertical bars indicate the standard error.

ratios of 50/50, 75/25, and 100/0 were 3.4%–12%, 0.1%–0.8%, and 0.1%–0.2%, respectively, of the value in the control soil. The amount of water-soluble organic carbon in the soil is shown in Figure 1(b). At the beginning of the incubation period, the amounts of water-soluble organic carbon in the soils treated with different AMC compositions were arranged in the following order: inorganic/organic = 0/100 > inorganic/organic = 25/75 > inorganic/organic = 50/50 > inorganic/organic = 75/25 > control > inorganic/organic = 100/0. The amounts of water-soluble organic carbon in the soil treated by AMC with inorganic-to-organic ratios of 50/50, 75/25, and 100/0 remained at their initial values during the incubation period. The amounts of water-soluble organic carbon in the soil treated by AMC with inorganic-to-organic ratios of 0/100 and 25/75 had rapidly decreased between days 0 and 7, after which gradual decreases were observed. The decreases in water-soluble organic carbon between days 0 and 184 were 1,071 and 615 mg kg^{-1} under inorganic-to-organic ratios of 0/100 and 25/75, respectively.

3.2. CO$_2$ Emission from Soil Amended with AMC. The cumulated CO$_2$ emission derived from AMC is shown in Figure 2. The level of CO$_2$ emission from the AMC with an inorganic-to-organic ratio of 0/100 increased during the early stage of incubation and then remained at approximately 500 mg-C kg^{-1}. The CO$_2$ emission trends from the AMC with inorganic-to-organic ratios of 50/50 and 100/0 were similar; they slightly increased at the beginning of incubation and then remained at approximately 100 mg-C kg^{-1}.

3.3. Lead Phases by Sequential Extraction in Lead-Contaminated Soil. Figure 3 shows the lead fraction results under sequential extraction. The average recovery, which was defined herein as the ratio of the sum total level of each fraction (Figure 3) to the total level of lead in the soil (Table 1), was 119 ± 20%. The composition of the control soil was as follows: 12.8% exchangeable fraction, 51.2% carbonate fraction, 20.7% Fe/Mn oxide fraction, 9.3% organic fraction, and 6.0% residual fraction. The percentage of organic fraction

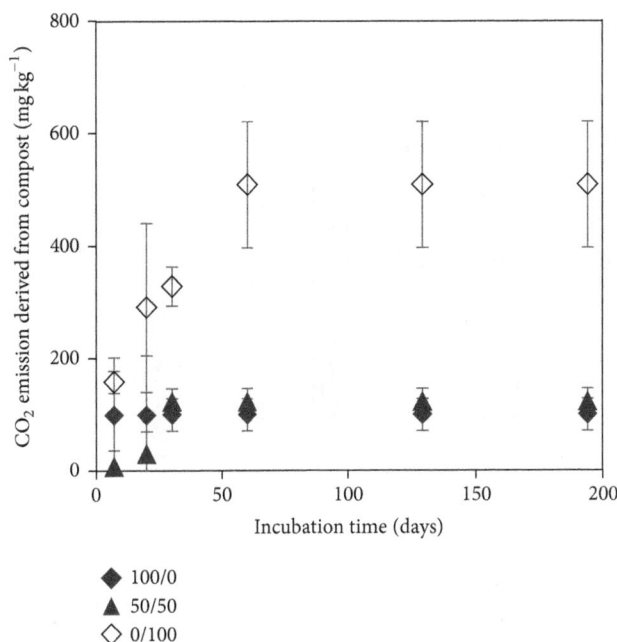

FIGURE 2: Cumulative CO_2 emission derived from AMC with inorganic/organic = 100/0, 50/50, and 0/100. Vertical bars indicate the standard error.

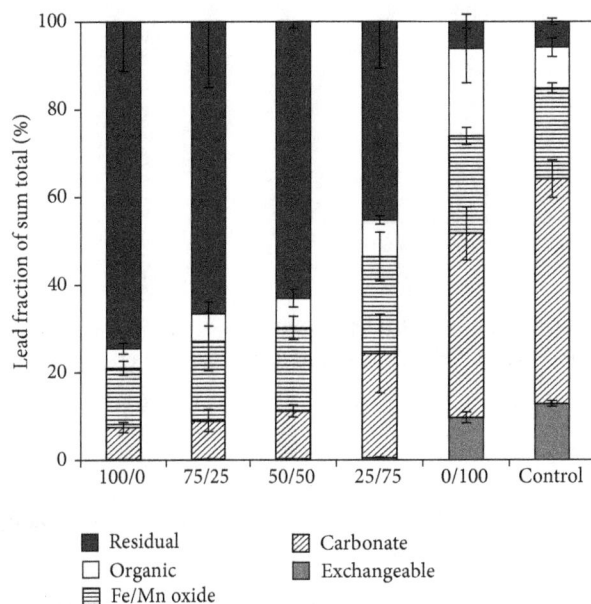

FIGURE 3: Sequential extraction of lead from soil amended with AMC of inorganic/organic = 100/0, 75/25, 50/50, 25/75, and 0/100 and without AMC (control) after 184 days of incubation. Vertical bars indicate the standard error.

by sequential extraction in the soil treated by AMC with an inorganic-to-organic ratio of 0/100 increased compared with that in the control soil, whereas that of carbonate decreased. The addition of AMC with a higher ratio of inorganic fraction resulted in a greater enhancement in the residual percentage. Furthermore, in soils with the AMCs containing an inorganic

fraction of 25% or more, the percentage of organic fraction was at the same level as that in the control soil.

3.4. Microbial Enzyme Activities in Lead-Contaminated Soil. The dehydrogenase and urease activities in the soil treated by AMC with inorganic-to-organic ratios of 0/100 and 25/75, respectively, were significantly higher than that in the control soil; however, those in the soil treated by AMC with inorganic-to-organic ratios of 75/25 and 100/0 were significantly lower or on the same level (Figure 4).

4. Discussion

4.1. Role of Organic Component in Animal Manure Compost Lead Immobilization and Rehabilitation of Microbial Activity. The water-soluble lead level in the soil treated by AMC with an inorganic-to-organic ratio of 0/100 did not become lower than that in the control soil during the incubation period (Figure 1(a)), whereas the lead phases in the soil by sequential extraction were altered to be more insoluble; the organic fraction was increased (Figure 3). Moreover, at the early stage of incubation, the water-soluble lead level remained at a higher level in the soil treated by AMC with an inorganic-to-organic ratio of 25/75. However, the water-soluble lead level decreased as the incubation time increased in soil with both composts. These observations would be explained by the higher level of water-soluble organic matter and its sorption and decomposition in the soil thereby increasing the incubation time. The water-soluble organic matter can easily form complexes with lead ions, and its complexes could enhance the lead mobility in the soil [17, 32, 33]. Therefore, at the early stage of incubation, the high level of water-soluble organic carbon would result in an enhancement of the lead mobility in the soil treated by AMC with inorganic-to-organic ratios of 0/100 and 25/75. The water-soluble organic matter was derived from the compost containing humic and fulvic acids with relatively low molecular weight [34, 35]. These organic compounds are also readily decomposable and sorbed on the soil surface [36, 37]. According to the result for CO_2 emission (Figure 2), the amount of CO_2 emission was lower than that of water-soluble organic carbon-decrease from day 0 to day 184, suggesting that some of the water-soluble organic matter derived from the compost was decomposed and some of it was sorbed in the soil with the increase in the incubation period. Thus, the level of water-soluble lead would decrease with that water-soluble organic carbon in the soil treated by AMC with inorganic-to-organic ratios of 0/100 and 25/75. It has been known that the compost amendment induced the very short-term leaching pulses of lead after the application [38], and lead was redistributed to the soil component by the decomposition of dissolved organic matter [39]. These results suggest that the organic component in the AMC does not considerably contribute to the immobilization of lead and suppression of lead mobility and bioavailability. This was comparable with the results of Schwab et al. [19] and Levonmäki et al. [40], who reported that water-soluble organic matter enhanced the lead mobility by the formation of complexation. However, all of the microbial enzyme activities

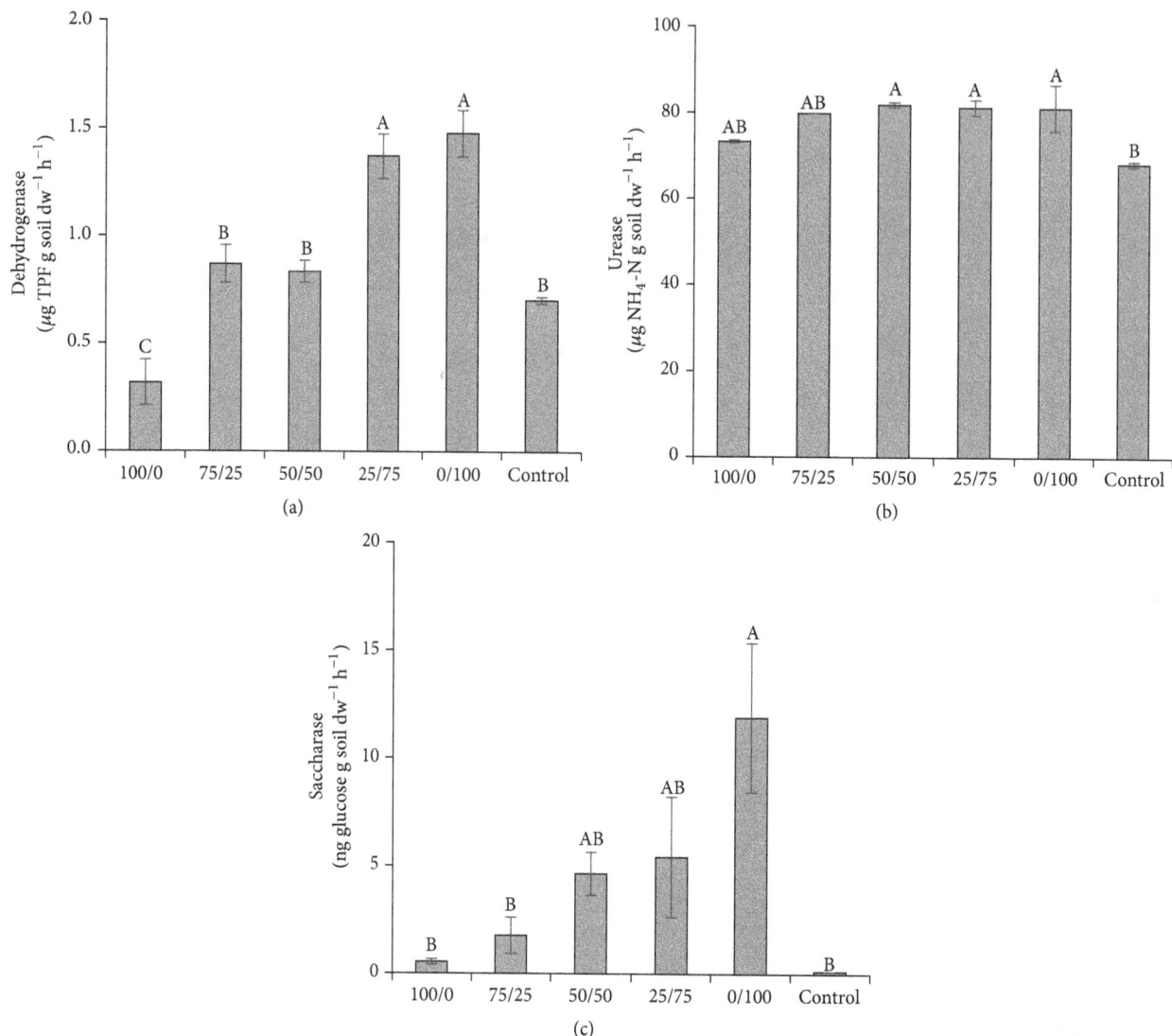

FIGURE 4: Enzyme activities of dehydrogenase (a), urease (b), and saccharase (c) in soil amended with AMC of inorganic/organic = 100/0, 75/25, 50/50, 25/75, and 0/100 and without AMC (control) after 184 days of incubation. Vertical bars indicate the standard error. Different letters indicate significant difference at $P < 0.05$.

measured herein increased in the soil treated by AMC with an inorganic-to-organic ratio of 0/100, despite the fact that the level of water-soluble lead was higher. The dehydrogenase activity, which is an indicator of the overall microbial activity [8, 9, 41–43], in the soil treated by AMC containing 50% or more inorganic fraction was the same level as that in the control soil, suggesting that the inorganic component in the AMC did not significantly contribute to the enhancement in the microbial activity, despite the fact that the lead mobility and bioavailability are greatly decreased by its addition. Farrell et al. [44] also showed that the microbial enzyme activities in the soil amended with the compost became higher than that in the inorganic material. The water-soluble organic matter derived from the compost is utilized as a nutrient source by microorganisms; thus, the addition of AMC with high organic matter content would induce the rehabilitation

of microbial activity owing to the large amount of readily decomposable organic matter.

4.2. Role of the Inorganic Component in Animal Manure Compost on Lead Immobilization. The addition of AMC containing 50% or more inorganic fraction reduced the water-soluble lead by over 88% during the incubation period (Figure 1(a)). Moreover, the addition of AMC containing 25% or more inorganic fraction could alter lead phases to be more insoluble. These results indicate that AMC containing 50% or more inorganic fraction could immobilize lead and reduce the lead bioavailability in the soil. The inorganic fraction used herein contained a considerable amount of phosphorus, which results in the precipitation of lead phosphate minerals such as pyromorphite [24]. Pyromorphite is thermodynamically stable with a solubility product of log K_{sp} =

−25.05 [45]. Therefore, the phosphorus in the AMC's inorganic component was responsible for the increase in the percentage of the residual fraction and decrease in the water-soluble lead level. This is supported by the results of Walker et al. [10], Liu et al. [11], and Clemente et al. [18], who suggested that the immobilization of lead and cadmium would be due to the precipitation of insoluble phosphate salts. The percentage of Fe/Mn oxide and organic fractions by sequential extraction in the soil with AMC containing 25% or more inorganic fraction was lesser than or approximately equal to that in the control soil, although the residual fraction increased with increase in the inorganic fraction ratio in AMC. Moreover, in the soil treated by AMC containing 50% or less inorganic fraction, the level of water-soluble organic carbon was higher than that in the control soil. These results suggest that the phosphorus in the inorganic component of AMC could immobilize lead-precipitating lead phosphate minerals even if the AMC contained more organic than inorganic components. This may be because the magnitude of lead immobilization by the inorganic component of AMC suppressed the magnitude of the facilitation of lead mobility by the organic component in AMC.

The level of water-soluble lead in the soil treated by AMC with an inorganic-to-organic ratio of 25/75 was higher than that in the control soil during the first 30 days of incubation, whereas the lead phases were altered to be more insoluble. This was attributed to incomplete immobilization. Scheckel et al. [46] demonstrated that not all lead in soil could be immobilized by phosphorus material, according to extended X-ray-absorption fine-structure analysis. The percentages of exchangeable and carbonate fractions by sequential extraction in the soil treated by AMC with an inorganic-to-organic ratio of 25/75 were higher than that in the soil treated by AMC that was 50% or more inorganic fraction. The levels of water-soluble organic carbon in the soil treated by AMC with an inorganic-to-organic ratio 25/75 were also higher, particularly during the first 30 days of incubation. The higher amount of water-soluble organic matter formed complexes with lead dissolved from the exchangeable and carbonate fractions, resulting in an enhancement in the level of water-soluble lead in the early stage of incubation. With the increasing incubation time, water-soluble organic matter was decomposed and the level of water-soluble lead decreased below that in the control soil.

4.3. Optimal Ratio of Inorganic and Organic Components in Animal Manure Compost for Lead Immobilization and the Rehabilitation of Microbial Activity. Various inorganic percentages are used in the AMC, ranging from 7.3 to 82.8% [25]. The ranges of the inorganic fraction of AMC in this study fall within this range. On the basis of this study's findings, an inorganic component with a high phosphorus content of 50% or more is required to alter lead phases to be more insoluble and reduce the water-soluble lead level to less than that in the soil without compost, although the microbial activities are not enhanced. The importance of an inorganic component for the immobilization of heavy metals including lead in soil is pointed out by other researchers [21, 47, 48].

In contrast, 25% inorganic AMC can alter lead phases to be more insoluble and simultaneously enhance microbial activities, whereas the water-soluble lead level is higher than that in the soil without compost during the first 30 days after the application. This is consistent with the result of Katoh et al. [17], who indicated that the application of 25% inorganic swine manure compost could alter lead phases to be more insoluble and improve plant growth and microbial enzyme activities; however, the level of water-soluble lead was higher than that in the cattle manure compost during the early stage of incubation (90 days after application). These findings suggest that AMCs cannot immobilize both lead and restore microbial activity. Therefore, to immobilize lead and suppress the lead mobility and bioavailability using AMC, AMC that is 50% inorganic or more and contains a large phosphorus component should be applied to contaminated sites. However, in the case of low precipitation over a month, for example, during a dry season, 25% inorganic AMC should be used to both immobilize lead and restore microbial activity.

5. Conclusion

The amount of water-soluble lead in the soil treated by 100% organic AMC remained higher than that in the control soil throughout the 184-day incubation period, although it tended to decrease with increasing incubation time. The amount of water-soluble lead in the soil treated by AMC with an inorganic-to-organic ratio of 25/75 was higher than that in the soil without compost during the early stage of incubation, but it reached a lower level than that in the soil without compost after 90 days. The amounts of water-soluble lead in the soil treated by AMC containing 50% or more inorganic fraction were suppressed by over 88% from the value in the soil without compost and remained low throughout the incubation period. The reduction in the level of water-soluble lead in the soil treated by AMC with inorganic-to-organic ratios of 0/100 and 25/75 would be explained by the decomposition of water-soluble organic matter. The ratio of the residual fraction after sequential extraction was enhanced and the readily soluble lead fractions (exchangeable and carbonate fractions) were reduced in soils treated by AMC containing 25% or more inorganic fraction. In soil treated with fully organic AMC, the readily soluble lead fraction in sequential extraction did not change in comparison with that in the control soil, whereas the organic fraction in sequential extraction was enhanced. The compost containing 25% or less inorganic fraction could improve the microbial activity, but the compost containing 50% or more inorganic fraction could not. These results indicate that the compost containing 50% or more inorganic fraction with a high phosphorus content is suitable for immobilizing lead and reducing lead mobility and bioavailability in the soil. However, to immobilize lead and improve the microbial activity at the same time, the AMC with 25% inorganic and 75% organic components should be used as a lead immobilization material, but note that higher lead mobility at the initial stage after the application should be considered.

Acknowledgments

The ICP-AES and CHN elemental analyzer instruments used for the chemical analysis in this study were made available by the Division of Instrumental Analysis at Gifu University. The authors are grateful to Professor F. Li and Professor T. Yamada (Gifu University) for allowing the use of the TOC analyzer. This study was supported by the Japan Society for the Promotion of Science (JSPS) KAKENHI [Grant no. 23710089].

References

[1] D. J. Walker, R. Clemente, and M. P. Bernal, "Contrasting effects of manure and compost on soil pH, heavy metal availability and growth of *Chenopodium album* L. in a soil contaminated by pyritic mine waste," *Chemosphere*, vol. 57, no. 3, pp. 215–224, 2004.

[2] J. H. Park, N. Bolan, M. Megharaj, and R. Naidu, "Comparative value of phosphate sources on the immobilization of lead, and leaching of lead and phosphorus in lead contaminated soils," *Science of the Total Environment*, vol. 409, no. 4, pp. 853–860, 2011.

[3] X. Cao, A. Wahbi, L. Ma, B. Li, and Y. Yang, "Immobilization of Zn, Cu, and Pb in contaminated soils using phosphate rock and phosphoric acid," *Journal of Hazardous Materials*, vol. 164, no. 2-3, pp. 555–564, 2009.

[4] X. Cao, D. Dermatas, X. Xu, and G. Shen, "Immobilization of lead in shooting range soils by means of cement, quicklime, and phosphate amendments," *Environmental Science and Pollution Research*, vol. 15, no. 2, pp. 120–127, 2008.

[5] G. M. Hettiarachchi, G. M. Pierzynski, and M. D. Ransom, "In situ stabilization of soil lead using phosphorus and manganese oxide," *Environmental Science and Technology*, vol. 34, no. 21, pp. 4614–4619, 2000.

[6] A. Davis, L. E. Eary, and S. Helgen, "Assessing the efficacy of lime amendment to geochemically stabilize mine tailings," *Environmental Science and Technology*, vol. 33, no. 15, pp. 2626–2632, 1999.

[7] E. Lombi, F.-J. Zhao, G. Zhang et al., "In situ fixation of metals in soils using bauxite residue: chemical assessment," *Environmental Pollution*, vol. 118, no. 3, pp. 435–443, 2002.

[8] P. Alvarenga, A. P. Gonçalves, R. M. Fernandes et al., "Evaluation of composts and liming materials in the phytostabilization of a mine soil using perennial ryegrass," *Science of the Total Environment*, vol. 406, no. 1-2, pp. 43–56, 2008.

[9] P. Alvarenga, A. P. Gonçalves, R. M. Fernandes et al., "Organic residue as immobilizing agents in aided phytostabilization: (I) Effects on soil chemical characteristics," *Chemosphere*, vol. 74, no. 10, pp. 1292–1300, 2009.

[10] D. J. Walker, R. Clemente, A. Roig, and M. P. Bernal, "The effects of soil amendments on heavy metal bioavailability in two contaminated Mediterranean soils," *Environmental Pollution*, vol. 122, no. 2, pp. 303–312, 2003.

[11] L. Liu, H. Chen, P. Cai, W. Liang, and Q. Huang, "Immobilization and phytotoxicity of Cd in contaminated soil amended with chicken manure compost," *Journal of Hazardous Materials*, vol. 163, no. 2-3, pp. 563–567, 2009.

[12] A. Sato, H. Takeda, W. Oyanagi, E. Nishihara, and M. Murakami, "Reduction of cadmium uptake in spinach (*Spinacia oleracea* L.) by soil amendment with animal waste compost," *Journal of Hazardous Materials*, vol. 181, no. 1-3, pp. 298–304, 2010.

[13] H.-S. Chen, Q.-Y. Huang, L.-N. Liu, P. Cai, W. Liang, and M. Li, "Poultry manure compost alleviates the phytotoxicity of soil cadmium: influence on growth of pakchoi (*Brassica chinensis* L.)," *Pedosphere*, vol. 20, no. 1, pp. 63–70, 2010.

[14] R. Clemente, C. Almela, and M. P. Bernal, "A remediation strategy based on active phytoremediation followed by natural attenuation in a soil contaminated by pyrite waste," *Environmental Pollution*, vol. 143, no. 3, pp. 397–406, 2006.

[15] E. Doelsch, A. Masion, G. Moussard, C. Chevassus-Rosset, and O. Wojciechowicz, "Impact of pig slurry and green waste compost application on heavy metal exchangeable fractions in tropical soils," *Geoderma*, vol. 155, no. 3-4, pp. 390–400, 2010.

[16] R. P. Narwal and B. R. Singh, "Effect of organic materials on partitioning, extractability and plant uptake of metals in an alum shale soil," *Water, Air, and Soil Pollution*, vol. 103, no. 1-4, pp. 405–421, 1998.

[17] M. Katoh, W. Kitahara, R. Yagi, and T. Sato, "Suitable chemical properties of animal manure compost to facilitate Pb immobilization in soil," *Soil and Sediment Contamination*, vol. 23, no. 5, pp. 523–539, 2014.

[18] R. Clemente, Á. Escolar, and M. P. Bernal, "Heavy metals fractionation and organic matter mineralisation in contaminated calcareous soil amended with organic materials," *Bioresource Technology*, vol. 97, no. 15, pp. 1894–1901, 2006.

[19] P. Schwab, D. Zhu, and M. K. Banks, "Heavy metal leaching from mine tailings as affected by organic amendments," *Bioresource Technology*, vol. 98, no. 15, pp. 2935–2941, 2007.

[20] R. Clemente and M. P. Bernal, "Fractionation of heavy metals and distribution of organic carbon in two contaminated soils amended with humic acids," *Chemosphere*, vol. 64, no. 8, pp. 1264–1273, 2006.

[21] A. Baghaie, A. H. Khoshgoftarmanesh, M. Afyuni, and R. Schulin, "The role of organic and inorganic fractions of cow manure and biosolids on lead sorption," *Soil Science and Plant Nutrition*, vol. 57, no. 1, pp. 11–18, 2011.

[22] S. Deiana, C. Gressa, B. Manunza, R. Rausa, and R. Seever, "Analytical and spectroscopic characterization of humic acids extracted from sewage sludge, manure, and worm compost," *Soil Science*, vol. 150, no. 1, pp. 419–424, 1990.

[23] M. A. A. Zaini, R. Okayama, and M. Machida, "Adsorption of aqueous metal ions on cattle-manure-compost based activated carbons," *Journal of Hazardous Materials*, vol. 170, no. 2-3, pp. 1119–1124, 2009.

[24] M. Katoh, W. Kitahara, and T. Sato, "Sorption of lead in animal manure compost: contributions of inorganic and organic fractions," *Water, Air, and Soil Pollution*, vol. 225, no. 1, article 1828, 2014.

[25] T. Yamaguchi, Y. Harada, and M. Tsuiki, "Basic data of animal waste composts," *Miscellaneous Publication of the National Agriculture Research Center*, vol. 41, pp. 1–178, 2000 (Japanese).

[26] G. W. Gee and J. M. Bauder, "Partical-size analysis," in *Methods of Soil Analysis, Part 1*, A. L. Page, R. H. Miller, and D. R. Keeney, Eds., pp. 383–411, American Society of Agronomy, Madison, Wis, USA, 1986.

[27] L. M. Shuman, "Fractionation method for soil microelements," *Soil Science*, vol. 140, no. 1, pp. 11–22, 1985.

[28] A. Tessier, P. G. C. Campbell, and M. Blsson, "Sequential extraction procedure for the speciation of particulate trace metals," *Analytical Chemistry*, vol. 51, no. 7, pp. 844–851, 1979.

[29] M. A. Tabatabai, "Soil enzymes," in *Methods of Soil Analysis, Part 2*, S. H. Mickelson and J. M. Bigham, Eds., pp. 77–83, American Society of Agronomy, Madison, Wis, USA, 1994.

[30] E. Kandeler and H. Gerber, "Short-term assay of soil urease activity using colorimetric determination of ammonium," *Biology and Fertility of Soils*, vol. 6, no. 1, pp. 68–72, 1988.

[31] W. T. Frankenberger and J. B. Johanson, "Method of measuring invertase activity in soils," *Plant and Soil*, vol. 74, no. 3, pp. 313–323, 1983.

[32] S. Sauvé, M. McBride, and W. Hendershot, "Soil solution speciation of lead (II): effects of organic matter and pH," *Soil Science Society of America Journal*, vol. 62, no. 3, pp. 618–621, 1998.

[33] H. Wang, X. Shan, T. Liu et al., "Organic acids enhance the uptake of lead by wheat roots," *Planta*, vol. 225, no. 6, pp. 1483–1494, 2007.

[34] M. Aoyama, "Fractionation of water-soluble organic substances formed during plant residue decomposition and high performance size exclusion chromatography of the fractions," *Soil Science and Plant Nutrition*, vol. 42, no. 1, pp. 31–40, 1996.

[35] M. Aoyama, "Properties of fine and water-soluble fractions of several composts II. Organic forms of nitrogen, neutral sugars, and muramic acid in fractions," *Soil Science and Plant Nutrition*, vol. 37, no. 4, pp. 629–637, 1991.

[36] T. Paré, H. Dinel, M. Schnitzer, and S. Dumontet, "Transformations of carbon and nitrogen during composting of animal manure and shredded paper," *Biology and Fertility of Soils*, vol. 26, no. 3, pp. 173–178, 1998.

[37] D. L. Jones and D. S. Brassington, "Sorption of organic acids in acid soils and its implications in the rhizosphere soil," *European Journal of Soil Science*, vol. 49, pp. 447–455, 1998.

[38] M. Farrell, W. T. Perkins, P. J. Hobbs, G. W. Griffith, and D. L. Jones, "Migration of heavy metals in soil as influenced by compost amendments," *Environmental Pollution*, vol. 158, no. 1, pp. 55–64, 2010.

[39] A. W. Schroth, B. C. Bostick, J. M. Kaste, and A. J. Friedland, "Lead sequestration and species redistribution during soil organic matter decomposition," *Environmental Science & Technology*, vol. 42, no. 10, pp. 3627–3633, 2008.

[40] M. Levonmäki, H. Hartikainen, and T. Kairesalo, "Effect of organic amendment and plant roots on the solubility and mobilization of lead in soils at a shooting range," *Journal of Environmental Quality*, vol. 35, no. 4, pp. 1026–1031, 2006.

[41] A. Pérez-de-Mora, P. Burgos, E. Madejón, F. Cabrera, P. Jaeckel, and M. Schloter, "Microbial community structure and function in a soil contaminated by heavy metals: effects of plant growth and different amendments," *Soil Biology and Biochemistry*, vol. 38, no. 2, pp. 327–341, 2006.

[42] S. Doni, C. MacCi, E. Peruzzi, M. Arenella, B. Ceccanti, and G. Masciandaro, "In situ phytoremediation of a soil historically contaminated by metals, hydrocarbons and polychlorobiphenyls," *Journal of Environmental Monitoring*, vol. 14, no. 5, pp. 1383–1390, 2012.

[43] J. Wyszkowska, J. Kucharski, and W. Lajszner, "The effects of copper on soil biochemical properties and its interaction with other heavy metals," *Polish Journal of Environmental Studies*, vol. 15, no. 6, pp. 927–934, 2006.

[44] M. Farrell, G. W. Griffith, P. J. Hobbs, W. T. Perkins, and D. L. Jones, "Microbial diversity and activity are increased by compost amendment of metal-contaminated soil," *FEMS Microbiology Ecology*, vol. 71, no. 1, pp. 94–105, 2010.

[45] P. Miretzky and A. Fernandez-Cirelli, "Phosphates for Pb immobilization in soils: a review," *Environmental Chemistry Letters*, vol. 6, no. 3, pp. 121–133, 2008.

[46] K. G. Scheckel, J. A. Ryan, D. Allen, and N. V. Lescano, "Determining speciation of Pb in phosphate-amended soils: method limitations," *Science of the Total Environment*, vol. 350, no. 1-3, pp. 261–272, 2005.

[47] P. Castaldi, L. Santona, and P. Melis, "Heavy metal immobilization by chemical amendments in a polluted soil and influence on white lupin growth," *Chemosphere*, vol. 60, no. 3, pp. 365–371, 2005.

[48] A. Ruttens, M. Mench, J. V. Colpaert, J. Boisson, R. Carleer, and J. Vangronsveld, "Phytostabilization of a metal contaminated sandy soil. I: influence of compost and/or inorganic metal immobilizing soil amendments on phytotoxicity and plant availability of metals," *Environmental Pollution*, vol. 144, no. 2, pp. 524–532, 2006.

15

Contamination of Soil with Pb and Sb at a Lead-Acid Battery Dumpsite and Their Potential Early Uptake by *Phragmites australis*

Abraham Jera, France Ncube, and Artwell Kanda

Department of Environmental Science, Bindura University of Science Education, P. Bag 1020, Bindura, Zimbabwe

Correspondence should be addressed to Artwell Kanda; alzkanda@gmail.com

Academic Editor: Marco Trevisan

Recycling of spent Lead-Acid Batteries (LABs) and disposal of process slag potentially contaminate soil with Pb and Sb. Total and available concentrations of Pb and Sb in three soil treatments and parts of *Phragmites australis* were determined by atomic absorption spectrophotometry. Soil with nonrecycled slag (NR) had higher total metal concentrations than that with recycled slag (RS). Low available fractions of Pb and Sb were found in the soil treatments before planting *P. australis*. After 16 weeks of growth of *P. australis*, the available fractions of Pb had no statistical difference from initial values ($p > 0.05$) while available Sb fractions were significantly lower when compared with their initial values ($p < 0.05$). Metal transfer factors showed that *P. australis* poorly accumulate Pb and Sb in roots and very poorly translocate them to leaves after growing for 8 and 16 weeks. It may be a poor phytoextractor of Pb and Sb in metal-contaminated soil at least for the 16 weeks of its initial growth. However, the plant established itself on the metalliferous site where all vegetation had been destroyed. This could be useful for potential ecological restoration. The long-term phytoextraction potential of *P. australis* in such environments as LABs may need further investigation.

1. Introduction

Soil is a geochemical sink of contaminants and natural buffer for controlling the transport of elements to the atmosphere, hydrosphere, and biota [1]. The mobility and bioavailability of these elements in soil are controlled by geochemical, climatic, and biological factors [2]. Anthropogenic activities may render soil unsuitable for various land uses. Processes for recycling of Lead-Acid Batteries (LABs) and dumping of resultant slag may be important routes for soil contamination. Some trace elements emitted to the biosphere have toxic effects on the environment and human health (i.e., exposure to Pb and Sb) [3, 4]. People living near or working at LAB recycling sites and dumpsites may be exposed to Pb and Sb through contact with contaminated water, runoff, and airborne particulate emissions. Various physicochemical clean-up technologies relying on intensive soil manipulation [5] or bioremediation [6] have been used for the reclamation of contaminated soil. However, most of these technologies are not cost-effective and environmentally friendly [1, 7]. Previous studies have focused on phytoremediation technologies [8–11] and available evidence suggests that the technologies are ecofriendly, innovative, and economical [12].

Human exposure to Pb and Sb may be occupational or nonoccupational. Lead has historically been used in water pipes, artifacts, LABs, and gasoline additive, tetraethyl-lead [13]. Antimony has been used as a fire retardant, in plastics, in coatings, in electronics, as a decoloring agent in glass, as alloys in LABs, and as a catalyst in the production of polyethylene terephthalate polymers [14]. The demand for energy and the subsequent widespread use of LABs by individuals, households, and industries result in large tonnage of spent batteries. Automobile LABs contain polypropylene, concentrated H_2SO_4, Pb electrodes with either PbO_2 paste cathode or Pb anode, and various metals such as Sb, As, Cd, Sn, and Cu [15]. Spent LABs are hazardous materials [14] that need appropriate handling and disposal in specially designed facilities and not in conventional landfills. Recycling of spent LABs may be more environmentally friendly than open dumping or incineration. According to Royer et al. [15], sites where LABs are recycled pose challenges for remediation due to a variety of soil contamination sources.

Phragmites australis (Cav.) Trin. ex Steud (common reed) is a widely distributed macrophyte throughout the world [16]. The terrestrial form of the plant has higher dry matter content and low growth rate and specific leaf surface area which makes it resist environmental stresses and adapt to adverse environments better than the aquatic form [17]. *Phragmites australis* can grow in environments of extreme pH and poor nutrient content [18]. Humans use the common reed to make woven mats, for fishing, as hunting traps, in musical instruments, or in baskets, among other applications. Animals browse the plant. *Phragmites australis* has reportedly been used for remediation of heavy-metal-contaminated aquatic ecosystems [19–21]. To the best of the authors' knowledge, limited information is available on the early uptake and accumulation of Pb and Sb by *P. australis* under field conditions, particularly in soil with recycled and nonrecycled slag from the recycling of automobile LABs.

The current study assessed the contamination of soil with LAB waste and the potential early uptake of Pb and Sb by *P. australis* under field conditions. The total and available concentrations of Pb and Sb in three soils (with recycled slag, nonrecycled slag, and reference) and total metal in three plant parts (root, stem, and leaf) of *P. australis* were determined. For remediation purposes, the study site required a plant species which could survive with limited water supply such as seasonal rainwater once established. We hypothesized that the common reed can adapt to such an environment and once introduced it could reestablish an ecosystem where other plants may grow. Then, it would take up Sb and Pb during its early stages of development. The identification of plant species for phytoremediation may help in ecological restoration of contaminated environments.

2. Materials and Methods

2.1. Description of the Study Site. An automobile LAB dumpsite located in Norton ($17°52'11''$S, $30°41'24''$E), a small town 46 km west of Harare, Zimbabwe, was studied. Lead-Acid Battery waste was discharged directly onto a one-hectare dumpsite, formerly a marshy area. It appears that the phytotoxic effects of the contaminants were severe since all indigenous vegetation had been destroyed (Figure 1(b)). Operations that generate LAB waste include automobile battery breaking (Figure 1(a)), smelting, and refining. Field work conducted on the dumpsite entailed planting vetiver grass and *P. australis*, of which the former subsequently died within three days (Figure 1(b)) but the latter survived (Figure 1(d)). The study site had fersialitic or chromic luvisol or rhodic paleustalf soil [22].

2.2. Sampling and Chemical Analyses. The experimental design consisted of three treatments which were replicated three times: soil with nonrecycled slag (NR), recycled slag (RS), and a reference site (RF). The reference site was 5 km away in the windward side. Three similar beds measuring

15 m² (5 m × 3 m) were prepared for each treatment. Each bed consisted of three rows with inter- and intrarow planting of 1 m × 1 m. It was raised to 30 cm in order to avoid ground or near-neighbor effects [23]. About 0.50 m was left from the edges. Three of the 15 established planting points (20%) from each bed were randomly selected for sampling the growth media (0.5 kg) to a depth of 30 cm using a soil auger. This was repeated for another two treatments. Split samples were used for the determination of selected physicochemical parameters. Remaining samples were oven-dried at 105°C for 1 h and then at 80°C for 24 h (Heraeus D6450) [24], ground, and sieved (<1 mm). These were used for the determination of other parameters including the total and available metal concentrations. The pH was determined using a glass pH meter (micropH 2002) in a 1:1 soil/water suspension. Electrical conductivity (EC) was measured using a conductivity meter (ERMA EC035). The NO_3-N concentration of soil was determined by extraction with 2 M KCl (1:10, m/v) and analyzed by colorimetry (UV-Vis spectrometer model GENESYS 10S, Thermo Scientific, Germany) [25]. Available phosphorus was determined as PO_4-P by extraction from soil with 0.5 M $NaHCO_3$ (42 g in 1 L) at pH 8.5 and then measured colorimetrically (UV-Vis spectrometer model GENESYS 10S, Thermo Scientific, Germany) using acidified blue ammonium molybdate [26]. The Walkley-Black method was used for determining organic matter [27]. In this procedure, carbon is oxidized by acidified dichromate and excess dichromate is back-titrated with ferrous iron with diphenylamine indicator. Soil textural composition was determined using the Bouyoucos hydrometer method [28]. Total Pb and Sb concentrations in the filtered digests were determined spectrometrically (Flame AAS; GBC-Savant).

Two split powdered growth media samples from each treatment (1 g and 0.5 g) were separately used for total *(aqua regia)* and available (ammonium acetate) Pb and Sb extraction. A 1 g sample was digested in an acid mixture (HCl and HNO_3, 20 ml, 1:3, v/v) on a hot plate (110°C, 3 h) until decomposition was complete. This was then evaporated to reduce volume to about 5 ml. Digests were filtered (Whatman filter paper No. 1), washed with deionized water, and transferred quantitatively to a 50 ml volumetric flask. Neutral ammonium acetate (1 M $CH_3CO_2NH_4$, pH 7) (10 ml) was added to the 0.5 g split sample of powdered growth media sample in a 250 ml beaker. The sample was shaken in a multishaker (Kahn Shaker, 140 rpm, 1 h) and filtered (Whatman filter paper No. 1) into a 50 ml volumetric flask which was completed to the mark with 1 M $CH_3CO_2NH_4$. This was replicated three times for each treatment. The concentrations of Pb and Sb in sample solutions were determined spectrophotometrically (Flame AAS; GBC-Savant) using appropriately prepared specific calibration curves. After 16 weeks of growth of *P. australis*, growth media in the rooting zone were sampled and analyzed for both total and available Pb and Sb. A reagent blank was run ten times. Metal recovery studies for growth media were carried out using a certified reference material (CRM) (channel sediment BCR 320R: 0083) [29].

Rhizomes (96) of *P. australis* (30–53 cm long) with some roots were taken from adult plants at a presumably unpolluted (with Sb and Pb) site. These were cut to approximately

FIGURE 1: (a) Battery breaking and (b) slag dumpsite with dead vetiver grass within 3 days of planting, at a Lead-Acid Battery recycling and smelting site in Norton, Zimbabwe. (c) *P. australis* rhizomes before preparation for planting and (d) *P. australis* growing in an experimental bed of soil with recycled slag.

25 cm each to give 157 rhizomes. Ten randomly selected cut rhizomes (6.4%) were repeatedly washed with stream water and then with deionized water. They were oven-dried (Heraeus D6450; 70°C, 24 h), ground, and sieved to <1 mm [12]. Ten 1.0 g split dried samples were separately decomposed in a muffle furnace (550°C, 6 h) and the ashes were dissolved in *aqua regia* (12 ml). These were filtered (Whatman filter paper No. 1) into 25 ml volumetric flasks which were completed to the mark with double-distilled water and analyzed for Pb and Sb using FAAS (GBC-Savant) [21]. The remaining cut rhizomes (145) were planted in nine beds (three beds for each treatment) at a depth of about 20 cm and watered with 40 L of tap water (20 L polyethylene bucket) per bed skipping a day or two in between watering events. No rainfall events were recorded during the study period. After eight weeks of planting, three whole plants of *P. australis* were randomly harvested from each bed within and across treatments. Composite plant part samples (leaves, stems, and roots) were made for each treatment. A garden hoe was used to uproot plants while a pair of secateurs was used to cut leaves. Plant tissues were separately put in open polypropylene bags, labeled, and sent to the laboratory. Each sample was washed with a jet of tap water and then with

distilled water. Oven-dried samples were ground to a fine powder and thoroughly mixed. Laboratory samples were then prepared as described above for the growth media in order to determine Pb and Sb concentrations. Bioaccumulation and translocation factors were calculated [12]. Lead and Sb were extracted from two sets of three replicate samples of the CRM by acid digestions and the other set was extracted by ammonium acetate and analyzed using similar procedures as the growth media. A procedural blank was run ten times for Pb and Sb analysis.

2.3. Quality Control and Statistical Procedures. During sampling, plant parts affected by herbivory or with signs of disease were avoided. Composite samples were used and replicated three times. Reagents blanks and calibration standards were run in between sample analyses. All apparatuses used were washed thoroughly before use and then rinsed thrice using deionized water. Analytical-grade reagents were used in all the analyses. A certified reference material was used to validate the analytical procedure. IBM SPSS statistical package version 21 was used for data analysis. Significant differences among physicochemical parameters for the three treatments were determined using One-Way ANOVA and

LSD post hoc test. All tests were considered significant at $p < 0.05$. A paired sample t-test was used to compare the concentrations of Pb and Sb in the growth media before planting and 16 weeks after planting.

3. Results and Discussion

3.1. Physicochemical Properties of Growth Media. Element recovery studies of a CRM gave 95.8% (81.43 ± 1.73 mg/kg) and satisfactory RSD (2.12%) for total Pb. Limits of detection (LODs) were 0.001 mg/L (0.10 mg/kg) for Pb and 0.003 mg/L (0.3 mg/kg) for Sb. Mean concentrations of Pb and Sb in 10 rhizomes before planting for the three treatments were less than the LOD. Table 1 shows the physicochemical parameters of the growth media before and after 16 weeks of planting *P. australis* in three soil treatments (NR, RS, and RF). Soil with recycled slag (RS) generally had significantly higher clay content and pH but lower organic matter (OM) and nutrients (NO_3^-, PO_4^{3-}) than soil with nonrecycled slag (NR) before and after 16 weeks of planting *P. australis* ($p < 0.05$). The concentrations of Pb and Sb in the soil treatments decreased in the order NRS > RS > RF. After 16 weeks of planting, the concentrations of NO_3^- (NRS and RS) and OM (RF) were significantly lower than their initial values before planting (t-test, $p < 0.05$). Not significantly different initial and final soil parameters after 16 weeks could be explained by the absence of rainfall events during the study period (except for occasional initial watering) which could promote leaching, overall insignificant element uptake by *P. australis* (for Sb), and additive effects of atmospheric deposition from the adjacent LAB recycling processes. However, leaching of Sb and Pb could not be excluded through watering but to a lesser extent as the two elements are strongly bound by soil clay minerals and organic matter [1]. High pH of slag soil treatments (10.7–12.1) may not favor mineralization of Pb and Sb. The observed significant decrease in soil Sb after 16 weeks in NR could be attributed to plant uptake, leaching, and other soil biochemical processes. There were no significant differences in the concentrations of Pb before planting and 16 weeks after planting of *P. australis* for both NRS and RS treatments ($p > 0.05$).

US EPA [5] reported 80,000 mg/kg Pb at LAB recycling sites. Soil that was 60 m away from an abandoned battery waste site had 41,890 mg/kg Pb [30]. At another abandoned scrap deposit site, soil had 104,000 mg/kg Pb [24]. This shows that LAB wastes are a very important route for the release of trace elements into the environment. People living near or working at sources of Sb and Pb such as smelters, coal fired plants, and refuse incinerators may be exposed to toxic elements in dust, soil, and vegetation. A study of the exposure of residents and children living in a battery recycling craft village in Vietnam showed Pb-contaminated hair, blood, and urine [31]. Similar observations of increased blood Pb levels were made in another study at a LAB recycling and manufacturing plant in Kenya [32]. Recycling processes may not be very efficient in recovering metals. Slag from recycling LABs still contains up to 5% Pb [13, 14]. Other than direct disposal of slag on the dumpsite, toxic elements may be added to the soil surface by atmospheric deposition of fugitive dust

and drainage or leachate from waste heaps. Soil naturally contains trace Sb at concentrations of less than 1 mg/kg (average: 0.48 mg/kg) but values of 109–2,550 mg/kg Sb from processing sites were reported [3]. An average background concentration of 0.67 mg/kg Sb was also reported [1].

Clay mineral content, organic matter, phosphorus, and moisture content of a soil are very important parameters that influence the bioavailability and uptake of trace elements by plants [1, 2, 33]. Results from the current study clearly show that soil at the LAB dumpsite was contaminated with Sb and Pb. Exposed populations may be encouraged to have Pb levels in their blood, hair, and urine monitored for possible adverse health effects.

3.2. Concentrations of Sb and Pb in Plant Parts of P. australis. Table 2 shows the concentrations of Sb and Pb in plant parts of *P. australis* grown in three different soil treatments (NR, RS, and RF) after 8 and 16 weeks of planting. The recycled slag treatment (RS) had lower concentrations of Pb and Sb for roots than the nonrecycled slag soil treatment (NR) ($p < 0.05$). No significant differences were observed for the concentrations of both elements between NR and RS treatments for leaves and stems after 8 and 16 weeks of planting *P. australis* ($p > 0.05$). A paired t-test to compare elemental concentrations within a soil treatment between 8- and 16-week growth periods showed significant differences ($p < 0.05$) for Pb and Sb in roots (NR and RS) only. Both metals were not detected in all plant parts from the RF treatment. Concentrations of Pb and Sb decreased in the order root > leaf > stem for NR and RS treatments. Results suggest that Pb and Sb were coming from LAB slag.

Table 3 shows that the Biological Absorption Factor (BAF) and Translocation Factor (TF) were less than unity for element transfer. These indicate that, after 8 and 16 weeks of planting, *P. australis* poorly accumulate Pb and Sb into roots and poorly translocate them to leaves. Plant roots appeared to take up more Sb than Pb in both NR and RS soil treatments after harvesting at 16 weeks. *Phragmites australis* appeared to take up more Pb after 16 weeks (3.3-fold) and Sb (5.5-fold) in the NR soil treatment and more Pb (5-fold) and Sb (4.5-fold) in the RS soil treatment than it did after 8 weeks of growth.

The above-ground : below-ground concentration ratios of Sb and Pb were very small (less than 0.15) in NR and RS soil treatments. This may suggest that the common reed (*P. australis*) is poor for phytoextraction of Pb and Sb from contaminated soil. A translocation factor well above one indicates a possible candidate for phytoextraction [10]. In its various remediation applications in contaminated aquatic ecosystems, *P. australis* showed that it is a poor phytoextractor of Pb and Sb but good for phytorhizofiltration [19, 34, 35]. One drawback in this field experimental setup, and thus phytoremediation of trace elements from contaminated soil, is the failure to control leaching of trace elements. Particulate fallout and foliar absorption of Pb cannot be excluded in this study since the element has been reported to be poorly translocated to leaves [1], yet Pb and Sb were recorded in appreciable amounts in the current study. Establishment of vegetation in an area where all native plants were destroyed was fascinating. This development may help filter and reduce

TABLE 1: Physicochemical characteristics of three soil treatments replicated three times (nonrecycled slag, recycled slag, and reference) at a LAB dumpsite in Norton, Zimbabwe. Parameters were determined before planting and 16 weeks after planting of *P. australis*. Values are expressed as mean ± SD of triplicate measurements.

Soil parameter	Before planting			16 weeks after planting		
	Nonrecycled slag treatment	Recycled slag treatment	Reference soil	Nonrecycled slag treatment	Recycled slag treatment	Reference soil
Clay%	6.33 ± 1.53	12.33 ± 1.53	18.67 ± 2.52	7.67 ± 1.63	10.63 ± 1.64	21.60 ± 2.95
Silt%	54.00 ± 2.65	61.00 ± 4.00	34.00 ± 3.00	55.10 ± 3.92	59.77 ± 1.53	34.60 ± 0.40
Fine sand%	39.67 ± 1.53	26.67 ± 2.52	47.33 ± 0.58	37.23 ± 4.52	29.60 ± 2.91	43.80 ± 3.17
Moisture content (g)	4.53 ± 0.22	6.76 ± 0.04	12.80 ± 0.30	—	—	—
pH (H_2O)	10.90 ± 0.36	12.10 ± 0.20*	7.23 ± 0.06	10.70 ± 0.10	10.93 ± 0.15*	7.17 ± 0.06
PO_4^{3-} (mg/kg)	0.85 ± 0.12	0.50 ± 0.02*	1.45 ± 0.03	0.74 ± 0.09	0.58 ± 0.03*	1.37 ± 0.17
NO_3^- (mg/kg)	0.61 ± 0.04*	ND	1.37 ± 0.05	0.55 ± 0.07*	ND	1.30 ± 0.08
EC ($\mu S/cm$)	144.50 ± 12.08	85.80 ± 8.35	41.03 ± 4.44	152.17 ± 8.92	108.40 ± 16.59	38.67 ± 1.82
Organic matter (%)	0.93 ± 0.06	0.31 ± 0.03	3.11 ± 0.18*	0.87 ± 0.09	0.38 ± 0.03	3.57 ± 0.05*
Total Pb (mg/kg)	48,840 ± 4,000	27,103 ± 3,869	0.51 ± 0.05	43860 ± 7,066	23,380 ± 2,495	ND
Total Sb (mg/kg)	3,460 ± 645*	2,583 ± 523	ND	2,343 ± 531*	2,163 ± 172	ND
Available Pb (mg/kg)	278.41 ± 20.50	111.54 ± 5.81	—	230.08 ± 42.33	95.34 ± 34.16	—
Available Sb (mg/kgl)	86.31 ± 13.08*	64.83 ± 8.32*	—	49.68 ± 16.36*	42.54 ± 8.24*	—

ND: not detected (below LOD). — (dash): parameter not determined. All parameters were significantly different across the three treatments before planting ($p < 0.05$). * Parameters significantly different before planting and after planting of *P. australis* for a given soil treatment (paired *t*-test; $p < 0.05$).

TABLE 2: Concentrations of Pb and Sb in tissues of *P. australis* harvested from different soil treatments after 8 and 16 weeks of planting. Values are expressed as mean ± SD of triplicate measurements in mg/kg, DW.

Element	Growth period (wks)	Nonrecycled slag treatment			Recycled slag treatment			Reference soil treatment		
		Root	Leaf	Stem	Root	Leaf	Stem	Root	Leaf	Stem
Pb	8	$136.57 \pm 18.91^{a*}$	8.77 ± 0.90^a	2.88 ± 0.27^a	$98.90 \pm 12.76^{b*}$	7.83 ± 1.74^a	2.63 ± 0.30^a	ND	ND	ND
	16	$590 \pm 36^{a*}$	8.61 ± 1.04^a	2.76 ± 0.38^a	$466.67 \pm 25.17^{b*}$	8.04 ± 0.23^a	2.96 ± 0.12^a	ND	ND	ND
Sb	8	$67.83 \pm 9.97^{a*}$	3.33 ± 0.36^a	1.45 ± 1.91^a	$47.84 \pm 3.86^{b*}$	2.89 ± 0.19^a	1.56 ± 0.23^a	ND	ND	ND
	16	$260 \pm 20^{a*}$	3.47 ± 0.32^a	1.73 ± 0.171^a	$205 \pm 18^{b*}$	2.94 ± 0.21^a	1.50 ± 0.16^a	ND	ND	ND

ND: not detected (<LOD). Different superscripts (a, b) in a row denote significantly different concentrations ($p < 0.05$) of a given element for a plant tissue across treatments; * in a column denotes significantly different concentrations (paired t-test, $p < 0.05$) for a given element for a specific plant tissue between 8- and 16-week growth periods within a soil treatment.

TABLE 3: Bioaccumulation and translocation factors of Pb and Sb for *P. australis* in different soil treatments after 8 and 16 weeks of planting.

Element	Growth period (wks)	Nonrecycled slag treatment			Recycled slag treatment			Reference soil treatment		
		[E]r/[E]s	[E]l/[E]r	([E]l + [E]st)/[E]r	[E]r/[E]s	[E]l/[E]r	([E]l + [E]st)/[E]r	[E]r/[E]s	[E]l/[E]r	([E]l + [E]st)/[E]r
Pb	8	0.003	0.06	0.09	0.004	0.08	0.11	—	—	—
	16	0.01	0.01	0.02	0.02	0.02	0.02	—	—	—
Sb	8	0.02	0.05	0.07	0.02	0.06	0.09	—	—	—
	16	0.11	0.01	0.02	0.09	0.01	0.02	—	—	—

[E]r: concentration of the element in roots of plant; [E]s: concentration of the element in soil on which the plant is growing; [E]l: concentration of the element in leaves of plant; [E]st: concentration of the element in the stem of plant.

surface runoff laden with contaminants. It will be interesting to find out the nature of other plants that would be established on the site with time and how *P. australis* may take up Pb and Sb.

4. Conclusions

Contamination of soil with LAB recycling wastes and potential early uptake of Pb and Sb by *P. australis* after 8 and 16 weeks of planting were studied at a dumpsite. Slag and LAB recycling processes introduced large quantities of Pb and Sb into soil, altered soil characteristics, and destroyed indigenous vegetation. Soil with recycled slag had lower nutrient (NO_3^-, PO_4^{3-}) content, OM, and trace elements (Pb, Sb) than soil with nonrecycled slag. Uptake of Pb and Sb by roots of *P. australis* appeared to increase over time although the poor accumulation in leaves and stems appeared to remain constant. The NR soil treatment appeared to promote root uptake of Sb compared to Pb. *Phragmites australis* poorly accumulates Pb and Sb into roots and poorly translocates them to leaves, and thus it is a poor candidate species for phytoextraction in contaminated field soil conditions. However, its ability to be established on a site where all vegetation had been destroyed could further be explored as a starting point for ecological restoration. Accumulation of Pb and Sb by *P. australis* in the long term and possibilities of whether other plant species would get established on this site may need further studies. Based on findings and results obtained in the current study, the authors recommend biological monitoring of Pb in urine, hair, and blood samples for LAB recycling workers and populations residing near LAB smelters and dumpsites.

References

[1] A. Kabata-Pendias, *Trace Elements in Soils and Plants*, Taylor and Francis, New York, NY, USA, 4th edition, 2011.

[2] A. Kabata-Pendias, "Soil-plant transfer of trace elements—an environmental issue," *Geoderma*, vol. 122, no. 2–4, pp. 143–149, 2004.

[3] Agency for Toxic Substances and Disease Registry (ATSDR), *Toxicological Profile for Antimony and Compounds*, US Public Health Service, Washington, DC, USA, 1992.

[4] Agency for Toxic Substances and Disease Registry (ATSDR), *Toxicological Profile for Lead*, US Public Health Service, Washington, DC, USA, 2007.

[5] US. Environmental Protection Agency (USEPA), "Innovative treatment technologies: semi-annual status report," EPA/540/2-91/014, US EPA, Washington, DC, USA, 1991.

[6] C. J. Rhodes, "Applications of bioremediation and phytoremediation," *Science Progress*, vol. 96, no. 4, pp. 417–427, 2013.

[7] H. Ali, E. Khan, and M. A. Sajad, "Phytoremediation of heavy metals—concepts and applications," *Chemosphere*, vol. 91, no. 7, pp. 869–881, 2013.

[8] M. M. Lasat, "Phytoextraction of toxic metals: a review of biological mechanisms," *Journal of Environmental Quality*, vol. 31, no. 1, pp. 109–120, 2002.

[9] N. Marmiroli, M. Marmiroli, and E. Maesti, "Phytoremediation and phytotechnologies: a review for the present and the future," in *Soil and Water pollution Monitoring, Protection and Remediation*, I. Twardowska et al., Ed., pp. 403–416, Springer, Berlin, Germany, 2006.

[10] M. Mench, N. Lepp, V. Bert et al., "Successes and limitations of phytotechnologies at field scale: outcomes, assessment and outlook from COST Action 859," *Journal of Soils and Sediments*, vol. 10, no. 6, pp. 1039–1070, 2010.

[11] P. Sharma and S. Pandey, "Status of phytoremediation in world scenario," *International Journal of Environmental Bioremediation and Biodegradation*, vol. 2, no. 4, pp. 178–191, 2014.

[12] N. Shabani and M. H. Sayadi, "Evaluation of heavy metals accumulation by two emergent macrophytes from the polluted soil: an experimental study," *Environmentalist*, vol. 32, no. 1, pp. 91–98, 2012.

[13] United Nations Environment Programme (UNEP), "Technical guidelines for the environmentally sound management of lead-acid battery wastes," UNEP/CHW/TWG/20/1, UNEP, Geneva, Switzerland, 2002.

[14] A. Smaniotto, A. Antunes, I. D. N. Filho et al., "Qualitative lead extraction from recycled lead-acid batteries slag," *Journal of Hazardous Materials*, vol. 172, no. 2-3, pp. 1677–1680, 2009.

[15] M. D. Royer, A. Selvakumar, and R. Gaire, "Control technologies for remediation of contaminated soil and waste deposits at super fund lead battery recycling sites," *Journal of the Air & Waste Management Association*, vol. 42, no. 7, pp. 970–980, 1992.

[16] J. Swearingen and K. Saltonstall, Phragmites Field Guide: Distinguishing Native and Exotic Forms of Common Reed (Phragmites australis) in the United States, Plant Conservation Alliance, Weeds Gone Wild, 2010 http://www.nps.gov/plants/alien/pubs/index.htm.

[17] P. Li, W. Han, N. Thevs et al., "A comparison of the functional traits of common reed (Phragmites australis) in Northern China: aquatic versus terrestrial ecotypes," *PLoS ONE*, vol. 9, no. 2, Article ID e89063, 2014.

[18] C. L. Gucker, Phragmites australis, in Fire Effects Information System, US. Department of Agriculture, Forest Service, Rocky Mountain Research Station, Fire Sciences Laboratory, 2008, http://www.fs.fed.us/database/feis/.

[19] F. Ghassemzadeh, H. Yousefzadeh, and M. H. Arbab-Zavar, "Removing arsenic and antimony by Phragmites australis: rhizofiltration technology," *Journal of Applied Sciences*, vol. 8, no. 9, pp. 1668–1675, 2008.

[20] N. A. Anjum, I. Ahmad, M. Válega et al., "Salt marsh macrophyte Phragmites australis strategies assessment for its dominance in mercury-contaminated coastal lagoon (Ria de Aveiro, Portugal)," *Environmental Science and Pollution Research*, vol. 19, no. 7, pp. 2879–2888, 2012.

[21] E. Grisey, X. Laffray, O. Contoz, E. Cavalli, J. Mudry, and L. Aleya, "The bioaccumulation performance of reeds and cattails in a constructed treatment wetland for removal of heavy metals in landfill leachate treatment (Etueffont, France)," *Water, Air, & Soil Pollution*, vol. 223, no. 4, pp. 1723–1741, 2012.

[22] J. Gotosa, K. Nezandonyi, A. Kanda, S. M. Mushiri, A. Kuhlande, and T. Nyamugure, "Effects of irrigating Eucalyptus grandis plantations with a mixture of domestic and pulp and paper mill effluent on soil quality at a site in northern Zimbabwe," *Journal of Sustainable Development in Africa*, vol. 13, no. 5, pp. 136–149, 2011.

[23] X. Song, X. Hu, P. Ji, Y. Li, G. Chi, and Y. Song, "Phytoremedi-ation of cadmium-contaminated farmland soil by the hyperac-cumulator beta vulgaris L. var. cicla," *Bulletin of Environmental Contamination and Toxicology*, vol. 88, no. 4, pp. 623–626, 2012.

[24] M. Wang, C. Zhang, Z. Zhang, F. Li, and G. Guo, "Distribution and integrated assessment of lead in an abandoned lead-acid battery site in Southwest China before redevelopment," *Ecotoxicology and Environmental Safety*, vol. 128, pp. 126–132, 2016.

[25] D. G. Maynard, Y. P. Kalra, and J. A. Crumbaugh, "Nitrate and exchangeable ammonium nitrogen," in *Soil Sampling and Methods of Analysis*, M. R. Carter and E. G. Gregorich, Eds., pp. 97–106, Taylor and Fracis, Boca Raton, Fla, USA, 2nd edition, 2006.

[26] L. P. van Reeuwijk, "Procedures for soil analysis," Tech. Rep. 9, ISRIC—World Soil Information, Wageningen, Netherlands, 2002.

[27] A. Walkley and I. Black, "An examination of the Degtjareff method for determining soil organic matter, and a proposed modification of the organic acid titration method," *Soil Science*, vol. 37, no. 1, pp. 29–38, 1934.

[28] G. W. Gee and J. W. Bauder, "Particle size analysis," in *Methods of Soil Analysis: Part 1—Physical and Mineralogical Methods*, A. Klute, Ed., Agronomy Monograph, American Society of Agronomy, Madison, Wis, USA, 2nd edition, 1986.

[29] Food and Agriculture Organisation (FAO), *Guidelines for Quality Management in Soil and Plant Lboratories: Quality of Analytical Procedures*, FAO and ISRIC, Rome, Italy, 1998.

[30] S. Oni, O. Ogunlaja, and O. Ladkun, "Wild plants in an abandoned battery waste site," *International Journal of Scientific and Engineering Research*, vol. 5, no. 4, pp. 991–994, 2014.

[31] T. Noguchi, T. Itai, N. M. Tue et al., "Exposure assessment of lead to workers and children in the battery recycling craft village, Dong Mai, Vietnam," *Journal of Material Cycles and Waste Management*, vol. 16, no. 1, pp. 46–51, 2014.

[32] F. H. Were, G. N. Kamau, P. M. Shiundu, G. A. Wafula, and C. M. Moturi, "Air and blood lead levels in lead acid battery recycling and manufacturing plants in Kenya," *Journal of Occupational and Environmental Hygiene*, vol. 9, no. 5, pp. 340–344, 2012.

[33] F. M. G. Tack, "Trace elements: general soil chemistry, principles and processes," in *Trace Elements in Soils*, P. Hooda, Ed., pp. 9–32, Blackwell, West Sussex, UK, 2010.

[34] W. A. Al-Taisan, "Suitability of using Phragmites australis and Tamarix aphylla as vegetation filters in industrial areas," *American Journal of Environmental Sciences*, vol. 5, no. 6, pp. 740–747, 2009.

[35] G. Bonanno, "Comparative performance of trace element bioaccumulation and biomonitoring in the plant species Typha domingensis, Phragmites australis and Arundo donax," *Ecotoxicology and Environmental Safety*, vol. 97, pp. 124–130, 2013.

16

Influence of Nanolime and Curing Period on Unconfined Compressive Strength of Soil

Panbarasi Govindasamy,[1] **Mohd Raihan Taha,**[1,2]
Jamal Alsharef,[1] **and Kowstubaa Ramalingam**[1]

[1]*Department of Civil and Structural Engineering, Universiti Kebangsaan Malaysia (UKM), 43600 Bangi, Selangor, Malaysia*
[2]*Institute for Environment and Development (LESTARI), Universiti Kebangsaan Malaysia (UKM), 43600 Bangi, Selangor, Malaysia*

Correspondence should be addressed to Panbarasi Govindasamy; gpam_21@yahoo.com

Academic Editor: Teodoro M. Miano

This paper presents the improvement of the unconfined compressive strength (UCS) of soil by mixing different percentages of nanolime and 5% lime with soil. The UCS of treated soil increased significantly over curing time with increasing percentage of nanolime. The optimum results were reached at only 0.5% nanolime admixtures which were much higher than 5% lime admixture. This may be due to higher ability of nanolime to flocculate and agglomerate the soil particles compared with the lime. In addition, the lime could fill only the micropores while nanolime could fill the micro- and nanopores as well. The strength gain is inversely proportional to the remolded moisture content and curing period. However, when the content of nanolime used is larger than 0.5%, nanolime particles are not uniformly dispersed. Therefore, a weak area in the form of voids is created, consequently the homogeneous hydrated microstructure cannot be formed, and finally the strength will decrease.

1. Introduction

In geotechnical engineering practice, soft clay is common and widely encountered. It forms a great challenge to geotechnical engineers, particularly in metropolitan areas due to its low strength and high compressibility. This type of ground at a construction site is not always suitable for supporting structures such as buildings, bridges, highways, and dams. In order to overcome this problem several methods have been employed worldwide to improve engineering characteristics of soils such as ground improvement techniques. These techniques are used to prepare soft soils for construction and they have become much more common nowadays.

In the 1970s and 1980s, soil stabilization by admixture was developed in Japan. Soil treated in such a manner was better than the original soil in terms of strength, reduced compressibility, and hydraulic conductivity [1]. The use of lime as admixture to improve soft ground has also long been used such as in ancient China and Egypt [2].

Lime has been found to be an effective stabilizer, which significantly changes the characteristics of a soil to produce long-term permanent strength and stability, particularly with respect to the action of water and frost [3]. However, previous work in stabilization of clayey soils with lime has shown that small amounts of lime considerably improve workability but contribute little to strength, whereas larger amounts of lime only improve the strength and bearing capacities of soil. In the results of unconfined compressive strength, kaolinitic and montmorillonitic clayey soils were found to be well stabilized with larger amount of lime, alone, whereas illitic-chloritic clayey soils require addition of fly ash to obtain significant gains in the strength [4].

2. Nanotechnology in Geotechnical Engineering

Nanotechnology is a rapidly emerging technology with vast potential to create new materials with unique properties and to produce new and improved products for numerous applications. At scales of nanometer range, materials can exhibit unique properties different from their bulk state. Nanotechnological achievements provided a modern approach in geotechnical engineering also. Nanotechnology can be applied in geotechnical engineering by studying the soil structure in nanometer scale to gain a better understanding of soil nature [5].

In recent years, application of nanomaterials in engineering studies has drawn interest by researchers from all over the world. Nanoparticles are referred to as nanocrystals due to its crystalline form. The transition from microparticles to nanoparticles can be attributed to positive changes in physical properties. This is because of the increase in ratio of surface area and changes in volume. Due to the large surface area of nanoparticles, a lot of interactions occur between intermixed materials such as nanocomposites which leads to an increase in strength of the materials [6].

Taha [7] has presented the laboratory experiments to study the fundamental geotechnical properties of mixtures of natural soils and its product after ball milling operation. The product after ball milling process was termed nanosoil herein. The plastic and liquid limits of soil mixtures consisting of 98% original soil and 2% nanosoil increased compared to the values of 100% original soil. Since the increase in the liquid limit was less than that of plastic limit, it reduced the plasticity index which is advantageous in many geotechnical constructions. These results showed that a small amount of these milled particles or nanosoil can provide significant improvement in the geotechnical properties of soil.

The materials in nanodimensions have specific chemical and physical properties, linked to the physical and chemistry properties of very small particles with respect to the massive materials. Comparably, nanosize is more reactive than its original size and act as a good catalyst. Materials with new applications could be produced due to the increase in total size of surface area. Thus, these materials become more reactive and suitable for improving the properties of clayey soil for various applications [7]. Increasing research and development is ongoing using nanoparticles as filler or additives for various desired effects [8].

Interesting applications for nanomaterials are being found within the conservation of cultural heritage. Nanolime is a recent type of nanoproducts. Nanolime is aimed at overcoming some of the limitations of traditional lime-based materials, such as the difficulty in achieving a complete carbonation, while maintaining their advantages [9]. This paper presents the improvement of UCS of original soil by mixing different percentages of nanolime and 5% lime with soil. Finally, it gives information about optimum nanolime content for the studied soil to improve the unconfined compressive strength at four different curing periods.

3. Materials

3.1. Properties of Original Soil.
The basic geotechnical properties and classification of the soil collected from the hilly area of Universiti Kebangsaan Malaysia (UKM) were listed in Table 1. Original soil was collected at shallow depth after removing the top 0.5 m depth of ground surface. The natural water content was 22.8% while the maximum dry density (γ_{dmax}) and optimum water content (ω_{opt}) were 1.87 g/cm^3 and 15.6%, respectively. According to the unified soil classification system (USCS), the tested soil was classified under

TABLE 1: Basic properties of original soil.

Property	Value
Natural water content, (ω_n) (%)	22.8
Maximum dry density, γ_{dmax} (g/cm^3)	1.87
Optimum water content, (ω_{opt}) (%)	15.6
Liquid limit (LL) (%)	31
Plastic limit (PL) (%)	18.7
Plasticity index PI (%)	12.3
USCS classification	CL
Specific gravity (ρ_s)	2.63
% gravel	0.0
% sand	57.5
% silt	22.5
% clay	20.0

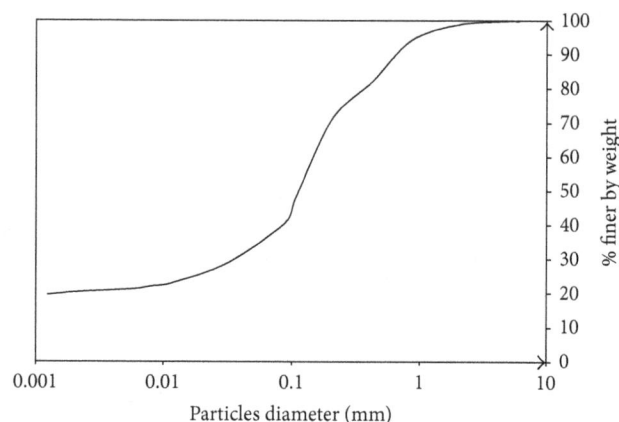

FIGURE 1: Particles sizes distribution curve of original soil.

group symbol of CL. The soil contained no gravel. Meanwhile, the fractions of sand, silt, and clay are 57.5, 22.5, and 20%, respectively. Figure 1 presents the distribution of particle size of original soil.

3.2. Chemical Characterization of Soil and Additives.
Lime and nanolime were the two additives used for this of soil improvement. Lime powder was obtained from local market. Nanolime powder was imported from Strem Chemicals, Inc., Mulliken Way, Newburyport, Massachusetts, United States. The results of X-ray fluorescence (XRF) test for lime used as additive are illustrated in Table 2 for both regular grades and nanogrades. It can be seen that the nanolime has greater purity than the lime. A large trace of CaO was available in both grades, while lime contained a noticeable amount of MgO (1.75%) compared to 1.24% which was available in nanolime. Table 3 presents the chemical composition of original soil, where the silica and alumina are the main components and then iron oxide comes as a third most abundant component.

X-ray diffraction (XRD) test was used to analyze the crystallite size of the particles. The results of this test in the form of relation graphs between intensity and 2 theta

TABLE 2: Chemical composition of lime additive.

	CaO	MgO	SO$_3$	SiO$_2$	Al$_2$O$_3$	Fe$_2$O$_3$	SrO	Cl	K$_2$O	ZnO	ZrO$_2$
Lime	95.21%	1.75%	0.56%	0.22%	0.13%	0.08%	0.03%	0.02%	0.01%	—	—
Nanolime	97.22%	1.24%	0.02%	0.47%	0.09%	0.07%	0.03%	0.01%	0.01%	48 ppm	12 ppm

TABLE 3: Chemical composition of original soil.

SiO$_2$	Al$_2$O$_3$	Fe$_2$O$_3$	TiO$_2$	K$_2$O	MgO	ZrO$_2$	SO$_3$	CaO	Others
60.35%	21.83%	4.36%	1.13%	0.51%	0.47%	0.09%	0.08%	0.05%	0.08%

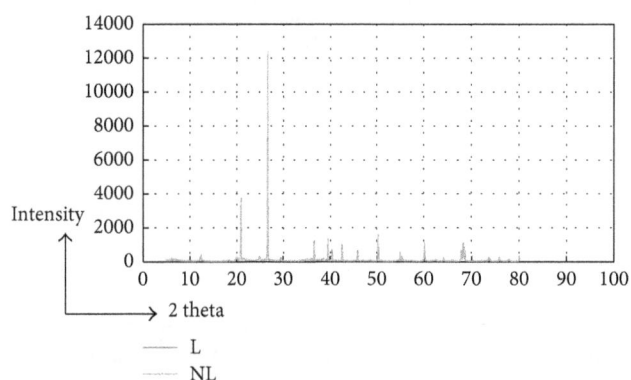

FIGURE 2: Relation graph between intensity and 2 theta for lime and nanolime material.

are shown in Figure 2. The values for the full width at half maximum (FWHM) and the theta were obtained. These numbers were analyzed using Scherer's formula to measure the crystallite size of lime and nanolime.

4. Methodology

4.1. Preparation of Test Samples.
Samples prepared for UCS tests are original soil, soil mixed with lime, and soil mixed with nanolime. The percentages of nanolime mixed with soil were 0.2%, 0.3%, 0.5%, 0.8%, and 1% each. 5% lime was added to soil. This amount was the value of Initial Lime Consumption (ILC) for the studied soil which was determined based on the pH test. The dry original soil was mixed with lime or nanolime percentages and then the water was added to the mixture. Contents of 12, 14, 16, and 18% of water from the dry weight of soil were added for every series of the mixture. In this method, mixing is carried out into two stages. Initially, the mixing was done through hand-mixing where the quantity of soil was divided into ten layers and each layer was sprayed with the required amount of additives. Each layer was mixed alone and then put in the pot and then the mixture is mixed again by horizontal cylindrical mixer for at least 3 hours [11]. The dry mixtures in the pot were sprayed with the required amount of water during the mixing process. The type of water used for preparing samples was distilled water. This procedure was found to be the best method to obtain homogeneous samples since homogeneous color was obtained after compaction.

4.2. UCS Test.
The UCS test was used to determine the strength of original soil and evaluate the strength evolution of soil-lime and soil-nanolime mixtures. Strain-controlled machine was used to perform this test and BS 1377-7:1990 was employed. In this test, 5% lime for soil-lime and different content of nanolime were used for soil-nanolime mixtures, that is, 0.2, 0.3, 0.5, 0.8, and 1.0% by weight of dry soil. The specimens were prepared with remolded water contents of 12, 14, 16, and 18%. The Proctor test was conducted directly after mixing process.

A cylindrical mould of diameter of 38 mm was inserted in the compacted soil. Thereafter, the specimen was ejected from this mould using a hydraulic extruder. The length of the specimen was trimmed to twice the diameter. A tolerance of 8% less and 12% more than the standard length is allowed according to BS standard. Plastic wrap was used to keep the cylindrical specimens of unconfined compression test for curing periods of 0, 7, 28, and 56 days.

4.3. X-Ray Fluorescence (XRF) Analysis.
XRF is a nondestructive method for the elemental analysis of solids and liquids. In this technique, the samples were irradiated by an intense x-ray beam causing the emission of fluorescent X-rays. The X-rays emitted were detected by energy dispersive or wavelength dispersive detector. The energies or wavelengths were used to identify the elements present in the samples, whereas concentrations (quantities) were determined by the intensity of the X-rays. XRF analysis was conducted for the native materials, that is, UKM soil, lime, and nanolime, to obtain their chemical compositions and to create a perception about the expected reactants using Bruker S4 Pioneer model. The soil was milled and sieved at 425 μm before XRF analyses were done.

4.4. X-Ray Diffraction (XRD) Analysis.
XRD analysis was conducted using Bruker AXS, Karlsruhe, Germany, model D8 Advanced. The source of X-ray was Cu-Kα with input voltage of 40 kV and current of 40 mA. The wave length and step size were 0.15406 nm and 0.025°/0.1 s, respectively.

XRD is a very reliable and dominant technique for minerals identification in soils and rocks [12]. This analysis was conducted to identify the chemical reactions and development of new products if any. When X-rays interact with a crystalline substance, a definite diffraction pattern is achieved. Every crystalline substance has a specific pattern and the same substance always produces the same pattern and in a mixture

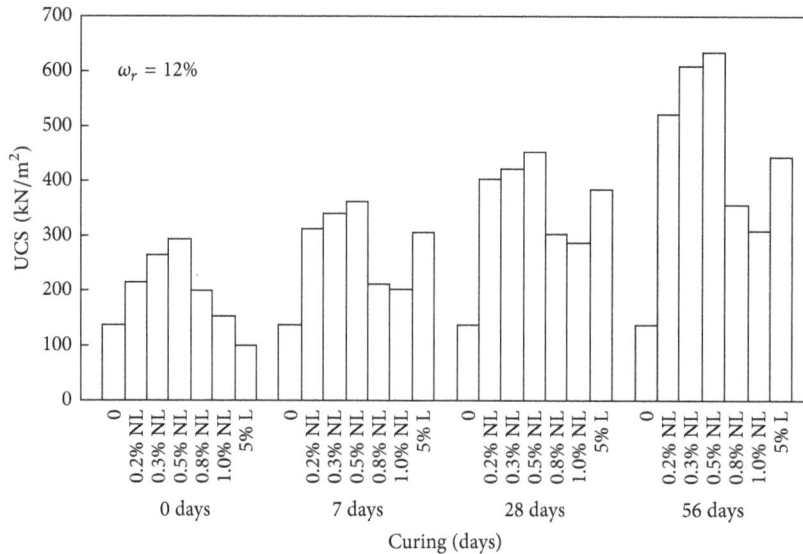

FIGURE 3: UCS development over curing periods for original soil, soil-nanolime mixtures, and soil-lime mixture at remolded water content of 12%.

of substances every substance gives its independent pattern from the others [13]. In this study, XRD analysis was used to determine the crystallite size of lime and nanolime particles using Scherer's formula.

4.5. Field-Emission Scanning Electron Microscope (FESEM). Supra 55VP Zeiss 2008 at CRIM, UKM, for FESEM test was employed to analyze the untreated and the treated soil samples to observe the change in the soil aggregation and formation of any new material. Initially the same percentages of lime and nanolime were hand mixed with water content according to their respective optimum moisture content level. The soil-lime and nanolime mixtures were kept in polyethylene bags and cured at room temperature for 28 days. Then the mixtures were oven dried for 24 hours at temperature of $100 \pm 1°C$ before they were milled and sieved at $425\,\mu m$. FESEM in this study was conducted to disclose the morphology of the native materials and to observe the morphology of the materials after mixing and curing. The change in the soil aggregation and formation of new material can be observed through this analysis. Since the microscope of FESEM uses electron in place of light, the objects that need to be observed by FESEM must be able to be conductive for current. Therefore, all the specimens in this study were coated by a very thin layer of gold. Besides that, mean particle sizes of lime and nanolime were also determined by analyzing the images provided by FESEM analysis.

5. Results and Discussions

5.1. Impact on UCS. Figures 3–6 show the strength gain over curing period for various contents of nanolime and 5% lime, respectively. The results of UCS test for the pure soil showed that there was a high sensitivity towards the remolded water contents (ω_r). Although the UCS for ω_r of ω_{opt} (16%) was $183.67\,kN/m^2$, the strength value started off at $138\,kN/m^2$

and increased to $157\,kN/m^2$ before dropping to $74\,kN/m^2$ at ω_r at 12, 14, and 18%, respectively. A dramatic increase occurred with the initial water content till optimum content of water. However, UCS value decreased drastically at 18% of water content. The reduction in soil strength as water content increase may be related to the effect of soil suction. The soil suction formed meniscus between neighboring particles which creates bonding attractive normal force between them. This is likely to be the case at optimum condition. As extra water is available (the condition of moisture content beyond optimum level), the attractive force by the soil suction induced water to form between soil particles leading to reduction of friction between them.

Moreover, a dramatic increase in soil strength with curing time was realized after mixing the tested soil with nanolime and lime. This strength gain was directly proportional to the curing periods. The results demonstrate that increasing the amount of additive increases the UCS. It can be seen more apparent with the mixture of nanolime. The UCS for mixtures of ω_r of 16% after 1-day curing periods were 184, 393, 405, 485, 325, and $241\,kN/m^2$ for original soil, soil-nanolime mixtures of 0.2, 0.3, 0.5, 0.8, and 1.0%, and soil-lime dose at 5%, respectively. These values of UCS grew, respectively, to 184, 615, 660, 685, 491, 289, and $559\,kN/m^2$ for curing period of 56 days.

In the same context, nanolime induced noticeable increase in UCS of soil-nanolime mixtures; that is, the strength of the mixtures with nanolime was higher than the strength of mixtures with lime. These may be due to higher ability of nanolime to flocculate and agglomerate the soil particles compared with the lime. Agglomeration and flocculation occured with the availability of Ca^{+2} cations. Ions exchange mechanism possibly took place between monovalent ions (Ca^{+2}) with divalent (Na^+, H^+) ions leading to reduction of repulsion forces and bridging the clayey particles of negative surface ions. This may again be related to the reactivity of

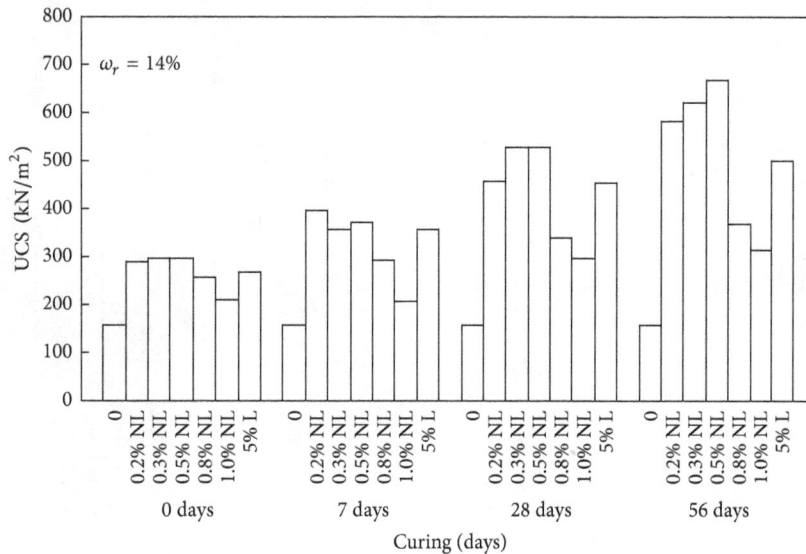

FIGURE 4: UCS development over curing periods for original soil, soil-nanolime mixtures, and soil-lime mixture at remolded water content of 14%.

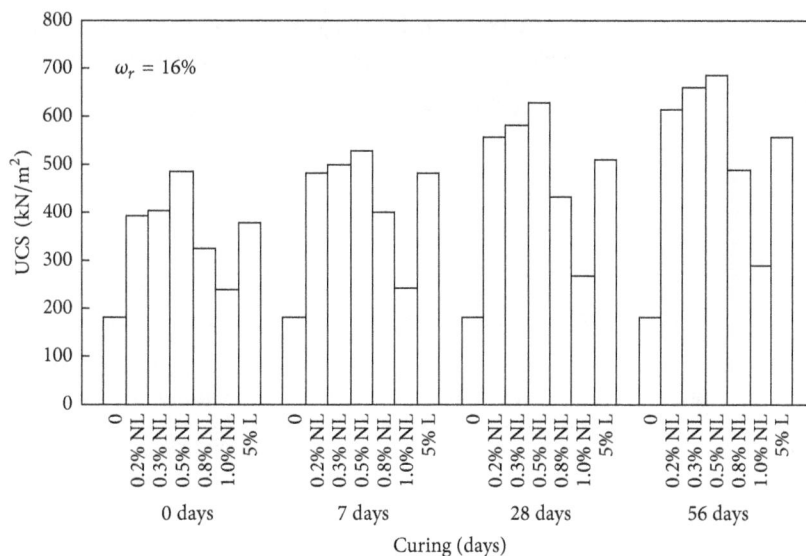

FIGURE 5: UCS development over curing periods for original soil, soil-nanolime mixtures, and soil-lime mixture at remolded water content of 16%.

nanolime and its ability to release more Ca^{+2} ions than lime. This is supported with the purity of nanolime which is higher than the lime as shown in Table 2.

Compared with natural soil, Table 4 presents the percentage of improvement in the UCS of soil-lime mixtures after 56-day curing periods. The improvement was more than 320% for lime mixtures and about 459% for nanolime mixtures depending on the remolded water contents and lime or nanolime contents. This is in comparison with 0, 7, and 14 days of curing period to 56 days of curing period which enhances the strength of the treated soil.

Although the UCS for all the soil-nanolime mixtures reportedly increased with nanolime percentages compared to the natural soil, the value slightly decreases among the nanolime percentages. It was found to decrease with the increase of nanolime from 0.8% to 1.0%. When the content of nanoparticles used is large, nanoparticles are not uniformly dispersed. Therefore, a weak area in the form of voids is created. Consequently, the homogeneous hydrated microstructure cannot be formed and finally the strength will decrease.

Arabani et al. [14] have studied the UCS test of soil cement mixed with different percentages of nanoclay at 7, 14, and 28 days of curing period. The compressive strength of soil cement mixture with 1% ratio of nanoclay was 67% higher than control mixture. The 14 days' and 28 days' compressive strength of nanoclay with soil cement mixture were found to decrease with the increase in nanoparticles ratio from 3

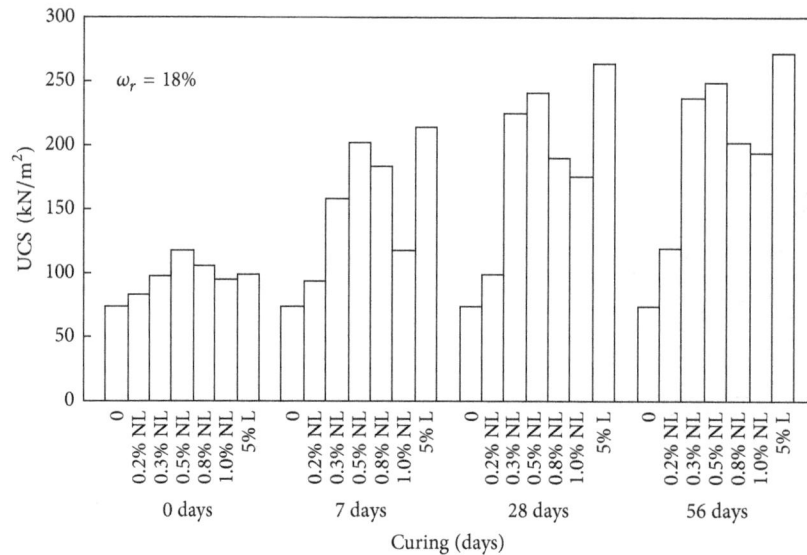

FIGURE 6: UCS development over curing periods for original soil, soil-nanolime mixtures, and soil-lime mixture at remolded water content of 18%.

TABLE 4: Improvement of UCS for soil-nanolime mixtures after 56 days' curing period.

Nanolime/lime (%) ω_r (%)	% increment					
	Soil-nanolime mixtures					Soil-lime mixtures
	0.2	0.3	0.5	0.8	1.0	5.0
12	386	441	459	258	224	322
14	371	396	426	235	200	318
16	334	359	372	267	157	304
18	162	322	336	273	262	327

TABLE 5: UCS (kN/m^2) value from Kassim and Chem [10] study.

Soil	ω_r (%)	Pure soil	Curing period		% increment	
			28 days	56 days	28 days	56 days
Tapah kaolin		24.4	145	275	580	1100
Pelepas marine	Optimum water content	24.5	120	150	480	600
Sg Buloh clay		35.5	80	80	228	228
Jerangau clay		150.1	255	405	170	270
Kulai clay		38.1	50	130	142	371

to 5%. When the content of nanoparticles used was large, nanoparticles were not uniformly dispersed. Therefore, a weak area in the form of voids was created, consequently the homogeneous hydrated microstructure cannot be formed, and finally the strength decreased. Reduction in compressive strength by adding 2% nanoparticles may be due to the excess silica leaching out.

The results of UCS can also be compared with the results of the study conducted by Kassim and Chem [10] in Table 5. In that study, residual acidic soils were collected from different locations in Malaysia. The obtained soils were treated by lime and cured for a different curing period. The results of Kassim and Chem [10] were for soils treated

with optimum lime content and cured for a period of 28 and 56 days. Jerangau clay was mixed with 3.0% lime and comparison was made with 0.5% nanolime and 5% lime mixtures. All samples were compacted at optimum water contents. The UCS of pure soils (i.e., Jerangau clay and studied original soil) was almost identical.

After 28 days and 56 days relatively, soil-nanolime mixtures at 0.2%, 0.3%, and 0.5% exhibit increment in the UCS value (Table 6). These values are higher than the increment by soil-lime mixture at 5%. The UCS increment was at 290% for soil-5% lime mixtures and 342% for soil-0.5% nanolime mixtures after 28 days' curing period. For 56 days' curing period, the UCS increment was at 327% for

TABLE 6: UCS (kN/m^2) value from current study.

Soil	ω_r (%)	Pure soil	Current study							
			Curing Period				% increment			
			28 days		56 days		28 days		56 days	
			0.5% nanolime	5.0% lime	0.5% nanolime	5.0% lime	0.5% nanolime	5.0% lime	0.5% nanolime	5.0% lime
UKM soil	12**	138	453	385	634	445	328	279	459	322
	14**	157	530	455	669	500	338	290	426	318
	16*	184	630	512	685	559	342	278	372	304
	18**	74	242	204	249	272	327	276	336	327

*Optimum water content of untreated soil.
**Remolded water content.

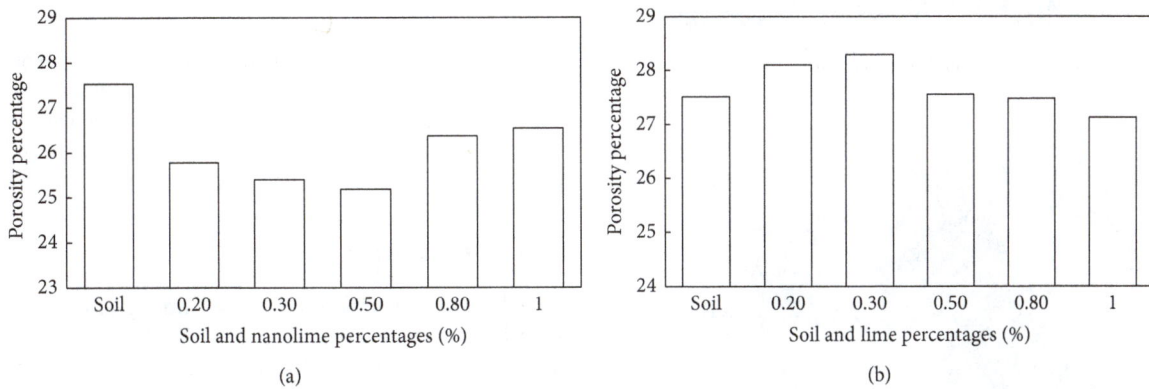

FIGURE 7: (a) Porosity percentage of soil and soil-nanolime mixtures. (b) Porosity percentage of soil and soil-lime mixtures.

TABLE 7: Average particle size and crystallite size of lime and nanolime.

	FESEM test: average particle size	XRD test: crystallite size
Lime	7.20 μm	75.44 nm
Nanolime	21.03 nm	19.36 nm

soil-5% lime mixtures and 459% for soil-0.5% nanolime mixtures.

5.2. Improvement Mechanism Investigation.

The porosity of original soil which is 28% has been reduced with the addition of nanolime which could be seen in the formation of flocculation and agglomeration of the soil particles in the SEM images below. Lime considerably reduced the porosity at 0.8% and 1% of lime addition. Figures 7(a) and 7(b) show the porosity percentages of soil and soil treated with additives: lime and nanolime.

In addition, the lime could fill only the micropores while nanolime could fill the micro- and nanopores. Table 7 shows the average particle size and crystallite size of lime and nanolime which was obtained from FESEM and XRD tests, respectively. The average particle size of lime was 7.20 μm while nanolime was much smaller at 21.03 nm as studied. Through XRD tests, data were calculated using Scherer's formula to find the crystallite size of lime which was 75.44 nm

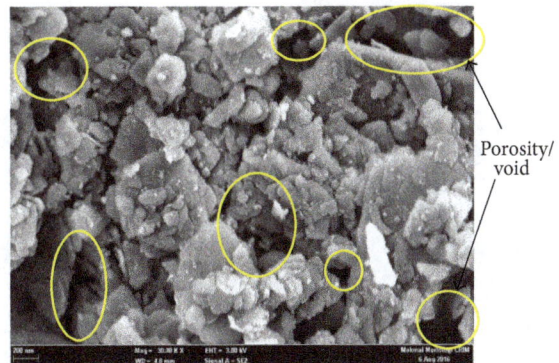

FIGURE 8: FESEM micrograph of original soil: Mag = 30.00 KX.

and that of nanolime was 19.36 nm. This also proved that the nanolime grains were much smaller compared to lime grains. This further supports the materials in nanodimensions which are very small particles with respect to the massive materials. Nanoparticles create positive changes in physical properties because of the increase in ratio of surface area and changes in volume. Due to the large surface area, nanoparticles become more reactive and suitable in improving the properties of clayey soil. This leads to an increase in strength of the soil.

5.3. FESEM Analysis.

The FESEM images of original soil were presented in Figure 8. The voids are seen clearly between the soil particles. Figures 9 and 10 show the micrograph

FIGURE 9: FESEM micrograph of soil-0.5% nanolime: Mag = 30.00 KX.

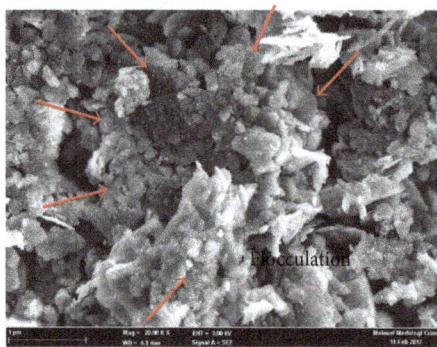

FIGURE 10: FESEM micrograph of soil-0.5% nanolime: Mag = 20.00 KX.

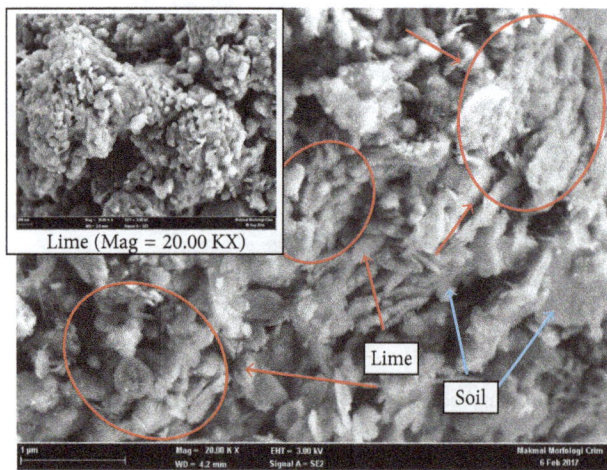

FIGURE 11: FESEM micrograph of soil-1% lime: Mag = 20.00 KX.

of soil mixed with 0.5% nanolime where it looks as a self-assembly or flocculation of gel-like particles, while soil mixed with 1% lime shown in Figures 11 and 12 looks like soft and aggregated particles in chain-like shape. It can be seen that the nanolime particles agglomerated more than that of lime. This is possibly because of van der Waal forces developed

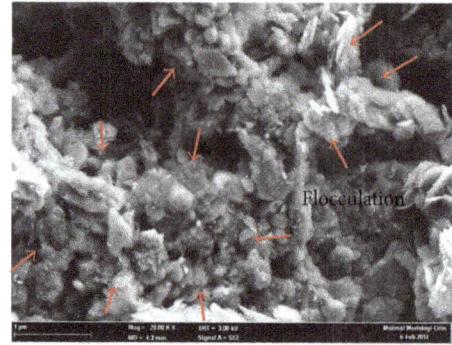

FIGURE 12: FESEM micrograph of soil-5% lime: Mag = 20.00 KX.

between the nanoparticles. This morphology can be observed more clearly in Figure 11 and Figure 12 where the particles of soil-lime seemed to be discrete small particles with voids in contrast to soil-nanolime particles which appeared to be fine particles agglomerated in ball-like structure.

6. Conclusion

Nanolime shows superiority in soil improvement compared with lime even at the dosage of 0.5%. Due to higher ability of nanolime to flocculate and agglomerate, the UCS of treated soil increased significantly over time with increasing percentage of nanolime from 0.2% until 0.5%. The strength gain is inversely proportional to the remolded moisture content and curing period.

Acknowledgments

The authors would like to acknowledge Ministry of Higher Education (MOHE), Malaysia, UKM Geotechnical Laboratory, Pusat Pengurusan Penyelidikan dan Instrumentasi (CRIM), and staffs for their full support and guidance in doing this paper.

References

[1] S. Kazemian and B. B. K. Huat, "Assessment of stabilization methods for soft soils by admixtures," in *Proceedings of the International Conference on Science and Social Research (CSSR '10)*, pp. 118–121, Kuala Lumpur, Malaysia, December 2010.

[2] C. McDowell, "Stabilization of soils with lime, lime-fly ash and other lime reactive materials," *High Research Board Bulletin*, vol. 231, pp. 60–66, 1959.

[3] National Lime Association, *Lime-Treated Soil Construction Manual: Lime Stabilization and Lime Modification*, 2004.

[4] G. H. Hilt and D. Davidson, "Lime fixation in clayey soils," *Highway Research Board Bulletin*, vol. 262, 1960.

[5] M. R. Taha and O. M. E. Taha, "Influence of nano-material on the expansive and shrinkage soil behavior," *Journal of Nanoparticle Research*, vol. 14, no. 10, article 1190, 13 pages, 2012.

[6] P. Holister, J. W. Weener, C. Román, and T. Harper, "Nanoparticles," Technology White Papers 3, 2003.

[7] M. R. Taha, "Geotechnical properties of soil-ball milled soil mixtures," in *Nanotechnology in Construction 3: Proceedings of the NICOM3*, pp. 377–382, Springer, Berlin, Germany, 2009.

[8] F. Uddin, "Clays, nanoclays, and montmorillonite minerals," *Metallurgical and Materials Transactions A: Physical Metallurgy and Materials Science*, vol. 39, no. 12, pp. 2804–2814, 2008.

[9] V. Daniele, G. Taglieri, and R. Quaresima, "The nanolimes in Cultural Heritage conservation: characterisation and analysis of the carbonatation process," *Journal of Cultural Heritage*, vol. 9, no. 3, pp. 294–301, 2008.

[10] K. A. Kassim and K. K. Chem, "Lime stabilized Malaysian cohesive soils," *Jurnal Kejuruteraan Awam*, vol. 16, no. 1, pp. 13–23, 2004.

[11] J. R. Jones, D. J. Parker, and J. Bridgwater, "Axial mixing in a ploughshare mixer," *Powder Technology*, vol. 178, no. 2, pp. 73–86, 2007.

[12] H. Willie and W. Norman, *Methods of Soil Analysis: Part 5. Mineralogical Methods*, Soil Science Society of America, 2007.

[13] A. W. Hull, "A new method of chemical analysis," *Journal of the American Chemical Society*, vol. 41, no. 8, pp. 1168–1175, 1919.

[14] M. Arabani, A. K. Haghi, A. Mohammadzade Sani, and N. Kamboozia, "Use of nanoclay for improvement of the microstructure and mechanical properties of soil stabilized by cement," in *Proceedings of the 4th International Conference on Nanostructures*, Kish Island, Iran, March 2012.

17

Previous Crop and Cultivar Effects on Methane Emissions from Drill-Seeded, Delayed-Flood Rice Grown on a Clay Soil

Alden D. Smartt,[1] Kristofor R. Brye,[1] Christopher W. Rogers,[2] Richard J. Norman,[1] Edward E. Gbur,[3] Jarrod T. Hardke,[4] and Trenton L. Roberts[1]

[1]Department of Crop, Soil, and Environmental Sciences, University of Arkansas, Fayetteville, AR 72701, USA
[2]Department of Plant, Soil, and Entomological Sciences, Aberdeen Research and Extension Center, University of Idaho, Aberdeen, ID 83210, USA
[3]Agricultural Statistics Laboratory, University of Arkansas, Fayetteville, AR 72701, USA
[4]Department of Crop, Soil, and Environmental Sciences, Rice Research and Extension Center, University of Arkansas, Stuttgart, AR 72160, USA

Correspondence should be addressed to Kristofor R. Brye; kbrye@uark.edu

Academic Editor: Amaresh K. Nayak

Due to anaerobic conditions that develop in soils under flooded-rice (*Oryza sativa* L.) production, along with the global extent of rice production, it is estimated that rice cultivation is responsible for 11% of global anthropogenic methane (CH_4) emissions. In order to adequately estimate CH_4 emissions, it is important to include data representing the range of environmental, climatic, and cultural factors occurring in rice production, particularly from Arkansas, the leading rice-producing state in the US, and from clay soils. The objective of this study was to determine the effects of previous crop (i.e., rice or soybean (*Glycine max* L.)) and cultivar (i.e., Cheniere (pure-line, semidwarf), CLXL745 (hybrid), and Taggart (pure-line, standard-stature)) on CH_4 fluxes and emissions from rice grown on a Sharkey clay (very-fine, smectitic, thermic Chromic Epiaquerts) in eastern Arkansas. Rice following rice as a previous crop generally had greater ($p < 0.01$) fluxes than rice following soybean, resulting in growing season emissions ($p < 0.01$) of 19.6 and 7.0 kg CH_4-C ha^{-1}, respectively. The resulting emissions from CLXL745 (10.2 kg CH_4-C ha^{-1}) were less ($p = 0.03$) than those from Cheniere or Taggart (15.5 and 14.2 kg CH_4-C ha^{-1}, resp.), which did not differ. Results of this study indicate that common Arkansas practices, such as growing rice in rotation with soybean and planting hybrid cultivars, may result in reduced CH_4 emissions relative to continuous rice rotations and pure-line cultivars, respectively.

1. Introduction

Agricultural practices around the globe are estimated to account for nearly half of anthropogenic methane (CH_4) emissions, and rice (*Oryza sativa* L.) cultivation is one of the leading agricultural sources of CH_4, accounting for 22% of global anthropogenic agricultural emissions, second only to enteric fermentation [1, 2]. Rice is the only major row crop grown under flooded-soil conditions and the anoxic environment leads to the production and emission of CH_4, a greenhouse gas with a global warming potential (GWP) 25 times stronger than carbon dioxide (CO_2) [3]. The GWP of rice cultivation has been estimated to be 2.7 and 5.7 times greater than that of maize (*Zea mays* L.) and wheat (*Triticum*

aestivum L.) systems, respectively, with CH_4 specifically contributing more than 90% to the GWP of rice systems [4, 5].

Methane production occurs in anaerobic soils as a specific group of *Archaea*, collectively known as methanogens, utilize acetate or hydrogen gas and CO_2, which are formed as fermentation products of a greater consortium of anaerobic bacteria, as substrates for methanogenesis [6]. A portion of the CH_4 produced during methanogenesis, however, is oxidized by a group of aerobic bacteria, known as methanotrophs, as CH_4 moves through oxidized portions of soil surrounding rice roots [7, 8] and near the soil surface [9, 10]. Studies have shown that up to 90% of CH_4 produced in the soil of rice systems is oxidized prior to entering the atmosphere, greatly reducing the proportion of produced CH_4 that is released from the soil [9, 11–15].

Studies have indicated that the majority of CH_4 released from rice fields occurs through the aerenchyma tissues of rice plants, with this plant-mediated transport mechanism accounting for about 90% of emissions, compared to around 8 and 2% of emissions from ebullition and diffusion through the floodwater, respectively [9, 12, 15, 16]. Furthermore, several studies have identified a positive correlation between CH_4 emissions and both aboveground and belowground dry matter accumulation [17–20], which may result from an increase in available substrate as root exudates have been correlated to biomass [21] or due to differences in methane transport capacity (MTC) between cultivars. Butterbach-Bahl et al. [15], for example, attributed a 24 to 31% difference in emissions between two pure-line cultivars to differences in MTC, as no differences were observed between CH_4 production and oxidation. Several studies have observed increased emissions from standard-stature relative to semidwarf cultivars, which is consistent with the positive effect of biomass on CH_4 emissions [22–24]. Cultivar differences extend beyond the impact of biomass on emissions, however, as Ma et al. [25] observed a reduction in emissions and soil CH_4 concentration accompanied by a 67% increase in CH_4 oxidation from a hybrid cultivar relative to pure-line *Indica* and *Japonica* cultivars. Additional studies have observed a similar reduction in emissions from hybrid cultivars [26–28]. With the exception of plant height through its general relationship to aboveground biomass, few other plant morphological characteristics, such as leaf area or photosynthetic activity, have been shown to be related to CH_4 emissions.

While the study of Rogers et al. [26] is the only known study that has directly compared emissions from rice following rice or soybean as previous crops and that observed a 31% reduction in emissions following soybean compared to following rice, several other studies have reported reductions in emissions when previous crop residue was burned [29–31] or when growing rice following a tillage-suppressed fallow period [17, 32]. Furthermore, it has been suggested that promoting aerobic decomposition of residues or growing rice in rotation with upland crops may provide a means of CH_4 mitigation as composted residues reportedly resulted in a sixfold decrease in available substrate for methanogenesis relative to rice straw or green manure [33, 34]. Although it has not been studied greatly, the impact of residue management and rotation with upland crops, such as soybean, has shown potential for mitigation of CH_4 emissions.

Currently, CH_4 emissions budgets in the US are calculated by summing contributions from identifiable homogeneous areas, such that average measured fluxes of representative factors are used in estimates on a regional or national basis [35, 36]. Based on limited studies with emissions ranging from 46 to 375 kg CH_4-C ha^{-1} season^{-1}, the United States Environmental Protection Agency (USEPA) is currently using one emission factor (178 kg CH_4-C ha^{-1} season^{-1}) for all non-California primary rice crops, while separate factors are used when ratooning or for winter-flooded and non-winter-flooded rice in California [37]. As more data become

available, CH_4 budgets and models can be further refined to account for factors such as soil texture, previous crop, and cultivar. This is particularly important for Arkansas, which accounts for nearly 50% of the total US rice production and contains a large portion of production following soybean (71%) and planted with hybrid cultivars (>40%), both of which have been shown to reduce emissions [26] and could be used to create more accurate and potentially lower CH_4 emission factors for midsouthern US rice production [38].

The impact of rice cultivation on greenhouse gas emissions coupled with the intense management involved in rice production allows for potential mitigation strategies based on various practices that are known to reduce CH_4 emissions, such as increasing the use of high-yielding cultivars that have shown potential for reduced emissions. Consequently, it is necessary to study the impacts of various practices on CH_4 emissions in a wide array of soils and climates in order to adequately understand the extent of the problem and to direct management practices toward mitigation of the greenhouse gas, while maintaining high yields and profitability.

While research on CH_4 emissions from rice has recently been conducted in Arkansas [26–28, 39–44], no study has examined the influence of previous crop and cultivar selection on direct-seeded, delayed-flood rice production on a clay soil in the midsouthern US. Direct measurements of CH_4 fluxes and emissions from field studies are necessary to further refine the USEPA emission factors. Therefore, the objective of this study was to assess the impact of previous crop (i.e., rice and soybean) and cultivar (i.e., standard-stature, semidwarf, and hybrid) on CH_4 fluxes and season-long emissions from drill-seeded, delayed-flood rice produced on a clay soil in eastern Arkansas. It was hypothesized that CH_4 fluxes and emissions would be greater when following rice as a previous crop due to the more recalcitrant nature of the rice straw residue, compared to the more labile soybean residue. It was also hypothesized that the hybrid cultivar would result in lower CH_4 fluxes and emissions than the two pure-line cultivars due to increased methanotrophic activity and CH_4 oxidation that has been observed in hybrid cultivars [25]. Furthermore, it was hypothesized that CH_4 fluxes and emissions would be less from the semidwarf cultivar than those from the standard-stature cultivar as was observed by Lindau et al. [22].

2. Materials and Methods

2.1. Site Description. Research was conducted during the 2013 growing season at the University of Arkansas System Division of Agriculture Northeast Research and Extension Center in Keiser, Mississippi County, Arkansas (35°40′N 90°05′W). Field plots were located on a Sharkey clay (very-fine, smectitic, thermic Chromic Epiaquerts), which makes up 31% of the Mississippi County soil survey area [45]. The study site is located within the Southern Mississippi River Alluvium Major Land Resource Area (MLRA 131A), which is located along the Mississippi River from the southern tip of Illinois to the Gulf Coast and is composed of approximately 70% cropland [46]. The location of the study has been cropped in a rice-soybean rotation for more than 25 years and crop residues are typically incorporated in the fall by disking to a depth of 15 cm. Mean annual precipitation at this site is 126 cm, ranging from an average of 6.8 cm in August to an average of 14.1 cm in May [47]. The mean annual air temperature is 15.5°C, while the mean minimum and maximum temperatures occur in January (−2.4°C) and July (33.3°C), respectively [47].

2.2. Treatments and Experimental Design. The purpose of this study was to examine the impacts of previous crop (rice or soybean) and rice cultivar (standard-stature, semidwarf, or hybrid) on CH_4 fluxes and season-long emissions from rice grown on a clay soil in Arkansas. Cultivars were selected in an attempt to represent rice commonly produced in Arkansas with various growth characteristics and breeding lines. The cultivar "Cheniere," developed at Louisiana State University [48], was selected as a pure-line, semidwarf cultivar. Cheniere is an early season, long-grain rice cultivar with an average height of 97 cm and average grain yield of 8.9 Mg ha^{-1} based on Arkansas Rice Performance trials [49]. The standard-stature, pure-line cultivar "Taggart," developed at the University of Arkansas [50], was also selected due to its high yield potential. Taggart is a midseason, long-grain cultivar with an average grain yield of 10.0 Mg ha^{-1} and an average height of 117 cm [49]. The final cultivar selected for use in this study was the hybrid "CLXL745" (RiceTec, Inc., Houston, TX), which is a very early season, long-grain cultivar averaging 114 cm in height and achieving an average yield of 10.1 Mg ha^{-1} in Arkansas [49]. The hybrid cultivar CLXL745 was the most popular cultivar in Arkansas in 2012 and 2013, accounting for 28 and 22% of total production, respectively, in those years [38].

Research plots were 1.6 m wide by 5 m long and arranged in a split-plot design. Previous crop was the whole-plot factor, which was arranged as a randomized complete block with four replicates of each previous crop. The split-plot factor was rice cultivar and each of the three cultivars was randomly located within each of the previous crop, whole-plot units. Therefore, there were a total of 12 field plots per previous crop. Sample date was treated as a repeated measure in analyzing CH_4 flux data.

2.3. Plot Management. Previous crop residues, which were left standing in the field following harvest, were incorporated one week prior to planting by disking to a depth of 15 cm. Research plots were independently seeded on 28 May 2013 with nine rows of rice drill-seeded using 18 cm row spacing. The two pure-line cultivars, Cheniere and Taggart, and the hybrid, CLXL745, were seeded at rates of 112 kg ha^{-1} and 34 kg ha^{-1}, respectively [51]. Levees were constructed following seeding and plots were irrigated with groundwater by flushing as necessary prior to permanent flood establishment, which occurred at the 4–6 leaf stages on 2 July 2013. According to University of Arkansas Cooperative Extension Service (UACES) guidelines [52], nitrogen (N) was applied as urea (46% N) in a split application with the pure-line cultivars and hybrid cultivar receiving 151 kg N ha^{-1} and 168 kg N ha^{-1}, respectively, one day prior to permanent flood establishment.

The second application of N occurred on 30 July 2013 at the beginning of internode elongation for the pure-line cultivars (50 kg N ha^{-1}) and at boot on 20 August 2013 for the hybrid cultivar (33 kg N ha^{-1}), amounting to a total of 201 kg N ha^{-1} for all cultivars [53]. A floodwater depth of 5 to 10 cm was maintained until grain maturity on 23 September 2013, after which the floodwater was released and plots were allowed to dry prior to harvest, which occurred on 24 October 2013. Plots were scouted regularly and managed to remain insect- and weed-free during the growing season according to UACES guidelines [54, 55].

2.4. Soil Sampling and Analyses. Composite soil samples from six, 2 cm diameter soil cores were collected from the top 10 cm of each plot prior to flooding and N fertilization. Composite samples were then oven-dried at 70°C for 48 hours and passed through a 2 mm mesh screen sieve prior to subsamples being analyzed for Mehlich-3 extractable nutrients (i.e., P, K, Ca, Mg, Fe, Mn, Na, S, Cu, and Zn) using inductively coupled plasma atomic emission spectroscopy (Spectro Analytical Instruments, Spectro Arcos ICP, Kleve, Germany) [56]. Additional dried, sieved subsamples were analyzed for total N (TN) and total C (TC) concentrations by high-temperature combustion using a VarioMax CN analyzer (Elementar Americas Inc., Mt. Laurel, NJ) [57] and analyzed for soil pH and electrical conductivity (EC) potentiometrically in a 1:2 (m:v) soil-to-water paste. Soil organic matter (OM) concentration was determined by weight-loss-on-ignition [58].

Additional soil samples were collected from the top 10 cm in each plot prior to flooding using a slide hammer and 4.7 cm diameter core chamber with a beveled core tip. Bulk densities were determined after samples were oven-dried for 48 hours at 70°C. Samples were then ground and sieved through a 2 mm mesh screen and analyzed for particle-size distribution using a modified 12-hour hydrometer method [59]. Bulk densities measured from each plot were then used in combination with measured TN, TC, and OM concentrations to determine total contents (Mg ha^{-1}) of each in the top 10 cm of soil.

2.5. Soil Redox Potential and Soil Temperature Monitoring. Soil oxidation/reduction (redox) potential (Eh) was monitored throughout the flooded portion of the growing season using redox potential sensors (Sensorex, Model S650KD-ORP, Garden Grove, CA) with Ag/AgCl reference solution and a built-in reference electrode installed to a soil depth of 7.5 cm immediately prior to flooding. Additionally, chromel-constantan thermocouples were installed immediately prior to flooding to a soil depth of 7.5 cm in order to monitor soil temperature. Due to equipment limitations, soil Eh and temperature readings were only conducted in two of the four replicates of Cheniere and CLXL745 following each of the previous crops. Soil redox potential and temperature measurements were recorded at 4-hour intervals using a datalogger (CR 1000, Campbell Scientific Inc., Logan, UT) contained within an environmental enclosure. Soil redox potential measurements were corrected to the standard hydrogen electrode by adding 199 mV [60].

2.6. Trace Gas Sampling and Analysis. Non-steady-state, enclosed headspace gas sampling chambers, similar to those used by Rogers et al. [26, 41, 42] and detailed by Livingston and Hutchinson [61], were used for collection of gas samples for CH_4 analysis in this study. This methodology is common in measuring trace gas fluxes [62] and involves installing permanent base collars into the soil and using various sized chamber extensions along with a vented cap in order to accommodate increasing plant growth throughout the season. Base collars, chamber extensions, and chamber caps were constructed using schedule 40 polyvinyl chloride (PVC) pipe with an inside diameter of 30 cm. Chamber base collars were cut to a length of 30 cm with one beveled edge for driving into the soil and four 12.5 mm holes placed 12 cm from the bottom to allow free movement of the floodwater. Chamber extensions cut to lengths of 40 and 60 cm, in order to accommodate growing plants while minimizing chamber headspace volume, and 10 cm caps were covered with reflective aluminum tape (CS Hyde, Mylar metalized tape, Lake Villa, IL) in order to reduce temperature elevation during sampling. Cross sections of tire inner tubes cut to a width of 10 cm were adhered to the bottom of chamber extensions and caps in order to seal the separate pieces together. Chamber caps also included a 15 cm section of 4.5 mm ID copper tubing as a vent to maintain atmospheric pressure, sampling and thermometer ports of gray butyl-rubber septa (Voigt Global, part number 73828A-RB, Lawrence, KS), and a 2.5 cm diameter, battery operated (9 V) fan (Sunon Inc., MagLev, Brea, CA) to mix air within the chamber during CH_4 sampling.

Boardwalks were established between plots prior to flooding in order to access chambers for sampling, while minimizing damage to plants and soil disturbance during sampling. Permanent base collars were installed within each plot to a depth of 11 cm, where the four holes were just above the soil surface, and were situated to contain 40 cm of row length in order to duplicate the plant density of the plots. Plants were carefully bundled with plant tie wire in order to deploy chamber extensions during each sampling event without damaging plants and ties were removed immediately after extension placement as to not affect the plants during sampling. Headspace gas samples were collected weekly for the duration of flooding (i.e., 7, 14, 28, 36, 42, 49, 56, 63, 71, and 77 days after flooding (DAF)), with the exception of the period during the third week after flooding when poor weather conditions did not permit sampling, and every other day following flood release (i.e., 1, 3, 5, and 7 days after flood release (DAFR)).

Chamber headspace gas sampling occurred between 0800 and 1000 hours, similar to previous studies [20, 26, 41, 42, 44], in order to reduce excessive chamber heating during sampling, while sampling during a time of near-average soil temperatures. Samples were collected at 20-minute intervals (i.e., 0, 20, 40, and 60 minutes after sealing) using 20 mL B-D syringes (Becton Dickinson and Co., Franklin Lakes, NJ) and immediately transferred to evacuated 10 mL, crimp-top glass vials (Agilent Technologies, part number 5182-0838, Santa Clara, CA). Chamber air temperature, 10 cm soil temperature, barometric pressure, and relative humidity were recorded throughout each sampling event and chamber volumes were calculated by measuring each chamber's height above the floodwater. Duplicate sets of CH_4 standards (i.e., 1, 2, 5, 10, 20, and 50 $\mu L\,L^{-1}$) were collected in the field into evacuated glass vials and an additional set of laboratory standards was again collected immediately prior to sample analysis in order to ensure that sample integrity was maintained as samples were transported from the field to the laboratory.

Field samples, field standards, and laboratory standards were analyzed within 48 hours after each sampling event using an Agilent 6890-N gas chromatograph with a 30 m-long by 0.53 mm-diameter HP-Plot-Q capillary column (Agilent Technologies, Santa Clara, CA) and equipped with a flame-ionization detector (FID). Methane concentrations of field samples were determined based on calibration curves for each sampling event created from peak-area responses from known sample concentrations. Methane fluxes ($\mu L\,CH_4\,m^{-2}\,min^{-1}$) were calculated for each chamber by using changes in headspace CH_4 concentration ($\mu L\,L^{-1}$, *y*-axis) regressed against time (min, *x*-axis) and multiplying the resulting best-fit line from that regression by chamber volume (L) and dividing by chamber surface area (m^2) as outlined by Parkin and Venterea [62]. Fluxes were then converted to mass-based units (i.e., mg $CH_4\,m^{-2}\,min^{-1}$) using the Ideal Gas Law. Based on details by Parkin et al. [63], minimum detection limits (MDLs) for CH_4 fluxes were calculated to be 0.03, 0.08, 0.11, and 0.16 mg CH_4-C $m^{-2}\,h^{-1}$ with the use of no extension or a 40, 60, or 100 cm extension, respectively. While MDLs were determined, measured fluxes below the MDLs were retained in calculating cumulative season-long emissions and for statistical analyses. Season-long total CH_4 emissions were determined for each chamber by linear interpolation between flux measurement dates.

2.7. Plant Sampling and Analyses. Plant samples were collected at physiological maturity in order to determine any impact of previous crop and cultivar on aboveground dry matter accumulation as well as to compare aboveground dry matter from within and outside the chambers to investigate the impact of the chamber on plant growth. All biomass from within each chamber and a 1 m row of rice from adjacent to each chamber were cut at the soil surface, dried at 60°C until no further moisture loss occurred, and weighed in order to determine total aboveground dry matter accumulation. A 4 m length of the center five rows of each plot was harvested at physiological maturity (24 October 2013) using a plot-scale combine. Grain samples were then weighed and analyzed for moisture content so that final grain yields could be reported at 120 g kg^{-1} grain moisture content.

2.8. Statistical Analyses. Initial soil chemical and physical properties were analyzed by analysis of variance (ANOVA) in SAS 9.2 (SAS Institute, Inc., Cary, NC) using PROC Mixed based on a split-plot design (i.e., the whole-plot factor was previous crop and the split-plot factor was cultivar) in order to determine any differences in soil properties among treatment combinations. Similarly, grain yield was analyzed by ANOVA based on the split-plot design in order to determine the impact of previous crop and cultivar on grain yields. An additional ANOVA was performed based on a split-split-plot design (i.e., the whole-plot factor was previous crop, the first split-plot factor was cultivar, and the second split-plot factor was sampling location) in order to compare total aboveground dry matter accumulation as affected by sampling location (i.e., in-chamber or in-plot), previous crop, and cultivar.

Methane flux data showed no indication of a nonnormal distribution based on an inspection for normality using normal probability plots of the studentized residuals. Consequently, an ANOVA was performed based on a split-plot, repeated measures design (i.e., previous crop was the whole-plot factor, cultivar was the split-plot factor, and time was a repeated measure) to evaluate the impact of previous crop, cultivar, and their interaction on CH_4 fluxes over time. Flux data were analyzed separately for the duration of flooding and following flood release due to differences in CH_4 transport mechanisms and sampling intervals. Seasonal total CH_4 emissions, calculated based on mass-per-area (area-scaled) and mass-per-grain-yield (yield-scaled), as well as post-flood-release emissions, on an area-scaled basis and as a percentage of total seasonal emissions, were analyzed by ANOVA based on a split-plot design (i.e., previous crop was the whole-plot factor and cultivar was the split-plot factor). When appropriate, means were separated at the 0.05 level using the Fisher protected least significant difference (LSD). Linear correlation and regression analyses were performed using Minitab (version 16, Minitab, Inc., State College, PA) in order to evaluate the relationships between sand and clay contents and growing season emissions and aboveground dry matter and growing season emissions.

3. Results and Discussion

3.1. Initial Soil Physical and Chemical Properties. Several initial soil physical and chemical properties in the top 10 cm differed based on previous crop ($p < 0.05$); however, initial soil properties did not differ based on cultivar (Table 1). The most notable differences were in soil particle-size distribution, where sand content was 4% greater and clay content was 4.7% less in the treatments where rice was the previous crop compared to soybean as a previous crop. The difference in particle-size distribution was likely due to natural spatial variability within the alluvial study site. While these differences likely had no agronomic significance, several studies have observed an inverse correlation between soil clay content and CH_4 emissions as well as a positive correlation between soil sand content and CH_4 emissions [32, 64, 65].

In addition to having a slightly greater clay content, the treatment combinations following soybean as a previous crop had greater extractable calcium (Ca), magnesium (Mg), and zinc (Zn), as well as a greater OM concentration (Table 1). The differences in Ca, Mg, and Zn concentrations, however, were minor in comparison to the relatively high concentrations of these nutrients in both previous crop treatments and likely posed no practical significance in this study. There were no differences in OM, TN, or TC contents among treatments (Table 1), indicating that available substrate for methanogenesis was similar among all treatment combinations prior to flooding. Extractable soil phosphorus (P) was

TABLE 1: Mean soil properties ($N = 12$ for each previous crop) prior to flood establishment from Sharkey clay during the 2013 growing season at the University of Arkansas Northeast Research and Extension Center in Keiser, Arkansas.

Soil property	Previous crop	
	Rice	Soybean
pH	7.06[a†]	7.13[a]
Electrical conductivity (dS m^{-1})	0.216[a]	0.202[a]
Sand (g g^{-1})	0.14[a]	0.10[b]
Silt (g g^{-1})	0.34[a]	0.33[a]
Clay (g g^{-1})	0.52[b]	0.57[a]
Bulk density (g cm^{-3})	1.17[a]	1.09[a]
Mehlich-3 extractable nutrients (mg kg^{-1})		
P	55.9[a]	46.3[b]
K	387[a]	390[a]
Ca	4147[b]	4570[a]
Mg	867[b]	919[a]
Fe	467[a]	445[a]
Mn	51.1[a]	53.9[a]
Na	54.5[a]	59.3[a]
S	17.6[a]	12.3[a]
Cu	4.0[a]	4.7[a]
Zn	3.4[b]	3.5[a]
Organic matter (g kg^{-1})	37.7[b]	39.6[a]
Organic matter (Mg ha^{-1})	44.0[a]	43.1[a]
Total N (g kg^{-1})	1.4[a]	1.4[a]
Total N (Mg ha^{-1})	1.6[a]	1.5[a]
Total C (g kg^{-1})	15.0[a]	14.8[a]
Total C (Mg ha^{-1})	17.4[a]	16.1[a]
C : N ratio	10.8[a]	10.5[a]

[†]Values in the same row followed by different letters are significantly different ($p < 0.05$).

greater ($p < 0.05$) following rice as a previous crop and was within the above-optimum level (≥ 51 mg kg^{-1}), while the P concentration following soybean was within the optimum level (36 to 50 mg kg^{-1}), indicating adequate native soil P in both previous crop treatments based on UACES recommendations [53]. Extractable soil potassium (K) was unaffected by previous crop and was within the above-optimum level (≥ 174 mg kg^{-1}). Extractable soil zinc (Zn) was also unaffected by previous crop and was within the medium level (2.6 to 4.0 mg kg^{-1}) recommended for rice, indicating adequate levels of both K and Zn for rice production [53].

3.2. Methane Fluxes from Flooding to Flood Release. Methane fluxes measured during the flooded portion of the 2013 growing season differed between previous crops over time ($p < 0.001$) and differed among cultivars over time ($p < 0.001$) (Table 2). Averaged across cultivar, CH_4 fluxes did not differ among previous crops on the first two or the final sampling dates (i.e., 7, 14, and 77 DAF), while fluxes were greater when the previous crop was rice for the remainder of the sampling dates (Figure 1). Both previous crop treatments exhibited the same trend, where fluxes generally increased from less

TABLE 2: Analysis of variance summary of the effects of previous crop, cultivar, time, and their interaction on methane (CH_4) fluxes from flooding to flood release and following flood release from a clay soil during the 2013 growing season at the Northeast Research and Extension Center in Keiser, Arkansas.

Source of variation	Measurement period	
	Flooding to flood release	Post-flood release
	p	
Previous crop	0.004	0.131
Cultivar	0.027	0.962
Previous crop × cultivar	0.099	0.770
Time	<0.001	0.002
Previous crop × time	<0.001	0.369
Cultivar × time	<0.001	0.270
Previous crop × cultivar × time	0.639	0.537

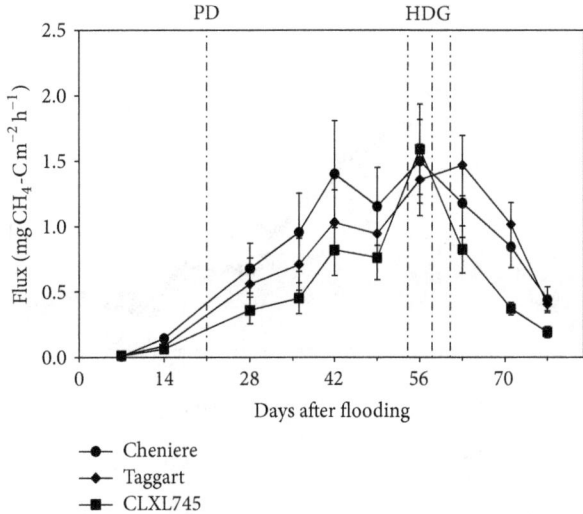

FIGURE 1: Methane fluxes over time throughout the flooded portion of the 2013 growing season from previous crop treatments averaged across cultivar at the Northeast Research and Extension Center in Keiser, Arkansas. The vertical dashed lines represent panicle differentiation (PD) and 50% heading (HDG) dates for CLXL745, Cheniere, and Taggart at 54, 58, and 61 days after flooding, respectively. Flood release occurred in 83 days after flooding. Least significant difference for the same previous crop treatment = 0.197 mg CH_4-C m^{-2} hr^{-1} and for different previous crop treatment = 0.309 mg CH_4-C m^{-2} hr^{-1}. Error bars indicate standard errors for the treatment means ($N = 12$).

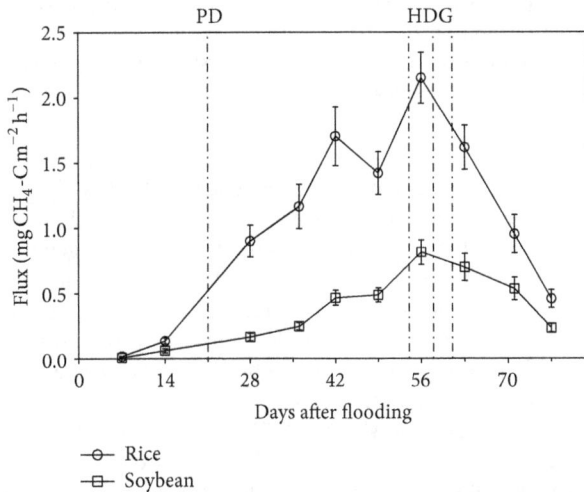

FIGURE 2: Methane fluxes over time throughout the flooded portion of the 2013 growing season from CLXL745, Cheniere, and Taggart averaged across previous crop treatment at the Northeast Research and Extension Center in Keiser, Arkansas. The vertical dashed lines represent panicle differentiation (PD) and 50% heading (HDG) dates for CLXL745, Cheniere, and Taggart at 54, 58, and 61 days after flooding, respectively. Flood release occurred in 83 days after flooding. Least significant difference for the same cultivar = 0.241 mg CH_4-C m^{-2} hr^{-1} and for different cultivars = 0.307 mg CH_4-C m^{-2} hr^{-1}. Error bars indicate standard errors for the treatment means ($N = 8$).

than 0.02 mg CH_4-C m^{-2} h^{-1} at 7 DAF to peaks of 2.15 and 0.81 mg CH_4-C m^{-2} h^{-1} that occurred at 56 DAF in the rice following rice and rice following soybean treatments, respectively. After peak CH_4 fluxes occurred near the time of 50% heading, fluxes decreased over time to 0.46 and 0.23 mg CH_4-C m^{-2} h^{-1} following rice and soybean as previous crops, respectively, at 77 DAF, the final sampling date of the flooded portion of the season. In the only other known study to have directly compared CH_4 fluxes from rice and soybean as previous crops, Rogers et al. [26] observed similar results on

a silt-loam soil, where fluxes were consistently lower when soybean was the previous crop and average peak fluxes of approximately 16.8 and 12.8 mg CH_4-C m^{-2} h^{-1} following rice and soybean, respectively, occurred near 50% heading.

Averaged across previous crop, CH_4 fluxes did not differ among cultivars on the first two sample dates (i.e., 7 and 14 DAF), where all fluxes were less than 0.15 mg CH_4-C m^{-2} h^{-1}, or on the final sample date, where all fluxes were less than 0.44 mg CH_4-C m^{-2} h^{-1} (i.e., 77 DAF) (Figure 2). All three cultivars exhibited a general increase in CH_4 fluxes up to a peak of approximately 1.5 mg CH_4-C m^{-2} h^{-1}, which occurred near 50% heading in all cultivars (i.e., 56 DAF for Cheniere and CLXL745 and 63 DAF for Taggart) and did not differ, followed by a decrease in fluxes up to the time of flood release. Methane fluxes from Cheniere and Taggart only differed at 42 DAF, where fluxes were greater from Cheniere, while the hybrid cultivar, CLXL745, had significantly lower fluxes than Cheniere on four of six sampling dates prior to heading and lower fluxes than Cheniere and Taggart on two of three dates following heading.

Reduced CH_4 fluxes from CLXL745 relative to Cheniere and Taggart were similarly observed by Rogers et al. [26], especially toward the end of the season where fluxes from CLXL745 decreased much more rapidly than those from the pure-line cultivars. Additional studies have also observed reduced fluxes from hybrid relative to pure-line cultivars [25, 27, 28]. Ma et al. [25] measured lower dissolved CH_4 concentrations and increased methanotrophic bacteria in the rhizosphere of hybrid rice accompanied by a 67% increase in

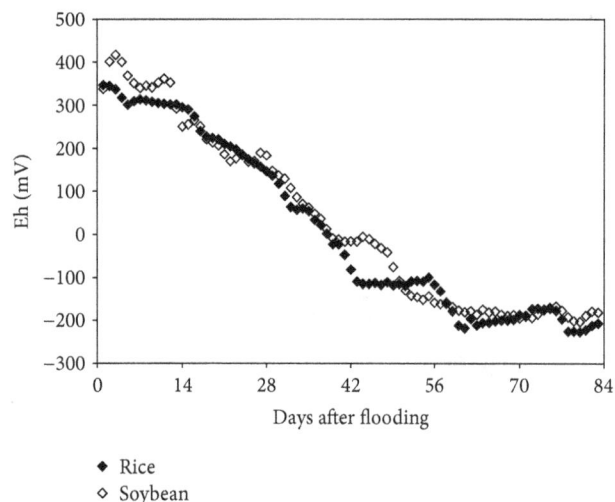

◆ Rice
◇ Soybean

FIGURE 3: Soil oxidation-reduction potential (Eh) at the 7.5 cm depth over the flooded portion of the 2013 growing season for CLXL745 and Cheniere at the Northeast Research and Extension Center in Keiser, Arkansas.

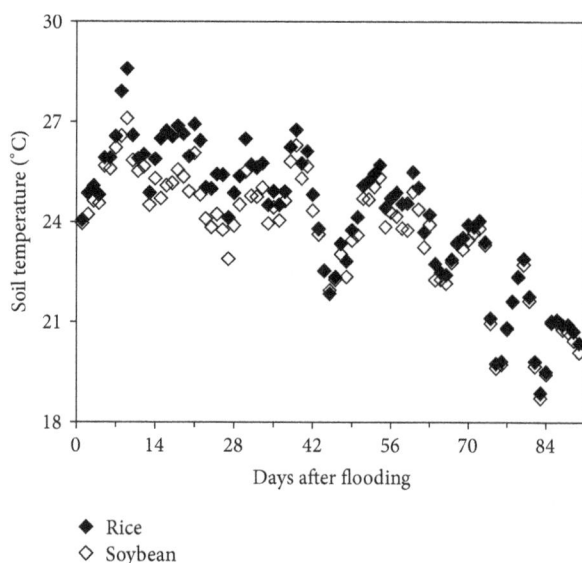

◆ Rice
◇ Soybean

FIGURE 4: Daily mean soil temperature at the 7.5 cm depth over the flooded portion of the 2013 growing season for rice and soybean previous crop treatments at the Northeast Research and Extension Center in Keiser, Arkansas.

CH_4 oxidation potential relative to pure-line cultivars, while all cultivars had similar CH_4 production potentials. This indicates that reduced CH_4 fluxes from hybrid cultivars may be due to differences in microbial community structure, where greater methanotrophic activity reduces CH_4 transport to the atmosphere by oxidizing a greater proportion of the produced CH_4. Similar to this study, Rogers et al. [26] observed only minor differences in CH_4 fluxes between Cheniere and Taggart, while previous studies have reported reduced fluxes from semidwarf relative to standard-stature cultivars [22–24].

While the relative treatment differences and flux trends measured in this study were consistent with previous studies, the magnitude of peak fluxes was on the low end of peaks measured from similar studies conducted on clay and clay-loam soils, which ranged from 2.1 to 25 mg CH_4-C m^{-2} h^{-1} [17, 29–31, 66]. On a silt-loam soil in Arkansas under similar management and methodology to this study, Rogers et al. [26] observed peak fluxes at heading ranging from 8.3 mg CH_4-C m^{-2} h^{-1} for CLXL745 following soybean to 18.7 mg CH_4-C m^{-2} h^{-1} for CLXL745 following rice. The low magnitude of fluxes observed in this study is likely large due to the high clay content, as several studies have indicated greater fluxes from coarse soils than those from fine-textured soils [23, 32, 43]. The low fluxes measured in this study may also be partially explained by soil Eh trends.

Soil Eh did not differ substantially between rice or soybean previous crops (data not shown); however, when averaged across previous crop, soil Eh was consistently lower from Cheniere than CLXL745 until approximately 60 DAF, after which soil Eh did not appear to differ much among cultivars and stabilized near −200 mV (Figure 3). Lower soil Eh prior to heading was indicative of more reduced conditions and greater potential for methanogenesis from Cheniere relative to CLXL745, which is consistent with CH_4 flux observations. The reason for the difference in soil Eh is not well understood but may be related to a difference in

root development and exudation as a degradable C supply for redox reactions or a difference in aerenchyma development or root structure that may result in greater oxygenation in the rhizosphere of CLXL745. Similar soil Eh at heading was consistent with observed fluxes; however, the greater reduction in fluxes from CLXL745 following heading was not reflected in soil Eh measurements but may have been a result of greater methanotrophic activity in the rhizosphere of the hybrid as observed by Ma et al. [25]. The general stabilization of soil Eh following heading may be a result of the decrease in root exudation observed in previous studies, which would limit the supply of degradable C and suppress a further drop in soil Eh as well as causing a reduction in methanogenesis.

In addition to soil Eh, low fluxes measured in this study may have also been partially the result of soil temperature variations. Average daily soil temperatures at the 7.5 cm depth ranged from 22 to 28.6°C for the first 10 weeks of flooding, before dropping as low as 18.7°C in the last few weeks of the growing season, and over the flooded portion of the growing season averaged 24.5 and 23.9°C following rice and soybean, respectively (Figure 4). While the seasonal trend in soil temperature did not appear to drive CH_4 flux trends, the unexpected decrease in fluxes at 49 DAF (Figures 1 and 2) may largely be due to uncharacteristically low temperatures in the week prior to 49 DAF. This period of low soil temperatures may have reduced fluxes at 49 DAF and likely caused a reduction in peak fluxes, which otherwise would have likely followed the preheading trend of continually increasing fluxes, as well as a reduction in fluxes following heading. Beyrouty et al. [67] observed an influence of soil temperature on root growth, suggesting that the low temperatures prior to 49 DAF may have slowed root growth and enhanced the effect of low temperatures reducing fluxes as observed in other studies [68, 69]. Average daily soil temperatures were consistently lower following soybean compared to rice, especially early in the season, as a result of faster canopy development in rice where soybean was the previous crop, which provided greater shading to the soil. The difference in canopy between rice following previous crops lessened over the growing season and had little impact by maturation. Greater soil temperatures following rice may partially explain greater fluxes following rice; however, the differences in temperature were small relative to the difference in fluxes, indicating that fluxes are more strongly influenced by the quantity and quality of previous crop inputs or by differences in microbial communities based on previous cropping system.

3.3. Methane Fluxes following Flood Release.

After the flood was released, CH_4 fluxes differed over time ($p = 0.002$), while previous crop and cultivar had no impact on the magnitude of fluxes (Table 2). Methane fluxes in all treatment combinations were below $0.46\,mg\,CH_4\text{-C}\,m^{-2}\,h^{-1}$ on the last sampling date prior to flood release and did not change substantially by 1 DAFR, averaging $0.29\,mg\,CH_4\text{-C}\,m^{-2}\,h^{-1}$. Averaged across all treatment combinations, post-flood-release CH_4 fluxes were the greatest at 3 DAFR, averaging $0.65\,mg\,CH_4\text{-}C\,m^{-2}\,h^{-1}$, while mean fluxes on the remaining sampling dates did not differ from each other but achieved a minimum value of $0.05\,mg\,CH_4\text{-C}\,m^{-2}\,h^{-1}$ on the last sampling date (7 DAFR). The post-flood-release CH_4 pulse observed in this study has been recorded by numerous other studies [26–28, 39, 42, 44], generally occurring from 3 to 6 DAFR, and is thought to result from the drying of soil macropores which release entrapped CH_4 [70, 71]. Similar to results obtained prior to flood release, CH_4 fluxes measured following flood release were lower than those observed in similar studies on silt-loam soils in Arkansas, which reported post-flood-release peak CH_4 fluxes ranging from 4.5 to $15\,mg\,CH_4\text{-C}\,m^{-2}\,h^{-1}$ [26, 42]. This difference is likely a result of soil textural differences as it has been demonstrated that clayey soils can result in increased CH_4 oxidation due to greater tortuosity and slower diffusion of gases through clay soils [23].

3.4. Aboveground Dry Matter and Grain Yield.

Sampling location (i.e., in-chamber or in-plot) had no effect ($p = 0.845$) on aboveground dry matter measured at the end of the growing season, indicating that the chambers did not adversely affect plant growth. Aboveground dry matter differed, however, between previous crops among cultivars ($p = 0.032$). Averaged across sampling location, aboveground dry matter following rice as the previous crop was greater from CLXL745 than that from Cheniere or Taggart, which did not differ and averaged about 20% lower than CLXL745 (Table 3). Following soybean as a previous crop, aboveground dry matter was also about 20% lower from Cheniere than that from CLXL745 and Taggart, which did not differ.

Similar to aboveground dry matter, grain yield differed between previous crops among cultivars ($p = 0.044$). Grain yields from CLXL745 and Taggart were greater when following soybean than those when following rice, while grain yields from Cheniere did not differ based on previous crop (Table 3). When rice was the previous crop, CLXL745 had a

TABLE 3: Mean aboveground dry matter and yields collected at harvest (24 October, 2013) from Cheniere, CLXL745, and Taggart following previous crops of rice and soybean at the Northeast Research and Extension Center in Keiser, Arkansas.

Plant property/previous crop	Cheniere	CLXL745	Taggart
Aboveground dry matter ($kg\,m^{-2}$)			
Rice	$1.76^{bA\dagger}$	2.17^{aA}	1.77^{bA}
Soybean	1.83^{bA}	2.23^{aA}	2.26^{aA}
Grain yield ($Mg\,ha^{-1}$)			
Rice	$9.3^{bA\dagger}$	9.9^{aB}	9.7^{abB}
Soybean	9.5^{cA}	11.0^{aA}	10.4^{bA}

†Different lowercase letters in the same previous crop for a plant property indicate differences among cultivars and different uppercase letters within a column for a plant property indicate differences between previous crops ($p < 0.05$).

greater grain yield than Cheniere, while Taggart did not differ from either. Similarly, following soybean, grain yields differed in all cultivars, with CLXL745 attaining the greatest yield and Cheniere the lowest. Grain yields achieved in this study, which ranged from $9.3\,Mg\,ha^{-1}$ for Cheniere following rice to $11.0\,Mg\,ha^{-1}$ for CLXL745 following soybean, were similar to 3-year means reported for Arkansas Rice Performance trials (i.e., $8.9\,Mg\,ha^{-1}$ for Cheniere (2010 to 2012), $9.5\,Mg\,ha^{-1}$ for CLXL745 (2011 to 2013), and $10.3\,Mg\,ha^{-1}$ for Taggart (2011 to 2013)) [49, 72].

3.5. Seasonal Methane Emissions.

As expected, based on CH_4 flux measurements, season-long area-scaled CH_4 emissions differed by both previous crop ($p = 0.003$) and cultivar ($p = 0.034$) (Table 4). Averaged across cultivar, area-scaled CH_4 emissions were greater following rice as a previous crop ($19.6\,kg\,CH_4\text{-C}\,ha^{-1}$ season^{-1}) than those following soybean ($7.0\,kg\,CH_4\text{-C}\,ha^{-1}$ season^{-1}) (Table 5). Many studies have reported an increase in CH_4 emissions with increasing additions of rice straw, indicating the importance of residue inputs on emissions [9, 66, 73–75]. Rogers et al. [26] reported a similar trend, although emissions were only reduced by 31% following soybean compared to those following rice, whereas a reduction of 64% was measured in this study. The observations made in this study are consistent with results indicating reduced fluxes resulting from a reduction in residue inputs prior to flooding. Reduced fluxes when following soybean as a previous crop are likely a result of lower residue inputs as well as increased decomposition of the more labile soybean residue under aerobic conditions.

Averaged across previous crop, season-long area-scaled CH_4 emissions from CLXL745 ($10.2\,kg\,CH_4\text{-C}\,ha^{-1}$ season^{-1}) were reduced by 31% relative to Cheniere and Taggart, which did not differ and averaged $14.9\,kg\,CH_4\text{-C}\,ha^{-1}$ season^{-1} (Table 5). The reduction in emissions from CLXL745 is consistent with previous studies that have reported a 37% reduction from CLXL745 relative to Cheniere and Taggart [26], a 25% reduction from CLXL745 relative to Francis and Jupiter [27], and a 30% average reduction from three hybrid cultivars (CLXL729, CLXL745, and XL753), which did not differ, relative to the standard-stature, pure-line cultivar Roy J [28].

TABLE 4: Summary of the effect of previous crop, cultivar, and their interaction on seasonal methane (CH_4) emissions from a clay soil during the 2013 growing season at the Northeast Research and Extension Center in Keiser, Arkansas.

Emissions property	Previous crop	Cultivar	Previous crop × cultivar
		p	
Area-scaled emissions (kg CH_4-C ha^{-1} $season^{-1}$)	0.003	0.034	0.122
Yield-scaled emissions (kg CH_4-C (Mg grain)$^{-1}$)	0.004	0.017	0.111
Postflood emissions (kg CH_4-C ha^{-1})	0.139	0.968	0.781
Postflood emissions (% total emissions)	0.367	0.877	0.841

TABLE 5: Season-long methane (CH_4) emissions as affected by previous crop and rice cultivar expressed on an area- and yield-scaled basis from the 2013 growing season on a clay soil at the Northeast Research and Extension Center in Keiser, Arkansas.

Property/effect	Cheniere	CLXL745	Taggart	Mean
Area-scaled emissions (kg CH_4-C ha^{-1} $season^{-1}$)				
Rice	23.9	14.6	20.4	19.6[A]
Soybean	7.0	5.8	8.1	7.0[B]
Mean	15.5[a†]	10.2[b]	14.2[a]	
Yield-scaled emissions (kg CH_4-C (Mg grain)$^{-1}$)				
Rice	2.59	1.45	2.11	2.05[A]
Soybean	0.74	0.53	0.78	0.68[B]
Mean	1.66[a†]	0.99[b]	1.45[a]	

[†]Different lowercase letters within a row for a measured property indicate differences among cultivars and different uppercase letters within a column for a measured property indicate differences between previous crop treatments ($p < 0.05$).

While little research has focused on determining how emissions are reduced from hybrid cultivars, Ma et al. [25] observed an increase in methanotrophic bacteria and CH_4 oxidation from hybrid rice relative to pure-line cultivars, which is consistent with greater redox potentials observed in the rhizosphere of CLXL745 in this study (Figure 3). Butterbach-Bahl et al. [15] attributed a 24 to 31% reduction in emissions from one cultivar relative to another (both pure-lines) to differences in transport capacity between the cultivars. Aulakh et al. [21] observed a positive correlation between total organic C from root exudates and CH_4 production potential, indicating the potential for cultivar differences in emissions based on variable root exudation rates. While the impacts of variable CH_4 oxidation rates, transport capacities, and root exudation rates are not well understood, evidence has consistently demonstrated a reduction in CH_4 emissions from hybrid cultivars grown in the US, particularly from CLXL745 grown in Arkansas [26–28].

Similar to this study, Rogers et al. [26] observed no difference in area-scaled emissions between Cheniere and Taggart, while several previous studies have reported reduced emissions from semidwarf relative to standard-stature cultivars [22–24]. The difference in fluxes between semidwarf and standard-stature cultivars in previous studies may be due to a positive correlation between plant biomass and C exudation rates from roots [21] or between aboveground dry matter and CH_4 emissions [17, 19, 20, 76]. While a reduction in emissions from semidwarf cultivars is oftentimes linked to reduced dry matter accumulation, this study, as well as Rogers et al.'s [26], observed a reduction in aboveground dry matter that was not accompanied by a reduction in emissions. Cultivars are being developed in the United States for reduced plant height and shorter growth duration, as breeding programs are focusing on the development of semidwarf and hybrid cultivars, but the aboveground morphological characteristics associated with new cultivars (i.e., shorter plants and presumably lower leaf area) may not relate to CH_4 emissions. Furthermore, using data from the Arkansas Rice Performance trials [72] for the cultivars used in this study, neither area- or yield-scaled emissions were correlated ($p > 0.05$) to plant height.

As was the case with area-scaled emissions, yield-scaled emissions varied based on both previous crop ($p = 0.004$) and cultivar ($p = 0.017$, Table 4). Yield-scaled emissions, averaged across cultivar, were reduced by 67% following soybean (0.7 kg CH_4-C (Mg grain)$^{-1}$) compared to those following rice as a previous crop (2.1 kg CH_4-C (Mg grain)$^{-1}$) and, averaged across previous crop, emissions from CLXL745 (1.0 kg CH_4-C (Mg grain)$^{-1}$) were reduced by 36% relative to Cheniere and Taggart, which did not differ and averaged 1.6 kg CH_4-C (Mg grain)$^{-1}$ (Table 5). While the difference in yield-scaled emissions following soybean compared to rice as a previous crop is greater than previously reported from a silt-loam soil (i.e., 31% reduction following soybean), the reduction in emissions from CLXL745 is consistent with an average reduction of 44% relative to Cheniere and Taggart reported by Rogers et al. [26]. Yield-scaled emissions measured in this study, however, were only about 10% of those reported by Rogers et al. [26], which ranged from 11.1 to 20.5 kg CH_4-C (Mg grain)$^{-1}$, indicating a strong suppression of CH_4 emissions from a clay relative to a silt-loam soil under similar management and production practices.

Methane emissions following flood release were unaffected ($p > 0.05$) by previous crop or cultivar both on an area-scaled basis and as a percentage of total emissions (Table 4). Averaged across all treatment combinations, postflood emissions amounted to 0.6 kg CH_4-C ha^{-1}, which was equivalent to an average of 4.5% of total season-long, area-scaled emissions. The proportion of CH_4 emitted following flood release in this study was much less than post-flood-release emissions of 10.5, 13, and 16% from CLXL745, Taggart, and Cheniere, respectively, reported by Rogers et al. [26] from a silt-loam soil, which may be a result of greater CH_4 oxidation in the clay soil reducing the amount of built-up CH_4 that is released upon soil drying. Post-flood-release CH_4 emissions observed in this study, however, were similar to the 5.1% reported by Smartt [40] and the 5.2% reported by Rogers et al. [42]. Additional studies have reported post-flood-release emissions ranging from 7 to 20% of total area-scaled emissions [34, 70, 77]. While the magnitude and fraction of post-flood-release emissions vary, it is apparent that, under certain conditions, CH_4 builds up in the soil and is rapidly released in a pulse as the soil dries and macropores become accessible for gas movement.

Season-long, area-scaled emissions measured in this study (Table 5) only amounted to 4 to 11% of the current USEPA emission factor for non-California, primary rice crops (178 kg CH_4-C ha^{-1}) and were substantially less than the lowest reported emissions used in calculating that factor (i.e., emissions ranged from 46 to 375 kg CH_4-C ha^{-1}), many of which were measured on clay soils in Texas [37]. Similarly, emissions measured in this study were only about 10% of those measured from a similar study on a silt-loam soil in eastern Arkansas [26]. Studies in California, however, have reported emissions of similar magnitudes (i.e., 6.7 to 14 kg CH_4-C ha^{-1}) from a Capay silty clay (48% clay) and from a Clear Lake clay (59% clay, 9.2 to 19 kg CH_4-C ha^{-1}), while also reporting emissions ranging from 58 to 69 kg CH_4-C ha^{-1} from another site with 47% clay [27, 44].

While emissions have been shown to be quite variable, even within studies on clay soils, it is likely that a textural effect is largely the cause for low emissions observed in this study. This is likely due to the impact of increasing clay content causing an increase in tortuosity and a decrease in diffusivity, effectively limiting CH_4 movement out of fine-textured soils [61, 78]. Multiple studies have observed an increase in CH_4 entrapment and decrease in emissions resulting from increasing clay contents [64, 70] and Sass and Fisher [23] attributed the reduction in CH_4 emissions from clay soils to the entrapment and slow movement of CH_4 that allows more CH_4 to be oxidized in aerated zones surrounding roots and at the soil surface. Wang et al. [64] observed varying degrees of CH_4 entrapment, even among soils with similar sand and clay contents, where the greatest entrapment (98.5%) was measured from a Sharkey clay soil compared to 80.6 and 67.8% entrapment from Beaumont clay and Sacramento clay, respectively. These results suggest that more than simple particle-size distribution affects CH_4 emissions and that the low emissions measured in this study likely reflect a large magnitude of CH_4 entrapment and oxidation in the Sharkey clay soil investigated.

Additional evidence suggesting large CH_4 oxidation rates in this study is provided by an examination of the soil Eh and temperature recorded in this study. Soil Eh decreased more slowly, while attaining a similar final Eh, in this study compared to a similar study conducted at the same site in 2012, which reported emissions of 35.6 kg CH_4-C ha^{-1} [40]. Similarly, Rogers et al. [42] and Bossio et al. [30] reported faster decreases and lower Eh values, even reaching as low as −275 mV, accompanied by greater emissions than those observed in this study. While lower soil Eh is likely to result in increased CH_4 production, Kludze et al. [79] also confirmed that a smaller proportion of CH_4 is oxidized by methanotrophs as soil Eh decreases, which supports greater oxidation rates in this study.

4. Conclusions

Emissions measured in this study only amounted to 4 to 11% of the current USEPA emission factor (178 kg CH_4-C ha^{-1}) and were lower than most previous studies. The large reduction in emissions here, relative to other studies, was likely a result of a large degree of CH_4 entrapment in the Sharkey clay resulting in a large degree of CH_4 oxidation by methanotrophs. Low emissions were also likely partially attributable to lower soil temperatures as emissions were reduced substantially from those reported for the same site and management of the previous season. Based on low emissions from clay soils, in combination with reductions when following soybean as a previous crop and from hybrid cultivars, it appears that emissions from rice in the mi-southern US may be less than current estimates.

Competing Interests

The authors declare that there are no competing interests regarding the publication of the paper.

Acknowledgments

The authors would like to acknowledge the Arkansas Rice Research and Promotion Board for supporting this research. Additional appreciation goes to Mike Duren, Donna Frizzell, Chuck Pipkins, Taylor Adams, Douglas Wolf, Anthony Fulford, and Chester Greub for their assistance in the field and laboratory.

References

[1] United States Environmental Protection Agency, *Global Anthropogenic Non-CO_2 Greenhouse Gas Emissions: 1990–2020*, 2006, http://nepis.epa.gov/Adobe/PDF/2000ZL5G.PDF.

[2] P. Smith, D. Martino, Z. Cai et al., "Agriculture," in *Climate Change 2007: The Physical Science Basis*, S. Solomon, D. Qin, M. Manning et al., Eds., Contribution of Working Group I to the Fourth Assessment Report of the Intergovernmental Panel on Climate Change, Cambridge University Press, Cambridge, UK, 2007.

[3] P. Forster, V. Ramaswamy, P. Artaxo et al., "Changes in atmospheric constituents and in radiative forcing," in *Climate Change 2007: The Physical Science Basis*, S. Solomon, D. Qin, M. Manning et al., Eds., Contribution of Working Group I to the Fourth Assessment Report of the Intergovernmental Panel on Climate Change, Cambridge University Press, Cambridge, UK, 2007.

[4] B. Linquist, K. J. van Groenigen, M. A. Adviento-Borbe, C. Pittelkow, and C. Van Kessel, "An agronomic assessment of greenhouse gas emissions from major cereal crops," *Global Change Biology*, vol. 18, no. 1, pp. 194–209, 2011.

[5] B. A. Linquist, M. A. Adviento-Borbe, C. M. Pittelkow, C. van Kessel, and K. J. van Groenigen, "Fertilizer management practices and greenhouse gas emissions from rice systems: a quantitative review and analysis," *Field Crops Research*, vol. 135, pp. 10–21, 2012.

[6] L. Nazaries, J. C. Murrell, P. Millard, L. Baggs, and B. K. Singh, "Methane, microbes and models: fundamental understanding of the soil methane cycle for future predictions," *Environmental Microbiology*, vol. 15, no. 9, pp. 2395–2417, 2013.

[7] W. Armstrong, "Radial oxygen losses from intact rice roots as affected by distance from the apex, respiration and waterlogging," *Physiologia Plantarum*, vol. 25, no. 2, pp. 192–197, 1971.

[8] R. Conrad and F. Rothfuss, "Methane oxidation in the soil surface layer of a flooded rice field and the effect of ammonium," *Biology and Fertility of Soils*, vol. 12, no. 1, pp. 28–32, 1991.

[9] H. Schütz, W. Seiler, and R. Conrad, "Processes involved in formation and emission of methane in rice paddies," *Biogeochemistry*, vol. 7, no. 1, pp. 33–53, 1989.

[10] F. Rothfuss and R. Conrad, "Effect of gas bubbles on the diffusive flux of methane in anoxic paddy soil," *Limnology and Oceanography*, vol. 43, no. 7, pp. 1511–1518, 1998.

[11] A. Holzapfel-Pschorn, R. Conrad, and W. Seiler, "Production, oxidation and emission of methane in rice paddies," *FEMS Microbiology Ecology*, vol. 1, no. 6, pp. 343–351, 1985.

[12] A. Holzapfel-Pschorn, R. Conrad, and W. Seiler, "Effects of vegetation on the emission of methane from submerged paddy soil," *Plant and Soil*, vol. 92, no. 2, pp. 223–233, 1986.

[13] R. L. Sass, F. M. Fisher, P. A. Harcombe, and F. T. Turner, "Methane production and emission in a Texas rice field," *Global Biogeochemical Cycles*, vol. 4, no. 1, pp. 47–68, 1990.

[14] R. L. Sass, F. M. Fisher, Y. B. Wang, F. T. Turner, and M. F. Jund, "Methane emission from rice fields: the effect of floodwater management," *Global Biogeochemical Cycles*, vol. 6, no. 3, pp. 249–262, 1992.

[15] K. Butterbach-Bahl, H. Papen, and H. Rennenberg, "Impact of gas transport through rice cultivars on methane emission from rice paddy fields," *Plant, Cell and Environment*, vol. 20, no. 9, pp. 1175–1183, 1997.

[16] I. Nouchi, S. Mariko, and K. Aoki, "Mechanism of methane transport from the rhizosphere to the atmosphere through rice plants," *Plant Physiology*, vol. 94, no. 1, pp. 59–66, 1990.

[17] R. L. Sass, F. M. Fisher, F. T. Turner, and M. F. Jund, "Methane emission from rice fields as influenced by solar radiation, temperature, and straw incorporation," *Global Biogeochemical Cycles*, vol. 5, no. 4, pp. 335–350, 1991.

[18] G. J. Whiting and J. P. Chanton, "Primary production control of methane emission from wetlands," *Nature*, vol. 364, no. 6440, pp. 794–795, 1993.

[19] Y. Huang, R. L. Sass, and F. M. Fisher, "Methane emission from Texas rice paddy soils. 2. Seasonal contribution of rice biomass production to CH_4 emission," *Global Change Biology*, vol. 3, no. 6, pp. 491–500, 1997.

[20] Q. Shang, X. Yang, C. Gao et al., "Net annual global warming potential and greenhouse gas intensity in Chinese double rice-cropping systems: a 3-year field measurement in long-term fertilizer experiments," *Global Change Biology*, vol. 17, no. 6, pp. 2196–2210, 2011.

[21] M. S. Aulakh, R. Wassmann, C. Bueno, J. Kreuzwieser, and H. Rennenberg, "Characterization of root exudates at different growth stages of ten rice (*Oryza sativa* L.) cultivars," *Plant Biology*, vol. 3, no. 2, pp. 139–148, 2001.

[22] C. W. Lindau, P. K. Bollich, and R. D. DeLaune, "Effect of rice variety on methane emission from Louisiana rice," *Agriculture, Ecosystems & Environment*, vol. 54, no. 1-2, pp. 109–114, 1995.

[23] R. L. Sass and F. M. Fisher Jr., "Methane emissions from rice paddies: a process study summary," *Nutrient Cycling in Agroecosystems*, vol. 49, no. 1-3, pp. 119–127, 1997.

[24] L. K. Sigren, G. T. Byrd, F. M. Fisher, and R. L. Sass, "Comparison of soil acetate concentratons and methane producton, transport, and emission in two rice cultivars," *Global Biogeochemical Cycles*, vol. 11, no. 1, pp. 1–14, 1997.

[25] K. Ma, Q. Qiu, and Y. Lu, "Microbial mechanism for rice variety control on methane emission from rice field soil," *Global Change Biology*, vol. 16, no. 11, pp. 3085–3095, 2010.

[26] C. W. Rogers, K. R. Brye, A. D. Smartt, R. J. Norman, E. E. Gbur, and M. A. Evans-White, "Cultivar and previous crop effects on methane emissions from drill-seeded, delayed-flood rice production on a silt-loam soil," *Soil Science*, vol. 179, pp. 28–36, 2014.

[27] M. B. Simmonds, M. Anders, M. A. Adviento-Borbe, C. van Kessel, A. McClung, and B. A. Linquist, "Seasonal methane and nitrous oxide emissions of several rice cultivars in direct-seeded systems," *Journal of Environmental Quality*, vol. 44, no. 1, pp. 103–114, 2015.

[28] A. D. Smartt, C. W. Rogers, K. R. Brye et al., "Growing-season methane fluxes and emissions from a silt-loam soil as influenced by rice cultivar," in *B.R. Wells Rice Research Studies, 2014*, R. J. Norman and K. A. K. Moldenhauer, Eds., vol. 626 of *Arkansas Agricultural Experiment Station Research Series*, pp. 289–297, Fayetteville, Ark, USA, 2015.

[29] R. J. Cicerone, C. C. Delwiche, S. C. Tyler, and P. R. Zimmerman, "Methane emissions from California rice paddies with varied treatments," *Global Biogeochemical Cycles*, vol. 6, no. 3, pp. 233–248, 1992.

[30] D. A. Bossio, W. R. Horwath, R. G. Mutters, and C. van Kessel, "Methane pool and flux dynamics in a rice field following straw incorporation," *Soil Biology & Biochemistry*, vol. 31, no. 9, pp. 1313–1322, 1999.

[31] G. J. Fitzgerald, K. M. Scow, and J. E. Hill, "Fallow season straw and water management effects on methane emissions in California rice," *Global Biogeochemical Cycles*, vol. 14, no. 3, pp. 767–776, 2000.

[32] R. L. Sass, F. M. Fisher, S. T. Lewis, F. T. Turner, and M. F. Jund, "Methane emissions from rice fields: effect of soil properties," *Global Biogeochemical Cycles*, vol. 8, no. 2, pp. 135–140, 1994.

[33] H. A. C. Denier van der Gon and H. U. Neue, "Influence of organic matter incorporation on the methane emission from a wetland rice field," *Global Biogeochemical Cycles*, vol. 9, no. 1, pp. 11–22, 1995.

[34] K. Yagi, H. Tsuruta, and K. Minami, "Possible options for mitigating methane emission from rice cultivation," *Nutrient Cycling in Agroecosystems*, vol. 49, no. 1-3, pp. 213–220, 1997.

[35] R. L. Sass, F. M. Fisher Jr., A. Ding, and Y. Huang, "Exchange of methane from rice fields: national, regional, and global

budgets," *Journal of Geophysical Research Atmospheres*, vol. 104, no. 21, pp. 26943–26951, 1999.

[36] Intergovernmental Panel on Climate Change, "Cropland," in *2006 IPCC Guidelines for National Greenhouse Gas Inventories*, vol. 4 of *Agriculture, Forestry and Other Land Use*, chapter 5, pp. 5.1–5.66, 2006, http://www.ipcc-nggip.iges.or.jp/public/2006gl/pdf/4_Volume4/V4_05_Ch5_Cropland.pdf.

[37] United States Environmental Protection Agency, *Inventory of U.S. Greenhouse Gas Emissions and Sinks: 1990–2013*, 2015, http://www.epa.gov/climatechange/Downloads/ghgemissions/US-GHG-Inventory-2015-Main-Text.pdf.

[38] J. T. Hardke, "Trends in Arkansas rice production," in *B.R. Wells Rice Research Studies, 2013*, R. J. Norman and K. A. K. Moldenhauer, Eds., vol. 617 of *Arkansas Agricultural Experiment Station Research Series*, pp. 13–23, Fayetteville, Ark, USA, 2014.

[39] A. D. Smartt, K. R. Brye, R. J. Norman, C. W. Rogers, and M. Duren, "Growing-season methane fluxes from direct-seeded, delayed-flood rice produced on a clay soil," in *B.R. Wells Rice Research Studies, 2012*, R. J. Norman and K. A. K. Moldenhauer, Eds., vol. 609 of *Arkansas Agricultural Experiment Station Research Series*, pp. 306–315, 2013.

[40] A. D. Smartt, *Influence of vegetation and chamber size on methane emissions from rice production on a clay soil in Arkansas [M.S. thesis]*, University of Arkansas, Fayetteville, Ark, USA, 2015.

[41] C. W. Rogers, K. R. Brye, R. J. Norman, T. Gasnier, D. Frizzell, and J. Branson, "Methane emissions from a silt-loam soil under direct-seeded, delayed-flood rice management," in *B. R. Wells Rice Research Studies, 2011*, R. J. Norman and K. A. K. Moldenhauer, Eds., vol. 600 of *Arkansas Agricultural Experiment Station Research Series*, pp. 240–247, 2012.

[42] C. W. Rogers, K. R. Brye, R. J. Norman et al., "Methane emissions from drill-seeded, delayed-flood rice production on a silt-loam soil in Arkansas," *Journal of Environmental Quality*, vol. 42, no. 4, pp. 1059–1069, 2013.

[43] K. R. Brye, C. W. Rogers, A. D. Smartt, and R. J. Norman, "Soil texture effects on methane emissions from direct-seeded, delayed-flood rice production in Arkansas," *Soil Science*, vol. 178, no. 10, pp. 519–529, 2013.

[44] M. A. Adviento-Borbe, C. M. Pittelkow, M. Anders et al., "Optimal fertilizer nitrogen rates and yield-scaled global warming potential in drill seeded rice," *Journal of Environmental Quality*, vol. 42, no. 6, pp. 1623–1634, 2013.

[45] Soil Survey Staff, Natural Resources Conservation Service, and United States Department of Agriculture, *Web Soil Survey*, 2012, http://websoilsurvey.sc.egov.usda.gov/App/HomePage.htm.

[46] Natural Resources Conservation Service and United States Department of Agriculture, *Land Resource Regions and Major Land Resource Areas of the United States, the Caribbean, and the Pacific Basin*, United States Department of Agriculture Handbook 296, 2006, http://www.nrcs.usda.gov/Internet/FSE_DOCUMENTS/nrcs142p2_050898.pdf.

[47] National Oceanic and Atmospheric Administration, "Climatography of the United States No. 81: Monthly station normals of temperature, precipitation, and heating and cooling degree days 1971–2000," 2002, http://www.ncdc.noaa.gov/climatenormals/clim81/ARnorm.pdf.

[48] S. D. Linscombe, X. Sha, K. Bearb et al., "Registration of 'Cheniere' rice," *Crop Science*, vol. 46, no. 4, pp. 1814–1815, 2006.

[49] J. T. Hardke, D. L. Frizzell, C. E. Wilson Jr. et al., "Arkansas rice performance trials," in *B.R. Wells Rice Research Studies, 2012*, R.

J. Norman and K. A. K. Moldenhauer, Eds., vol. 609 of *Arkansas Agricultural Eexperiment Station Research Series*, pp. 222–231, Fayetteville, Ark, USA, 2013.

[50] K. A. K. Moldenhauer, J. W. Gibbons, F. N. Lee et al., "Taggart, high yielding large kernel long-grain rice variety," in *B. R. Wells Rice Research Studies, 2008*, R. J. Norman and K. A. K. Moldenhauer, Eds., vol. 571, pp. 68–73, Arkansas Agricultural Experiment Station Research Series, Fayetteville, Ark, USA, 2008.

[51] J. T. Hardke, "RICESEED," 2014, http://riceseed.uaex.edu/Opt-1menu.asp.

[52] J. T. Hardke, Ed., *Arkansas Rice Production Handbook*, University of Arkansas, Division of Agriculture, Cooperative Extension Service MP192, Little Rock, AK, USA, 2013.

[53] R. Norman, N. Slaton, and T. Roberts, "Soil fertility," in *Arkansas Rice Production Handbook*, J. T. Hardke, Ed., pp. 69–102, University of Arkansas, Division of Agriculture, Cooperative Extension Service MP192, Little Rock, AK, USA, 2013.

[54] G. Lorenz and J. T. Hardke, "Insect management in rice," in *Arkansas Rice Production Handbook*, J. T. Hardke, Ed., pp. 139–162, University of Arkansas, Division of Agriculture, Cooperative Extension Service MP192, Little Rock, Ark, USA, 2013.

[55] B. Scott, J. Norsworthy, T. Barber, and J. Hardke, "Rice weed control," in *Arkansas Rice Production Handbook*, J. T. Hardke, Ed., pp. 53–62, University of Arkansas, Division of Agriculture, Cooperative Extension Service MP 192, Little Rock, AK, USA, 2013.

[56] M. R. Tucker, "Determination of phosphorus by Mehlich 3 extraction," in *Soil and Media Diagnostic Procedures for the Southern Region of the United States*, S. J. Donohue, Ed., vol. 374, pp. 6–8, Virginia Agricultural Experiment Station Bullitin, Blacksburg, Va, USA, 1992.

[57] D. W. Nelson and L. E. Sommers, "Total carbon, organic carbon, and organic matter," in *Methods of Soil Analysis. Part 3: Chemical Analysis*, D. L. Sparks, A. L. Page, P. A. Helmke et al., Eds., pp. 961–1010, Soil Science Society of America, Madison, Wis, USA, 3rd edition, 1996.

[58] E. E. Schulte and B. G. Hopkins, "Estimation of organic matter by weight loss-on-ignition," in *Soil Organic Matter: Analysis and Interpretation*, F. R. Magdoff, M. A. Tabatabai, and E. A. Hanlon Jr., Eds., pp. 21–31, Soil Science Society of America Special, Madison, Wis, USA, 1996.

[59] G. W. Gee and D. Or, "Particle-size analysis," in *Methods of Soil Analysis. Part 4: Physical Methods*, J. H. Dane and G. C. Topp, Eds., pp. 255–293, Soil Science Society of America, Madison, Wis, USA, 1st edition, 2002.

[60] W. H. Patrick, R. P. Gambrell, and S. P. Faulkner, "Redox measurements of soil," in *Methods of Soil Analysis. Part 3: Chemical Analysis*, D. L. Sparks, A. L. Page, P. A. Helmke et al., Eds., pp. 1255–1273, Soil Science Society of America, Madison, Wis, USA, 3rd edition, 1996.

[61] G. Livingston and G. Hutchinson, "Enclosure-based measurement of trace gas exchange: applications and sources of error," in *Biogenic Trace Gases: Measuring Emissions from Soil and Water*, P. A. Matson and R. C. Harriss, Eds., pp. 14–51, Blackwell Sciences, Oxford, UK, 1995.

[62] T. Parkin and R. Venterea, "Chamber-based trace gas flux measurements," in *GRACEnet Sampling Protocols*, R. Follett, Ed., 2010, http://www.ars.usda.gov/SP2UserFiles/Program/212/Chapter%203.%20GRACEnet%20Trace%20Gas%20Sampling%20Protocols.pdf.

[63] T. B. Parkin, R. T. Venterea, and S. K. Hargreaves, "Calculating the detection limits of chamber-based soil greenhouse gas flux measurements," *Journal of Environmental Quality*, vol. 41, no. 3, pp. 705–715, 2012.

[64] Z. P. Wang, C. W. Lindau, R. D. Delaune, and W. H. Patrick Jr., "Methane emission and entrapment in flooded rice soils as affected by soil properties," *Biology and Fertility of Soils*, vol. 16, no. 3, pp. 163–168, 1993.

[65] A. Watanabe and M. Kimura, "Influence of chemical properties of soils on methane emission from rice paddies," *Communications in Soil Science and Plant Analysis*, vol. 30, no. 17-18, pp. 2449–2463, 1999.

[66] R. L. Sass, F. M. Fisher, F. T. Turner, and M. F. Jund, "Methane emission from rice fields as influenced by solar radiation, temperature, and straw incorporation," *Global Biogeochemical Cycles*, vol. 5, no. 4, pp. 335–350, 1991.

[67] C. A. Beyrouty, R. J. Norman, B. R. Wells et al., "A decade of rice root characterization studies," in *Arkansas Rice Research Studies 1995*, R. J. Norman and B. R. Wells, Eds., vol. 453, pp. 9–20, Arkansas Agricultural Experiment Station Research Series, Fayetteville, Ark, USA, 1996.

[68] T. Hosono and I. Nouchi, "The dependence of methane transport in rice plants on the root zone temperature," *Plant and Soil*, vol. 191, no. 2, pp. 233–240, 1997.

[69] B. Wang, H. U. Neue, and H. P. Samonte, "The effect of controlled soil temperature on diel CH_4 emission variation," *Chemosphere*, vol. 35, no. 9, pp. 2083–2092, 1997.

[70] H. A. C. Denier van der Gon, N. van Breemen, H.-U. Neue et al., "Release of entrapped methane from wetland rice fields upon soil drying," *Global Biogeochemical Cycles*, vol. 10, no. 1, pp. 1–7, 1996.

[71] H. U. Neue, R. Wassmann, H. K. Kludze, B. Wang, and R. S. Lantin, "Factors and processes controlling methane emissions from rice fields," *Nutrient Cycling in Agroecosystems*, vol. 49, no. 1–3, pp. 111–117, 1997.

[72] J. T. Hardke, D. L. Frizzell, E. Castaneda-Gonzalez et al., "Arkansas rice performance trials," in *B.R. Wells Rice Research Studies, 2013*, R. J. Norman and K. A. K. Moldenhauer, Eds., vol. 617 of *Arkansas Agricultural Experiment Station Research Series*, pp. 265–273, Arkansas Agricultural Experiment Station, Fayetteville, Ark, USA, 2014.

[73] K. Yagi and K. Minami, "Effect of organic matter application on methane emission from some Japanese paddy fields," *Soil Science and Plant Nutrition*, vol. 36, no. 4, pp. 599–610, 1990.

[74] K. F. Bronson, H.-U. Neue, U. Singh, and E. B. Abao Jr., "Automated chamber measurements of methane and nitrous oxide flux in a flooded rice soil: I. Residue, nitrogen, and water management," *Soil Science Society of America Journal*, vol. 61, no. 3, pp. 981–987, 1997.

[75] J. Ma, X. L. Li, H. Xu, Y. Han, Z. C. Cai, and K. Yagi, "Effects of nitrogen fertiliser and wheat straw application on CH_4 and N_2O emissions from a paddy rice field," *Australian Journal of Soil Research*, vol. 45, no. 5, pp. 359–367, 2007.

[76] R. J. Cicerone and J. D. Shetter, "Sources of atmospheric methane: measurements in rice paddies and a discussion," *Journal of Geophysical Research*, vol. 86, no. 8, pp. 7203–7209, 1981.

[77] R. Wassmann, H. U. Neue, R. S. Lantin et al., "Temporal patterns of methane emissions from wetland rice fields treated by different modes of N application," *Journal of Geophysical Research*, vol. 99, no. 8, pp. 16457–16462, 1994.

[78] W. W. Nazaroff, "Radon transport from soil to air," *Reviews of Geophysics*, vol. 30, no. 2, pp. 137–160, 1992.

[79] H. K. Kludze, R. D. DeLaune, and W. H. Patrick, "Aerenchyma formation and methane and oxygen exchange in rice," *Soil Science Society of America Journal*, vol. 57, no. 2, pp. 386–391, 1993.

18

Dissolution of Metals from Biosolid-Treated Soils by Organic Acid Mixtures

Won-Pyo Park,[1] Bon-Jun Koo,[1,2] Andrew C. Chang,[2] Thomas E. Ferko,[1] Jonathan R. Parker,[1] Tracy H. Ward,[1] Stephanie V. Lara,[1] and Chau M. Nguyen[1]

[1]Department of Natural and Mathematical Sciences, California Baptist University, Riverside, CA 92504-3297, USA
[2]Department of Environmental Sciences, University of California, Riverside, CA 92521-0001, USA

Correspondence should be addressed to Bon-Jun Koo; bonjunkoo@calbaptist.edu

Academic Editor: Bernardino Chiaia

Results for the solubilization of metals from biosolid- (BSL-) treated soils by simulated organic acid-based synthetic root exudates (OA mixtures) of differing composition and concentrations are presented. This study used two BSL-treated Romona soils and a BSL-free Romona soil control that were collected from experimental plots of a long-term BSL land application experiment. Results indicate that the solubility of metals in a BSL-treated soil with 0.01 and 0.1 M OA mixtures was significantly higher than that of 0.001 M concentrations. Differences in composition of OAs caused by BSL treatment and the length of growing periods did not affect the solubility of metals. There were no significant differences in organic composition and metals extracted for plants grown at 2, 4, 8, 12, and 16 weeks. The amount of metals extracted tended to decrease with the increase of the pH. Results of metal dissolution kinetics indicate two-stage metal dissolution. A rapid dissolution of metals occurred in the first 15 minutes. For Cd, Cu, Ni, and Zn, approximately 60–70% of the metals were released in the first 15 minutes while the initial releases for Cr and Pb were approximately 30% of the total. It was then followed by a slow but steady release of additional metals over 48 hours.

1. Introduction

Organic acids (OAs) provide attractive options for extracting agents not only because they are biodegradable [1, 2], but also because they are able to extract metal contaminants from soils at mildly acidic conditions (pH 3–5). OAs found in root exudates, such as citric, oxalic, tartaric, and acetic acids, are capable of forming complexes with Cu, Zn, Pb, and Co ions in solutions [3–8] and enhancing their mobilization and uptake by plants [9–12]. Römheld and Awad [13] showed that plants grown on low Fe nutrient mediums in contaminated soil had a higher uptake of Fe, Zn, Ni, and Cd (up to 200%) than control plants in adequate Fe nutrient mediums [14]. Various OAs and/or their salts were tested against two metal chelating agents (EDTA and DTPA) for their potential effects in the remediation of loam and sandy clay loam polluted by heavy metals [15, 16]. Experimental evidence shows that OAs added to humic macromolecules induced the release of adsorbed metal ions [17, 18]. In near neutral and neutral

aqueous solutions, OAs are readily dissociated. The negatively charged OA ligands are capable of forming complexes with metals (e.g., Mn, Fe, and Zn) in solutions, increasing their availability to plants [11, 19, 20].

The extent of complexation depends on the characteristics of the OAs involved (number and proximity of carboxylic groups), their concentrations, types of metal, and the pH of the soil [21]. Organic acids with only one carboxyl group, such as acetic, formic, and lactic acids, have less metal complexing ability than malic and oxalic acids, which are frequently found OAs in soils that have a high affinity for complexing metals [14, 22, 23]. The ability of OAs to complex metals is also dependent on pH [24]. For instance, the complexation of Fe by malic and oxalic acids is highly dependent on soil pH, with little or no complexation at pH > 7.0 [25]. In addition, malic and oxalic acids have the tendency to precipitate in the presence of Ca^{2+}, thus reducing their potential to complex with other metals [26]. When compared with rainwater alone, OAs are able to double or even quadruple mineral dissolution rates. The extent of the chemical reactions, however, is dependent on mineral type, pH, and OA type [22]. Jones and Kochian [23] reported that the presence of OAs increases the dissolution of Fe and Al oxyhydroxides.

OAs may be adsorbed onto the hydroxyaluminum-montmorillonite (HyA-Mt) complex. Cambier and Sposito [27] concluded that the HyA-Mt complex is stable at $4 < \text{pH} < 5.5$, and only external HyA polycations could react with citrate. Janssen et al. [28] noted that citrate did not appear to be adsorbed on the Al-OH groups of the HyA-Mt complex; instead it is adsorbed at the edge of the clay [29]. Sakurai and Huang [30] studied the effect of oxalate on the adsorption of Cd by montmorillonite (Mt) and HyA-Mt complex at pH 5. The reaction was very rapid and virtually completed within 10 minutes. The presence of oxalate markedly interfered with Cd adsorption on clays, especially on the montmorillonite. Taniguchi et al. [31] investigated the adsorption phenomena of Cd on hydroxyaluminosilicate- (HAS-) montmorillonite (Mt) and HyA-Mt complexes as influenced by oxalate and citrate. They concluded that the optimal concentrations of oxalate and citrate for Cd adsorption depended on the form of Cd ions in the solution.

Early research by Eaton [32] showed that plant roots absorbed P and Fe from "water insoluble" P and Fe containing minerals present in the growth medium. Jenny and Overstreet [33] proposed that reactions between root and solid phase minerals that were in direct contact or in close proximity would facilitate plant absorption of sparingly soluble mineral nutrients in soils. Recent studies have demonstrated that root exudates, which contain OAs, play a significant role in the solubilization of metals in soil [5, 11, 34]. In BSL, metals are present almost entirely in solid phases. Upon land application, BSL-borne metals remain largely in their original solid phases in BSL-treated soils [35]. Mobilization of metals by OAs in root exudates would be the most significant pathway through which plant absorption of metals from BSL-treated soils occurs [36]. The total amount of bioavailable metals in BSL-treated soils can be estimated by extracting the soils using OAs found in the rhizosphere of plants grown on BSL-treated medium. It follows that the rate of metal availability for plants would be in proportion to the rate of metal dissolution in the OA mixture [37].

The pH of the rhizosphere is also important in determining metal and nutrient mobilization and uptake. It also affects microbial activity in the vicinity of the root. Root induced pH change in the rhizosphere is a known phenomenon [38, 39] and has an effect on the availability or solubility of nutrients such as P, Fe, Mn, Zn, Cu, and Al [40]. Rhizosphere pH may differ from the bulk soil pH by more than 2 units [41]. Buffer capacities of soil and root activity are the main factors influencing pH at the soil-root interface [42]. Several hypotheses have been postulated to explain the different abilities of plant species in affecting rhizosphere pH: these include differences in root exudation and respiration patterns and differences in cation/anion uptake rates [43]. Since plant roots acquire most mineral nutrients and metals as ions, imbalances between the absorption of cations and anions result in roots' excretion of compensating H^+ or OH^- ions into the soil in rhizosphere, to prevent changes

TABLE 1: Chemical properties of the soil used for the experiment.

Biosolid treatment[†]	pH[‡]	Total concentration (mg kg^{-1})					
		Cd	Cr	Cu	Ni	Pb	Zn
Control	7.7	0.5	37	105	24	25	95
135 Mg ha^{-1}	6.9	11	243	188	88	120	559
1,080 Mg ha^{-1}	6.1	26	596	478	215	396	1466

[†]Obtained at Moreno Field Station of the University of California, Riverside, CA. From 1976 through 1981, composted biosolids were applied at dry weight rates of 0 (control), 22.5, and 180 Mg ha^{-1} yr^{-1}, respectively.
[‡]1 : 1 w/v ratio.

in the electroneutrality of the root tissues [44]. Gollany and Schumacher [45] conducted a growth chamber study to characterize patterns of pH change within the rhizosphere of plants and the pH at different root zones was measured by a microelectrode at 1, 2, 3, and 4 mm distances from the root surface. They reported that the pH decreased to 4.82 and 4.95 in the rhizosphere around elongation and meristematic zones, respectively, compared to the control (pH = 7.6) without plants.

The objectives of this study were the following:

(1) To use an OA mixture as a substitute for actual OAs in root exudates to solubilize metals in BSL-treated soils.

(2) To test the solubilization of metals in BSL-treated soils by the different concentrations of the OA mixture of corn (*Zea mays* L.).

In order to meet these objectives, the researchers needed to assess OA mixture-specific metal solubility and dissolution rate constants of BSL-treated soils.

2. Materials and Methods

2.1. Chemical Properties of the Soils. Two BSL-treated Romona soils and the BSL-free Romona soil control from the field plots of a long-term BSL land application experiment were used [46]. These experiment plots were established in 1976 on a Romona sandy loam soil (fine-loamy, mixed, thermic Typic Haploxeralf) located in the Moreno Field Station of the University of California, Riverside. The Nu-earth BSL used throughout the experiment contained an average of 40, 600, 475, 250, and 3,547 mg kg^{-1} of Cd, Cr, Cu, Ni, and Zn, respectively. From 1976 through 1981, composted BSL were applied at a rate of 0 (control), 22.5, 45, 90, and 180 Mg ha^{-1} yr^{-1} dry weight. The entire experimental fields were cultivated for 10 years (1982–1991) following the termination of BSL application. Soil collected in 1991 from the control, 22.5, and 180 Mg ha^{-1} yr^{-1} treatments was used. Soil samples were air-dried and ground to pass through a 2 mm sieve, homogenized, and stored for subsequent analysis. The chemical properties of the selected soils used in the study are presented in Table 1. For metal determination, aliquots of soil samples were digested in Teflon Parr bombs by a HNO_3 microwave digestion procedure (0.3 g soil with a mixture of 1.0 mL H_2O + 4.5 mL concentrated HNO_3 + 1.5 mL concentrated HCl, in a 120 mL Teflon digestion vessel for 20 minutes and with maximum pressure of 484 kPa) [47].

2.2. Release and Analysis of the Metals. One gram of the soils was mixed with 10 mL of OA mixtures in 50 mL Teflon test tubes. The contents were shaken and allowed to equilibrate at 298°K using a rotary mixer, SA-12 Motor Speed Control (B & B Motor and Control Corp., Long Island City, NY), which rotated the capped test tubes head to tail at approximately 1 rpm for 48 hours. The speed of rotation was maintained constant in all treatments. One mL of chloroform was added to each test tube to control microbial activity and prevent decomposition of OA mixtures during equilibration. The pH and EC of the system in the beginning and at the end of the reaction period were monitored and attempts were made to keep the pH constant. Three OA mixture concentrations 0.001, 0.01, and 0.1 M in 13.5 mM $Ca(NO_3)_2$ along with a 13.5 mM $Ca(NO_3)_2$ blank were tested. Each treatment combination was replicated two times. After equilibration, the soil suspensions were centrifuged for 20 minutes at 8,000 rpm to separate the solution and solid phases. The solution phase was passed through a 0.45 μm filter paper into 25 mL volumetric flasks. The filtrates were acidified with 0.25 mL of concentrated HNO_3. The metal contents of the supernatants were determined using Inductively Coupled Plasma Optical Emission Spectroscopy (ICP-OES) and Atomic Absorption Spectrophotometry (AAS) [47]. The ICP-OES system was OPTIMA 3000 V (Perkin-Elmer, Norwalk, CT, USA) with AS-91 Autosampler and WinLab™ software for the optima family of ICP-OES. Another instrument utilized was the Perkin-Elmer Analyst 800 Atomic Absorption Spectrometer (Perkin-Elmer, Bodenseewerk, Germany) with AS-800 Autosampler. The same experimental setup was used for the batch metal dissolution kinetics. The metal concentrations at 0, 0.25, 0.5, 1, 2, 4, 8, 24, and 48 hours were determined for the kinetic studies.

2.3. Statistical Analysis. All experiments were repeated. Between-group differences were determined by one-way analysis of variance (ANOVA), followed by Student-Newman-Keuls test using a probability level of $P < 0.05$ in all cases. Tests were performed with SigmaStat 4.01 Software.

3. Results and Discussion

3.1. Formulation of Organic Acid Mixtures. Organic acids in rhizosphere are difficult to collect because the volume produced is limited and the components of the root exudates are readily biodegradable. To evaluate the OAs' ability in the rhizosphere to solubilize metals in the soils, a large amount of root exudates must be collected. Because of the difficulty in collecting and preserving root exudates, it is imperative that an OA-based synthetic root exudate (OA mixture) be formulated.

The OA mixture should contain the primary chemical components responsible for metal complex formation and should be in the concentration and pH ranges commonly observed in the rhizosphere. In addition, the OA mixtures should also be prepared under the same background chemical matrix of the soil solution. In this manner, the OA mixture would exhibit comparable ability of reacting with metals as the actual root exudates.

TABLE 2: Amounts of Cd extracted by organic acid (OA) mixtures of various compositions and concentrations[†].

OA concentration	OA composition[‡]	Cd concentration (mg kg^{-1})	
		Biosolid	Standard
0.1 M	2nd	2.11[Ba]	1.86[Ba]
	4th	2.36[Aa]	2.58[Aa]
	8th	1.98[Ba]	1.63[Ba]
	12th	1.79[Ba]	1.60[Ba]
	16th	1.71[Ba]	1.62[Ba]
0.01 M	2nd	0.39[Aa]	0.31[Aa]
	4th	0.41[Aa]	0.43[Aa]
	8th	0.44[Aa]	0.35[Aa]
	12th	0.47[Aa]	0.37[Aa]
	16th	0.51[Aa]	0.43[Aa]
0.001 M	2nd	0.29[Aa]	0.20[Aa]
	4th	0.32[Aa]	0.24[Aa]
	8th	0.36[Aa]	0.32[Aa]
	12th	0.29[Aa]	0.23[Aa]
	16th	0.30[Aa]	0.29[Aa]
13.5 mM $Ca(NO_3)_2$	Not applicable	0.10	

[†]Values represent means of four replicates. The differences of Cd concentrations among the OA compositions of growth periods were tested by one-way ANOVA. In each column of each organic concentration, values followed by the same uppercase letter were not significantly different at $P < 0.05$. The differences of the Cd concentrations between standard (STD) and biosolid (BSL) treatments were tested by Student-Newman-Keuls test. Each pair of Cd concentrations followed by the same lowercase letter was not significantly different at $P < 0.05$.
[‡]Compositions corresponding to OAs recovered from STD and BSL-treated medium of corn at 2nd, 4th, 8th, 12th, and 16th weeks of growth.

To simulate OAs in root exudates, an OA mixture of similar composition to the root exudate composition was formulated. A series of experiments were conducted to test the effect of OA compositions, the total concentration of metals, and the concentration of OAs in the dissolution of metals in the BSL-treated soils. The amount of metals extracted by the OA mixtures, representing compositions of OAs recovered from the rhizosphere of corn grown on BSL-treated medium and the standard (STD) sand medium [34], was illustrated by the Cd extractions summarized in Table 2. The differences of metals extracted due to the compositions representing OAs at 2, 4, 8, 12, and 16 weeks of plant growth were not significant at $P < 0.05$ and the differences due to OAs of STD and BSL-treated medium were not significant at $P < 0.05$. The trends were similar for other metals extracted. The data were then pooled to test the differences in metal extractions due to the concentrations of OA mixtures. They were significant at $P < 0.05$. Based on the results, the OA mixture was formulated.

3.1.1. Composition and Concentration of OAs. Although the OA compositions of the root exudates recovered from plants were grown for various lengths of time from different plant species, they did not vary significantly under our system

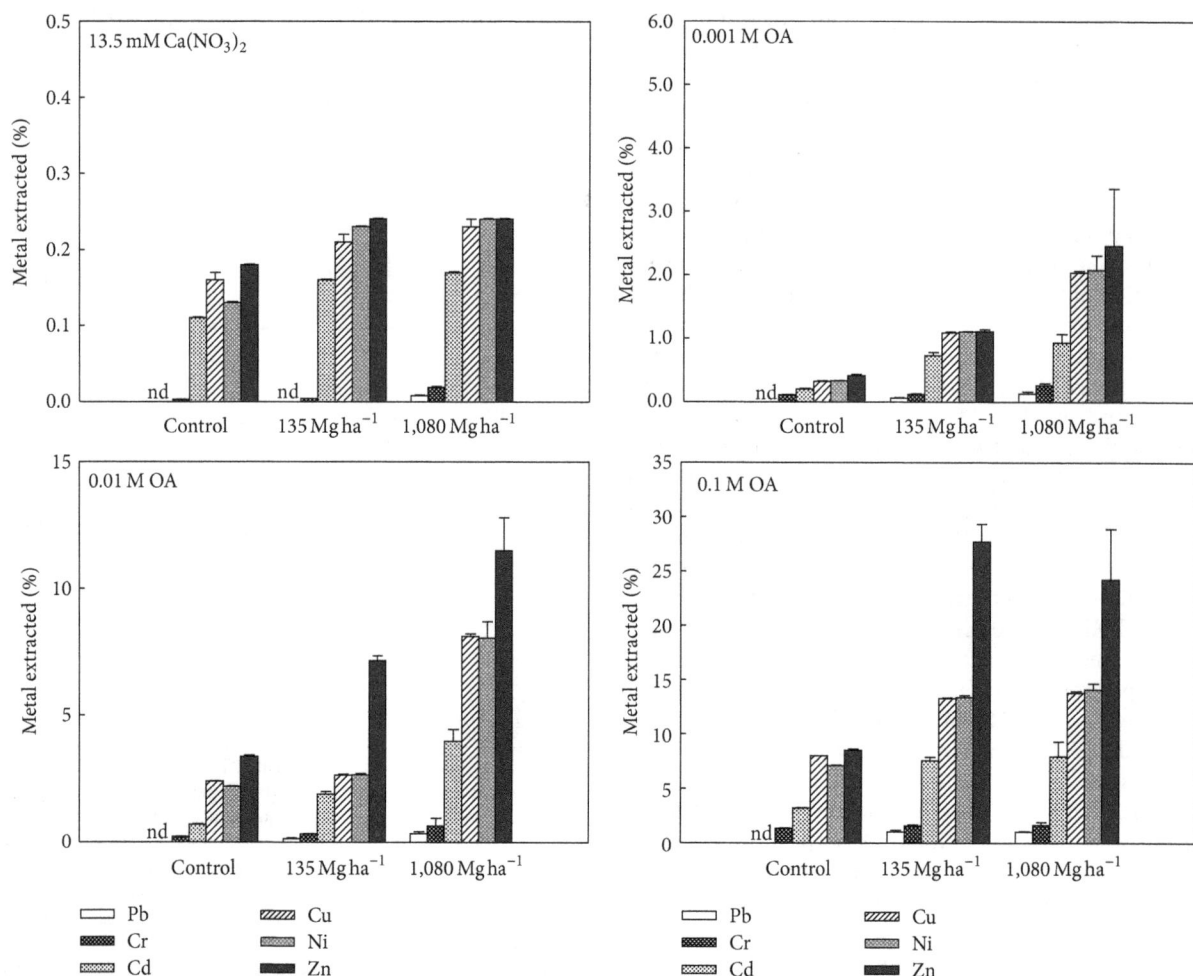

FIGURE 1: Percentages of metal extracted (means $\pm$ SD where $n = 4$) from control and biosolid-treated Romona soil by the organic acid mixtures. Obtained at Moreno Field Station of the University of California, Riverside, CA. From 1976 through 1981, composted biosolids were applied at dry weight rates of 0 (control), 22.5, and 180 $Mg\,ha^{-1}\,yr^{-1}$, respectively. Some of the observations were below detection limits (nd) of the AAS for Pb = 0.001 $mg\,kg^{-1}$.

[34]. The OA mixture composition was taken as the average composition of OAs recovered from the rhizosphere of corn grown on BSL-treated medium for 16 weeks and summarized in Table 3.

Figure 1 summarizes the amounts of metals extracted by OA mixtures at concentrations of 0.001, 0.01, and 0.1 M. The experimental results indicate that the concentrations of the OA mixtures significantly affect the amount of metals extracted. The amount of metals extracted by the OA mixtures was considerably greater than that of the neutral electrolyte solution, 13.5 mM of $Ca(NO_3)_2$, which was used to represent the background chemical matrix of the soil solution. Based on the moisture content of the rhizosphere and the amount of OAs recovered from the rhizosphere [34], the concentrations of OAs in the rhizosphere were estimated to be from 0.001 to 0.013 M (Table 4). As the OAs are closely associated with root exudates, their concentrations are expected to be considerably higher near the root and soil interface and could approach 0.1 M.

TABLE 3: Mole fraction of organic acids collected in root exudates of corn.

Organic acid	Molecular weight	Mole fraction ratio
Acetic	60.05	0.287
Butyric	88.11	0.209
Glutaric	132.12	0.004
Lactic	90.08	0.366
Maleic	116.07	0.042
Oxalic	90.04	0.043
Propionic	74.08	0.010
Pyruvic	88.06	0.0004
Succinic	118.09	0.006
Tartaric	150.09	0.032
Valeric	102.13	0.001

Quantitative measurements of root exudates from plant roots indicate that OA concentrations as high as 50 mM

TABLE 4: Estimated solution concentrations (16-week average) of organic acid mixtures in root exudates of corn[†].

Treatment	Estimated concentration (mM)			
	Standard (control)		Biosolid-treated	
	Mean	SD	Mean	SD
Blank	2.05	0.63	3.41	0.87
Planted	5.23	1.21	12.9	2.04

[†]All experiments performed in four replicates for each 2, 4, 8, 12, and 16 weeks. Values represent means and standard deviation of 20 replicates.

were found within 1 mm from the root surface [53], with typical concentrations in roots at 10–20 mM. For example, in corn, organic solutes present in root cells existed primarily as amino acids (10–20 mM) and sugars (90 mM) [54]. Generally, higher concentrations of OAs are expected in the rhizosphere soil compared to those in the bulk soil [34, 55]. Soil microorganisms do not utilize the carboxylic acids, which play a significant role in complexing metal ions in soils as rapidly as the carbohydrates [3].

3.1.2. pH of the Rhizosphere Soils.

The rhizosphere typically extends 1–5 mm outward from the interface of the root and soil but the pH measurement is complicated. Instead of directly measuring the pH of the rhizosphere, it was deduced and estimated from data found in literature. Studies of the rhizosphere changes in pH along roots of crops and pH changes at different distances from roots are listed in Table 5.

Zhang and Pang [50] demonstrated that while pH was essentially uniform in unvegetated soil, pH was lower near the root tips than at other locations around the root in vegetated soil. The pH varied from 6.2 near the surface (0.5 cm depth) to 4.99 at 6 cm below the soil surface.

It is apparent that pH at or near the root of living plants was altered by root exudation. It varied from pH = 4-5 at the soil-root interface and gradually varied to the level comparable to the pH of bulk soils over approximately 5 mm of distance. The pH = 4.8 was chosen as the pH for the synthetic root exudates based on the results of effects of pH on the dissolution of metals in the BSL-treated soils.

3.2. Effects of pH on the Dissolution of Metals in the Biosolid-Treated Soils.

The pH of the rhizosphere may vary from 4.0 to 8.0 (Table 5) and may affect the solubility of metals in the soil. The effects of pH on the dissolution of metals in BSL-treated soils by the OA mixture were determined. Based on the outcome of this experiment, the pH of the OA mixture was set at 4.8. A BSL-treated Romona soil (135 Mg ha^{-1}) was extracted by the 0.01 M OA mixture and 13.5 mM Ca(NO$_3$)$_2$ that were adjusted to pH = 4.5, 5.5, 6.5, and 7.5.

The amounts of metals extracted tended to decrease with the increase of the pH and, at the same pH level, the OA mixture typically extracted more metals than the 13.5 mM Ca(NO$_3$)$_2$ electrolyte solution (Figure 2). The extent of pH-induced changes in metal solubility was not substantial. Cadmium and zinc experienced by far the biggest decrease, from 0.04 to 0.02 and 2.3 to 1.1 mg kg^{-1}, respectively, when pH of the extracting OA mixture increased from 4.5 to 7.5. The

change in extractable Cd and Zn was larger from pH = 4.5 to 5.5 than from 6.5 to 7.5. Under the same circumstance, the Cr extracted decreased from approximately 0.1 to 0.08 mg kg^{-1}. Statistical tests indicated that metals extracted at pH = 4.5 were not significantly different from the corresponding metals extracted at pH = 5.5 (Figure 2).

In general, the holding capacity of soils for metals increases with increasing pH. Exceptions are Cr and Mo, which are commonly more mobile under alkaline conditions. Accordingly, a decrease in plant uptake of Cu, Mn, and Zn was observed when soil pH was increased [56, 57]. The pH can be issued as the main driving factor of all the factors because it can affect the surface charge of layer silicate clays, OM, and oxides of Fe and Al. In addition to the effect on the sorption of cations, which increases with increasing pH, Fernández-Ramos et al. [58] reported that the adsorption of certain trace metals onto hydrous ferric oxide depends on pH. The results correspond to our findings.

3.3. Metal Solubility by Organic Acid Mixtures.

Metals in the soils were not readily extractable by the neutral electrolyte solution blank that contained 13.5 mM Ca(NO$_3$)$_2$ and, in general, less than 0.25% of any metals were solubilized (Figure 1).

For Pb, the concentrations in the extracts of blank were below limits of quantification of AAS (<0.001 mg kg^{-1}). For the control soil, the total metal concentrations and percent of metals extracted were considerably lower than those in BSL-treated soils.

In BSL-treated soils, the amount of metals extracted was proportional to the amount present in soils and the percentage of the total metals extracted from BSL-treated soils increased with the concentration of OA mixtures; however, the percentage of extraction did not appear to change significantly with the BSL loading when soils were extracted by the 0.1 M OA mixture. In general, Cd, Cu, Ni, and Zn were more readily extractable by the OA mixtures (Figure 1) than Cr and Pb. The least extractable Pb and Cr from the BSL-treated soils follow the general trend reported in the previous paper [34] because of their strong complexation with OM compounds [56] and more specifically the effective role of OM of BSL to serve as electron donors for the reduction of Cr(VI) to Cr(III) [59]. Metals extracted from BSL-treated soils were almost all from BSL. For instance, the amount of Cd extracted from control Romona soil by 13.5 mM Ca(NO$_3$)$_2$ was 0.0008 mg kg^{-1} (Figure 1) and the amount of Cd extracted from BSL-treated soil with the same electrolyte solution blank was 0.023 mg kg^{-1}. We calculated that the amount of Cd extracted from BSL-treated soils is about 97% of the total extracted Cd.

3.4. Metal Dissolution Kinetics of Biosolid-Treated Soils by Organic Acid Mixtures.

The batch equilibrium method was used to study the kinetics of metal dissolution in BSL-treated soils. The experimental procedures were similar to those previously described in the metal solubility study, with the exception that samples were equilibrated for time periods ranging from 15 minutes to 48 hours. When the BSL-treated soils (1,080 Mg ha^{-1}) were equilibrated with 0.1 M OA

TABLE 5: Reported pH ranges of rhizosphere.

Plant	Genotype	pH range	Reference
Barley	Bowman	6.2–7.6	Gollany and Schumacher [45]
	Primus II	6.0–7.8	
	Dorirumugi	4.8–7.1	Youssef and Chino [42]
Corn	Pioneer-3737	5.2–7.6	Gollany and Schumacher [45]
	Pioneer-3732	5.2–7.6	
	CM-37	6.0–7.6	
	—	4.8–6.7	Fischer et al. [48]
Clover	Trikkala	6.2–7.1	Hinsinger and Gilkes [49]
Daisy fleabane	—	5.1–6.3	Zhang and Pang [50]
Nectarine tree	Maxim	5.3–8.2	Tagliavini et al. [36]
Oat	Hytest	6.0–7.6	Gollany and Schumacher [45]
	SD 84104	6.2–7.6	
Rape	—	5.7–6.4	Ruiz and Arvieu [25]
Rye	Standard	5.6–7.1	Hinsinger and Gilkes [49]
Sordan	S-757	6.0–7.6	Gollany and Schumacher [45]
	S-333	6.1–7.6	
Sorghum	SC-33-8-9EY	6.3–7.6	Gollany and Schumacher [45]
	SC-118-15E	6.6–7.6	
Soybean	Hawkeye	5.5–7.1	Römheld and Marschner [51]
		4.8–7.6	Gollany and Schumacher [45]
	PI-54169	5.8–7.6	Gollany and Schumacher [45]
	Toyosuzu	5.1–7.0	Youssef and Chino [42]
	—	4.7–7.1	Riley and Barber [52]

mixture, the amount of BSL-borne metals solubilized in the OA mixtures increased from 1.67 to 2.48, 1.46 to 5.04, 20.3 to 33.4, 16.1 to 24.1, 0.77 to 2.33, and 205 to 282 $mg\,kg^{-1}$ for Cd, Cr, Cu, Ni, Pb, and Zn, respectively, when equilibration time increased from 15 minutes to 48 hours (Tables 6–8).

The percentages of total BSL-borne metals extracted were 13.8, 1.63, 7.95, 14.1, 1.03, and 24.2% for Cd, Cr, Cu, Ni, Pb, and Zn, respectively. If OAs are responsible for converting metals in solid phases into plant available forms, the amount and rate of the metals' solubilization would be indicative of the metals' availability to plants.

The metal dissolution kinetics data was plotted as a fraction of the total dissolved metals (C_t/C_{48}, based on 48 hours' equilibration, C_{48}) that was found at time t (C_t) (Figure 3). The patterns for dissolution in the OA mixtures were essentially the same for all metals. When the BSL-treated soils were equilibrated with the 0.1 M OA mixture, there appeared to be an immediate and rapid release of metals from the soil. The amount of Cd, Cu, Ni, and Zn solubilized in the first 15 minutes ($C_{0.25}$) accounted for approximately 60–75% of the total dissolved metals. For Cr and Pb, approximately 8 hours was needed to dissolve 60–70% of the total soluble metals. After 15 minutes, the dissolution of metals slowly reached the steady state over a period of 48 hours. When the concentrations of the OA mixtures were 0.001 M and 0.01 M, the dissolution behavior of metals in BSL-treated soils

exhibited the same pattern as those of 0.1 M OA mixture. The amount of metals extracted by 0.001 M and 0.01 M OA mixtures was considerably less than the amount extracted by the 0.1 M OA mixture. The percentages of the total soluble metals that entered into the solution phase in the first 15 minutes, however, were essentially identical for 0.001, 0.01, and 0.1 M OA mixtures (Table 9).

A variety of chemical reactions occur in soils and reactions often take place simultaneously. Reaction time may vary from millisecond scale for ion exchange reactions to days (or months or years) for sorption/desorption reactions to reach equilibrium [60]. Metals present in BSL-treated soils may be present in different forms and therefore dissolve at different rates [61]. In the rhizosphere, the biosolids-borne metals are mainly present in solid forms and are not readily available to the growing plants. The dissolution by the root exudates is a significant pathway through which the plant absorbs metals from biosolid-treated soils. We hypothesized that the phytoavailable metals in biosolid-treated soils can be determined by amount of metals dissolved by root exudate derived organic acids in the rhizosphere. The metal update by plants is determined by the kinetics of metal released into solution by organic acids. In this manner, the phytoavailability of biosolid-borne metals may be defined in terms of capacity factor (i.e., organic acids extractable metals in soils), which describes the plant available metal concentration in

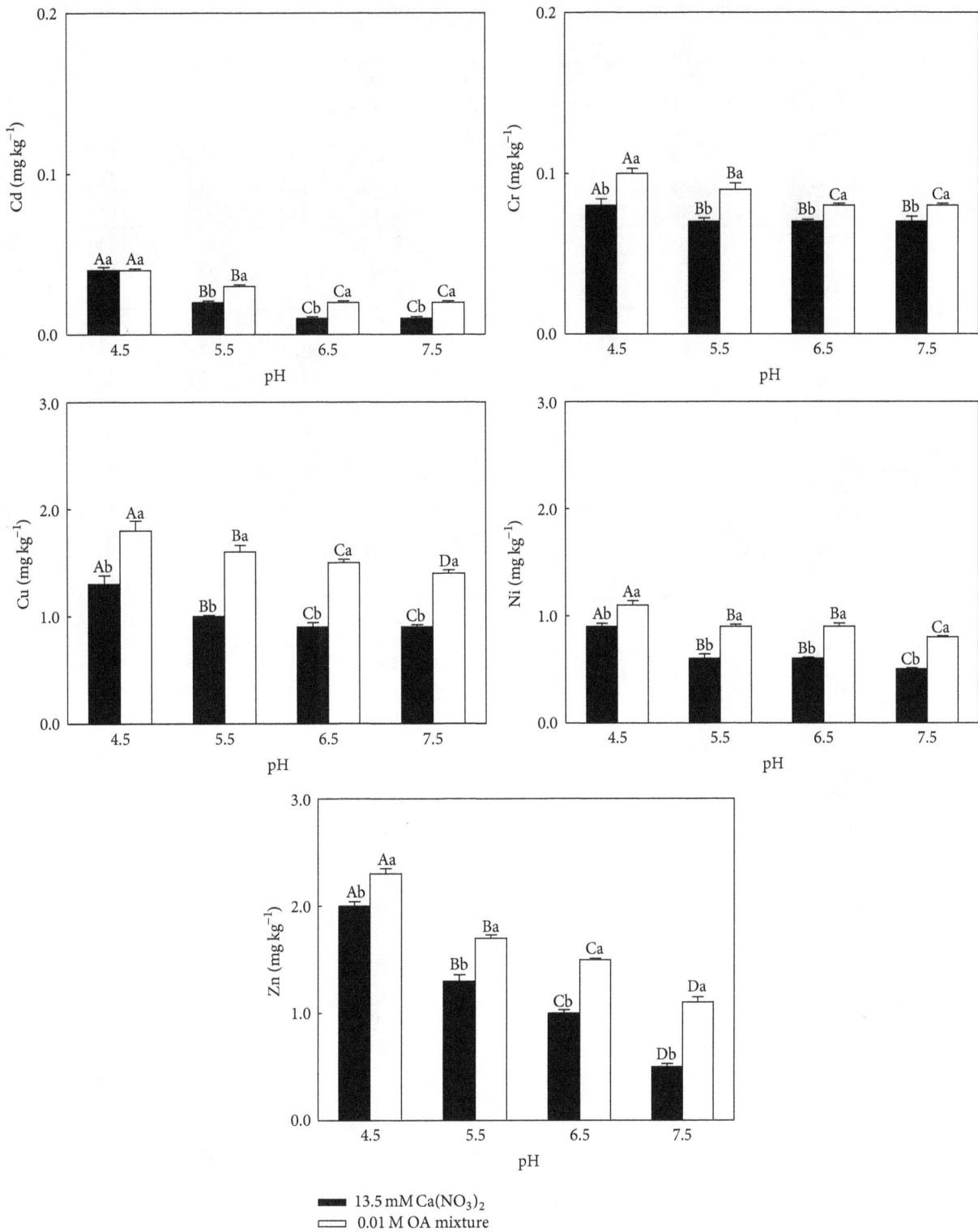

FIGURE 2: Effects of pH on the extraction of metals (four replicates) in biosolid-treated Romona soil (135 Mg ha^{-1}) by 13.5 mM Ca(NO$_3$)$_2$ electrolyte solution and 0.01 M organic acid mixtures. The differences of metal concentrations among the pH values were tested by one-way ANOVA. Values followed by the same uppercase letter were not significantly different at $P < 0.05$. The differences of the metal concentrations between 13.5 mM Ca(NO$_3$)$_2$ and 0.01 M OA mixture were tested by Student-Newman-Keuls test. Values followed by the same lowercase letter were not significantly different at $P < 0.05$.

TABLE 6: Metals release kinetics of biosolid-treated Romona soil (1,080 Mg ha^{-1}) in different concentrations of organic acid (OA) mixtures.

Element	OA concentration	Metal concentration (mg kg^{-1})							
		15 min	30 min	1 hr	2 hrs	4 hrs	8 hrs	24 hrs	48 hrs
Cd	0.001 M	0.31	0.32	0.33	0.35	0.37	0.40	0.46	0.47
	0.01 M	1.16	1.21	1.24	1.28	1.35	1.49	1.73	1.75
	0.1 M	1.67	1.70	1.75	1.82	1.90	2.03	2.40	2.48
Cr	0.001 M	0.34	0.40	0.44	0.49	0.57	0.73	1.09	1.12
	0.01 M	0.85	0.95	1.03	1.16	1.39	1.67	2.49	2.57
	0.1 M	1.46	1.78	1.91	2.17	2.53	3.17	4.79	5.04
Cu	0.001 M	2.41	2.48	2.63	2.78	2.88	3.14	3.81	3.89
	0.01 M	10.7	11.1	11.7	12.1	12.8	13.8	16.8	17.5
	0.1 M	20.3	21.2	22.5	23.7	25.2	26.4	32.1	33.4
Ni	0.001 M	2.53	2.64	2.83	2.94	3.13	3.28	3.72	3.79
	0.01 M	9.80	10.2	10.9	11.4	12.1	12.5	14.5	14.8
	0.1 M	16.1	16.8	18.0	18.8	19.5	20.8	23.3	24.1
Pb	0.001 M	0.13	0.14	0.15	0.18	0.22	0.29	0.42	0.46
	0.01 M	0.34	0.36	0.38	0.45	0.54	0.67	0.99	1.03
	0.1 M	0.77	0.82	0.87	1.02	1.09	1.89	2.01	2.33
Zn	0.001 M	25	26	28	29	31	32	36	38
	0.01 M	104	108	111	115	123	129	145	148
	0.1 M	205	211	220	227	235	245	275	282

TABLE 7: Metals release kinetics of biosolid-treated Romona soil (135 Mg ha^{-1}) in different concentrations of organic acid (OA) mixtures.

Element	OA concentration	Metal concentration (mg kg^{-1})							
		15 min	30 min	1 hr	2 hrs	4 hrs	8 hrs	24 hrs	48 hrs
Cd	0.001 M	0.055	0.059	0.063	0.066	0.068	0.073	0.089	0.091
	0.01 M	0.18	0.19	0.20	0.21	0.23	0.25	0.29	0.30
	0.1 M	0.79	0.83	0.88	0.92	1.01	1.10	1.26	1.29
Cr	0.001 M	0.07	0.08	0.09	0.10	0.13	0.16	0.22	0.23
	0.01 M	0.23	0.27	0.30	0.34	0.38	0.46	0.69	0.71
	0.1 M	0.91	1.07	1.13	1.23	1.47	1.80	2.63	2.68
Cu	0.001 M	0.86	0.88	0.92	0.97	1.05	1.15	1.38	1.42
	0.01 M	2.21	2.35	2.49	2.63	2.80	2.93	3.59	3.66
	0.1 M	7.6	7.8	8.3	8.9	9.4	10.1	12.6	12.7
Ni	0.001 M	0.53	0.55	0.59	0.62	0.66	0.68	0.79	0.81
	0.01 M	1.57	1.67	1.80	1.93	2.01	2.13	2.52	2.57
	0.1 M	6.8	7.3	7.8	8.4	8.9	9.5	11.0	11.3
Pb	0.001 M	0.021	0.023	0.025	0.030	0.035	0.044	0.063	0.065
	0.01 M	0.06	0.07	0.08	0.09	0.11	0.13	0.19	0.20
	0.1 M	0.32	0.34	0.38	0.45	0.53	0.62	0.90	0.94
Zn	0.001 M	3.96	4.19	4.35	4.48	4.69	4.91	5.60	5.66
	0.01 M	21.4	22.6	23.3	24.8	25.7	27.1	30.7	31.0
	0.1 M	100	103	108	114	117	121	134	137

biosolid-treated soil and intensity factor (i.e., the rate at which metals may be dissolved by organic acids), which indicates the rate at which metals will be absorbed by plants. The data presented in Tables 6–8 and Figure 3 may fit the two-site bicontinuum model. Under this conception, there appeared to be two dissolution reactions, a fast reaction, which quickly solubilized this component of the metals, followed by a slow reaction of the remaining components that may continue for a long period of time. These two reactions may occur in series or in parallel. Location and chemical bonding of metals in soil and solution versus metal ratio might be the effects for two pools of metal [62]. The rapid release of metals in the first 15 minutes was fitted into a zero-order kinetics model that

$$C_t = k_0 \times t \quad \text{for } t = 0 \text{ to } t = 0.25, \tag{1}$$

TABLE 8: Metals release kinetics of control Romona soil in different concentrations of organic acid (OA) mixtures.

Element	OA concentration	Metal concentration (mg kg^{-1})							
		15 min	30 min	1 hr	2 hrs	4 hrs	8 hrs	24 hrs	48 hrs
Cd	0.001 M	0.0013	0.0014	0.0015	0.0015	0.0016	0.0017	0.002	0.002
	0.01 M	0.0084	0.0088	0.0091	0.0097	0.0102	0.0115	0.0137	0.014
	0.1 M	0.020	0.021	0.022	0.023	0.025	0.027	0.031	0.032
Cr	0.001 M	0.010	0.012	0.013	0.016	0.019	0.023	0.034	0.035
	0.01 M	0.015	0.018	0.020	0.023	0.028	0.035	0.052	0.053
	0.1 M	0.11	0.13	0.14	0.16	0.20	0.23	0.35	0.36
Cu	0.001 M	0.1071	0.1139	0.1207	0.1275	0.1343	0.1433	0.170	0.177
	0.01 M	0.35	0.36	0.39	0.41	0.44	0.47	0.56	0.58
	0.1 M	1.97	2.03	2.12	2.21	2.34	2.51	3.11	3.17
Ni	0.001 M	0.0603	0.063	0.0675	0.0702	0.0737	0.0773	0.0891	0.090
	0.01 M	0.39	0.42	0.47	0.48	0.51	0.53	0.63	0.64
	0.1 M	1.07	1.12	1.18	1.24	1.33	1.42	1.64	1.67
Pb	0.001 M	n.d.[†]	n.d.[†]	n.d.[†]	n.d.[†]	n.d.[†]	n.d.[†]	n.d.[†]	n.d.[†]
	0.01 M	n.d.[†]	n.d.[†]	n.d.[†]	n.d.[†]	n.d.[†]	n.d.[†]	n.d.[†]	n.d.[†]
	0.1 M	0.014	0.016	0.017	0.019	0.024	0.030	0.042	0.044
Zn	0.001 M	0.30	0.32	0.34	0.35	0.37	0.38	0.44	0.45
	0.01 M	1.95	2.06	2.17	2.26	2.34	2.44	2.77	2.79
	0.1 M	5.3	5.6	6.0	6.2	6.5	6.8	7.8	7.9

[†]Concentration was below detection limits of the AAS for Pb = 0.001 mg kg^{-1}.

where C_t (mg kg^{-1}) is the metal concentration at time t (hour) and k_0 (mg kg^{-1} hr^{-1}) is the zero-order kinetics constant. The remainder portion of the metal dissolution reaction follows a first-order kinetics model that

$$C_t = (0.25 \times k_0) + \left\{ C_s \times \left[1 - e^{-k_1 \times (t-0.25)} \right] \right\}; \quad \text{for } t > 0.25, \quad (2)$$

where C_s represents the ultimate metal release due to the first-order dissolution reaction (mg kg^{-1}) and k_1 is the first-order kinetics constant (hr^{-1}). The kinetics models depicting the metal dissolution reactions are summarized in Table 10. The concentrations of OA mixtures and the amount of BSL-borne metals in the soils determined the metals' dissolution and their availability to plants. For the first-order slow metal release, the rate constants for a metal (k_1) did not vary significantly. k_1 ranged from 0.07 to 0.13, 0.08 to 0.09, 0.07 to 0.09, 0.12 to 0.14, 0.08 to 0.09, and 0.1 to 0.13 for Cd, Cr, Cu, Ni, Pb, and Zn, respectively.

For the zero-order rapid metal release, the metal dissolution of the soil increased with the concentration of OA mixtures and the amount of metals dissolved by the OA mixture of a given concentration increased with the amount of BSL-borne metals in the soils. When the concentrations of OA mixtures varied from 0.001 to 0.1 M, k_0 for a given soil generally increased by 5 to 10 times. At the same OA mixture concentration, k_0 of metals varied by 20–100, 15–60, and 10–60 times for control, soil receiving 135 Mg ha^{-1}, and soil receiving 1,080 Mg ha^{-1} of BSL, respectively.

The total metals that were extractable by the OA mixtures were calculated as the sum of metal rapidly released by the zero-order reaction ($0.25 \times k_0$) and the metal slowly released through the subsequent first-order reaction (C_s). The total

OA extractable metals would indicate the amount of metals possibly available to plants. The amount of metals that were extractable from BSL-treated soil (1,080 Mg ha^{-1}) by 0.1 M OA mixture is presented in Table 11.

Under this metal release model, plants grown on BSL-treated soils would be expected to absorb metals from the rapid release pool first. As the metal release followed a zero-order reaction, one would expect the plant uptake, and therefore tissue concentration, of metals to remain essentially the same until metals in this pool are exhausted, at which time the plant uptake of metals, and therefore the tissue concentration, is expected to decrease as metals in the slow release pool would be available to plants at a much slower rate.

The data presented in Tables 6–8 and Figure 3 also may fit the first-order kinetics model. Under this conception, the slope of the curve approached 0 as the time increased (Table 12). The trend on the cumulative metal extracted with respect to time was fitted into a first-order kinetics model and can be expressed as

$$C_t = C_0 \times \left(1 - e^{-k_1 \times t} \right), \quad (3)$$

where C_t (mg kg^{-1}) is the metal concentration at time t (hour) and k_1 (hr^{-1}) is the first-order kinetics constant and C_0 represents the ultimate metal release due to the first-order dissolution reaction (mg kg^{-1}). The concentrations of OA mixtures and the amount of BSL-borne metals in the soils determined the metals' dissolution and their availability to plants. For the first-order metal release, the rate constants for a metal (k_1) ranged from 3.88 to 6.25, 0.30 to 0.51, 3.81 to 4.88, 3.79 to 5.53, 0.24 to 0.50, and 4.85 to 6.80 for Cd, Cr, Cu, Ni,

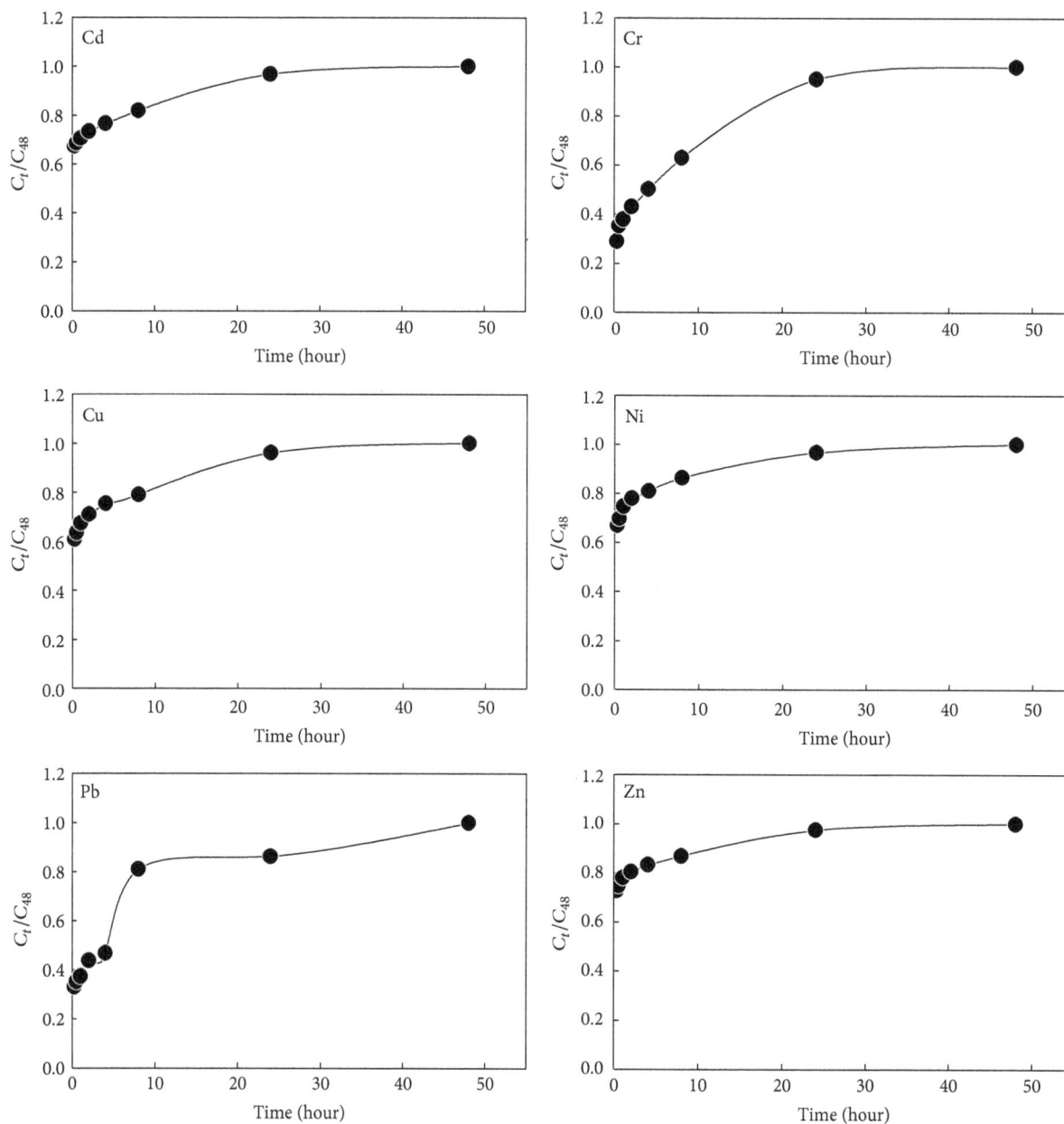

FIGURE 3: Time-dependent Cd, Cr, Cu, Ni, Pb, and Zn dissolutions of the biosolid-treated Romona soil (1,080 Mg ha^{-1}) in 0.1 M organic acid mixtures (C_t denotes concentration at time t and C_{48} denotes concentration at 48 hours).

Pb, and Zn, respectively (Table 13). The total metals (C_0) that were extractable by the OA mixtures were the same as the total released amounts in Table 11. The total OA extractable metals would indicate the amount of metals possibly available to plants.

4. Conclusion

(1) Metals present in BSL-treated soils are more extractable by an OA mixture than indigenous metals of the soil. In BSL-treated soil, more than 90% of metals extracted may be attributed to BSL-borne metals.

(2) In general, the amount of metals extracted decreased with the increase of the pH, and at the same (4.8) pH level, the OA mixture extracted more metals than the 13.5 mM Ca(NO$_3$)$_2$ electrolyte solution.

(3) In general, Cd, Cu, Ni, and Zn were more readily extractable by the OA mixtures and readily absorbed by plants grown on BSL-treated soils than Cr and Pb.

(4) The amount of metals extracted was a function of concentration of OA mixtures. Higher concentrations of OA mixture resulted in greater extraction of metals from the BSL-treated soils.

TABLE 9: Metals of control and biosolid-treated soils solubilized by organic acid (OA) mixtures in the first 15 minutes of equilibration.

Biosolid treatment[†]	Concentration of OA mixtures	Metals dissolved in 15 minutes (% of total dissolved)					
		Cd	Cr	Cu	Ni	Pb	Zn
Control	0.001 M	63	29	61	67	n.d.[‡]	67
	0.01 M	60	28	60	61	n.d.[‡]	70
	0.1 M	63	31	62	64	32	67
135 Mg ha^{-1}	0.001 M	60	30	61	66	32	70
	0.01 M	60	32	60	61	30	69
	0.1 M	62	34	60	60	34	73
1,080 Mg ha^{-1}	0.001 M	66	30	62	67	28	66
	0.01 M	66	33	61	66	33	70
	0.1 M	67	29	61	67	33	73

[†]Obtained at Moreno Field Station of the University of California, Riverside, CA. From 1976 through 1981, composted biosolids were applied at dry weight rates of 0 (control), 22.5, and 180 Mg ha^{-1} yr^{-1}, respectively.
[‡]Concentration was below detection limits of the AAS for Pb = 0.001 mg kg^{-1}.

TABLE 10: Kinetics constant for metal dissolution reaction extracted by organic acid (OA) mixtures according to (1) and (2)[†].

Biosolid treatment[‡] (Mg ha^{-1})	Metal extracted by OAs	Zero- and first-order kinetics constant											
		Cd		Cr		Cu		Ni		Pb		Zn	
		k_0	k_1	k_0	k_1	k_0	k_1	k_0	k_1	k_0	k_1	k_0	k_1
Control	0.001 M	0.01	0.11[a]	0.04	0.09	0.43	0.09	0.24	0.12	n.d.[§]	n.d.[§]	1.2	0.12
	0.01 M	0.03	0.10[a]	0.06	0.09	1.40	0.09	1.56	0.14	n.d.[§]	n.d.[§]	7.8	0.13
	0.1 M	0.08	0.12[a]	0.44	0.08	7.88	0.07	4.28	0.12	0.06	0.09	21.2	0.13
135	0.001 M	0.22	0.09[a]	0.28	0.09	3.44	0.09	2.12	0.12	0.84	0.09	15.8	0.11
	0.01 M	0.72	0.12[a]	0.92	0.08	8.84	0.09	6.28	0.12	0.24	0.09	85.6	0.13
	0.1 M	3.16	0.13[a]	3.64	0.09	30.4	0.08	26.8	0.14	1.28	0.08	400	0.13
1,080	0.001 M	1.24	0.10[a]	1.36	0.09	9.64	0.08	10.1	0.14	0.52	0.08	100	0.12
	0.01 M	4.64	0.10[a]	3.40	0.08	42.8	0.07	39.2	0.12	1.36	0.08	416	0.12
	0.1 M	6.68	0.07[a]	5.84	0.08	81.2	0.08	64.4	0.12	3.08	0.08	820	0.10

[†]Values represent means of two replicates. The differences of first-order kinetics constant for Cd among the treatments were tested by one-way ANOVA. In each column, values followed by the same lowercase letter were not significantly different at $P < 0.05$.
[‡]Obtained at Moreno Field Station of the University of California, Riverside, CA. From 1976 through 1981, composted biosolids were applied at dry weight rates of 0 (control), 22.5, and 180 Mg ha^{-1} yr^{-1}, respectively.
[§]Concentration was below detection limits of the AAS for Pb = 0.001 mg kg^{-1}.

TABLE 11: Amounts of metals that were extractable from the biosolid-treated Romona soil (1,080 Mg ha^{-1}) by 0.1 M organic acid mixture.

Element	Rapid release (mg kg^{-1})	Slow release (mg kg^{-1})	Total release (mg kg^{-1})
Cd	1.67	0.86	2.53
Cr	1.46	3.74	5.20
Cu	20.3	13.7	34.0
Ni	16.1	8.60	24.7
Pb	0.77	1.73	2.50
Zn	205	80.0	285

TABLE 12: The slope[†] of metal extracted from the biosolid-treated Romona soil (1,080 Mg ha^{-1}) by 0.1 M organic acid mixtures according to (3).

Element	k_1	k_2	k_3	k_4	k_5	k_6	k_7	k_8
Cd	6.68	3.40	1.75	0.91	0.48	0.25	0.10	0.05
Cr	5.84	3.56	1.91	1.09	0.63	0.40	0.20	0.11
Cu	81.2	42.4	22.5	11.9	6.30	3.30	1.34	0.70
Ni	64.4	33.6	18.0	9.40	4.88	2.60	0.97	0.50
Pb	3.08	1.64	0.87	0.51	0.27	0.24	0.08	0.05
Zn	820	422	220	113	58.8	30.6	11.5	5.88

[†]Values of $k_1, k_2, k_3, k_4, k_5, k_6, k_7$ or k_8 indicate the slopes of metal extracted at 15 min., 30 min., 1 hr, 2 hrs., 4 hrs., 8 hrs., 24 hrs., and 48 hrs.

(5) The percentages of total BSL-borne metals extracted were 13.8, 1.63, 7.95, 14.1, 1.03, and 24.2% for Cd, Cr, Cu, Ni, Pb, and Zn, respectively. If OAs were responsible for converting metals in solid phases into plant available forms, the amount and rate of metals' solubilization would be indicative of metals' availability to plants.

TABLE 13: First-order kinetics constant for metal dissolution reaction extracted by organic acid (OA) mixtures according to (3).

Biosolid treatment[†]	Metal extracted by OAs	First-order kinetics constant (k_1)					
		Cd	Cr	Cu	Ni	Pb	Zn
Control	0.001 M	4.71	0.34	4.11	4.87	n.d.[‡]	5.00
	0.01 M	3.99	0.30	3.90	3.93	n.d.[‡]	5.42
	0.1 M	4.30	0.37	4.49	4.46	0.36	4.92
135 Mg ha^{-1}	0.001 M	4.13	0.38	4.05	4.65	0.38	5.63
	0.01 M	3.88	0.51	4.03	3.95	0.40	5.33
	0.1 M	3.96	0.48	3.81	3.79	0.50	6.06
1,080 Mg ha^{-1}	0.001 M	4.93	0.32	4.30	4.81	0.24	4.85
	0.01 M	5.16	0.40	4.41	4.81	0.32	5.75
	0.1 M	6.25	0.37	4.88	5.53	0.43	6.80

[†]Obtained at Moreno Field Station of the University of California, Riverside, CA. From 1976 through 1981 composted BSL were applied at dry weight rates of 0 (control), 22.5, and 180 Mg ha^{-1} yr^{-1}, respectively.
[‡]Concentration was below detection limits of the AAS for Pb = 0.001 mg kg^{-1}.

(6) A rapid dissolution of metals occurred in the first 15 minutes of mixture. For Cd, Cu, Ni, and Zn, approximately 60–70% of the metals were released. For Cr and Pb, the initial releases were approximately 30% of the total.

(7) The data of the metal dissolution kinetics in BSL-treated soils may fit either the two-site bicontinuum model in which significant amounts of the soluble metals were dissolved rapidly, following a zero-order dissolution kinetics and the remaining soluble metals released slowly over a long period of time, following a first-order dissolution kinetics, or first-order dissolution kinetics alone.

Competing Interests

The authors declare that they have no competing interests.

Acknowledgments

The authors gratefully acknowledge Mr. D. Thomason, Mr. W. Smith, and Ms. N. J. Krage for technical assistance. This research was supported by the Water Environmental Research Foundation (WERF-97-REM-5) and California Baptist University's Microgrant.

References

[1] A. H. M. Veeken and H. V. M. Hamelers, "Removal of heavy metals from sewage sludge by extraction with organic acids," *Water Science and Technology*, vol. 40, no. 1, pp. 129–136, 1999.

[2] D. del Mundo Dacera and S. Babel, "Use of citric acid for heavy metals extraction from contaminated sewage sludge for land application," *Water Science and Technology*, vol. 54, no. 9, pp. 129–135, 2006.

[3] W. P. Inskeep and S. D. Comfort, "Thermodynamic predictions for the effects of root exudates on metal speciation in the rhizosphere," *Journal of Plant Nutrition*, vol. 9, no. 3–7, pp. 567–586, 1986.

[4] M. Mench, J. L. Morel, A. Guckert, and B. Guillet, "Metal binding with root exudates of low molecular weight," *Journal of Soil Science*, vol. 39, no. 4, pp. 521–527, 1988.

[5] G. S. R. Krishnamurti, G. Cieslinski, P. M. Huang, and K. C. J. Van Rees, "Kinetics of cadmium release from soils as influenced by organic acids: implication in cadmium availability," *Journal of Environmental Quality*, vol. 26, no. 1, pp. 271–277, 1997.

[6] J. Kumpiene, A. Lagerkvist, and C. Maurice, "Stabilization of As, Cr, Cu, Pb and Zn in soil using amendments—a review," *Waste Management*, vol. 28, no. 1, pp. 215–225, 2008.

[7] M. C. Hernandez-Soriano and J. C. Jimenez-Lopez, "Effects of soil water content and organic matter addition on the speciation and bioavailability of heavy metals," *Science of the Total Environment*, vol. 423, pp. 55–61, 2012.

[8] A. C. Chang, A. L. Page, and B.-J. Koo, "Biogeochemistry of phosphorus, iron, and trace elements in soils as influenced by soil-plant microbial interactions," *Developments in Soil Science*, vol. 28, no. 2, pp. 43–57, 2002.

[9] M. Treeby, H. Marschner, and V. Römheld, "Mobilization of iron and other micronutrient cations from a calcareous soil by plant-borne, microbial, and synthetic metal chelators," *Plant and Soil*, vol. 114, no. 2, pp. 217–226, 1989.

[10] F. Awad and V. Römheld, "Mobilization of heavy metals from a contaminated calcareous soil by plant borne and synthetic chelators and their uptake by wheat plants," *Journal of Plant Nutrition*, vol. 23, no. 11-12, pp. 1847–1855, 2000.

[11] B.-J. Koo, A. C. Chang, A. L. Page, D. E. Crowley, and A. Taylor, "Availability and plant uptake of biosolid-borne metals," *Applied and Environmental Soil Science*, vol. 2013, Article ID 892036, 10 pages, 2013.

[12] T. Mimmo, M. Ghizzi, C. Marzadori, and C. E. Gessa, "Organic acid extraction from rhizosphere soil: effect of field-moist, dried and frozen samples," *Plant and Soil*, vol. 312, no. 1-2, pp. 175–184, 2008.

[13] V. Römheld and F. Awad, "Significance of root exudates in acquisition of heavy metals from a contaminated calcareous soil by graminaceous species," *Journal of Plant Nutrition*, vol. 23, no. 11-12, pp. 1857–1866, 2000.

[14] L. Ström, A. G. Owen, D. L. Godbold, and D. L. Jones, "Organic acid behaviour in a calcareous soil implications for rhizosphere nutrient cycling," *Soil Biology and Biochemistry*, vol. 37, no. 11, pp. 2046–2054, 2005.

[15] S. A. Wasay, S. F. Barrington, and S. Tokunaga, "Remediation of soils polluted by heavy metals using salts of organic acids and chelating agents," *Environmental Technology*, vol. 19, no. 4, pp. 369–379, 1998.

[16] L. Di Palma and R. Mecozzi, "Heavy metals mobilization from harbour sediments using EDTA and citric acid as chelating agents," *Journal of Hazardous Materials*, vol. 147, no. 3, pp. 768–775, 2007.

[17] A. Piccolo, S. Nardi, and G. Concheri, "Structural characteristics of humic substances as related to nitrate uptake and growth regulation in plant systems," *Soil Biology and Biochemistry*, vol. 24, no. 4, pp. 373–380, 1992.

[18] S. Nardi, F. Reniero, and G. Concheri, "Soil organic matter mobilization by root exudates of three maize hybrids," *Chemosphere*, vol. 35, no. 10, pp. 2237–2244, 1997.

[19] B. Dinkelaker, V. Römheld, and H. Marschner, "Citric acid excretion and precipitation of calcium citrate in the rhizosphere of white lupin (*Lupinus albus* L.)," *Plant, Cell and Environment*, vol. 12, no. 3, pp. 285–292, 1989.

[20] Z. Rengel and V. Römheld, "Root exudation and Fe uptake and transport in wheat genotypes differing in tolerance to Zn deficiency," *Plant and Soil*, vol. 222, no. 1-2, pp. 25–34, 2000.

[21] H. Nietfeld and J. Prenzel, "Modeling the reactive ion dynamics in the rhizosphere of tree roots growing in acid soils. I. Rhizospheric distribution patterns and root uptake of M_b cations as affected by root-induced pH and Al dynamics," *Ecological Modelling*, vol. 307, pp. 48–65, 2015.

[22] A. A. Pohlman and J. G. McColl, "Kinetics of metal dissolution from forest soils by soluble organic acids," *Journal of Environmental Quality*, vol. 15, no. 1, pp. 86–92, 1986.

[23] D. L. Jones and L. V. Kochian, "Aluminium-organic acid interactions in acid soils: I. Effect of root-derived organic acids on the kinetics of Al dissolution," *Plant and Soil*, vol. 182, no. 2, pp. 221–228, 1996.

[24] M. Mench and E. Martin, "Mobilization of cadmium and other metals from two soils by root exudates of *Zea mays* L., *Nicotiana tabacum* L. and *Nicotiana rustica* L.," *Plant and Soil*, vol. 132, no. 2, pp. 187–196, 1991.

[25] L. Ruiz and J. C. Arvieu, "Measurement of pH gradients in the rhizosphere," *Symbiosis*, vol. 9, no. 1–3, pp. 71–75, 1990.

[26] K. Fujii, C. Hayakawa, P. A. W. Van Hees, S. Funakawa, and T. Kosaki, "Biodegradation of low molecular weight organic compounds and their contribution to heterotrophic soil respiration in three Japanese forest soils," *Plant and Soil*, vol. 334, no. 1, pp. 475–489, 2010.

[27] P. Cambier and G. Sposito, "Interactions of citric acid and synthetic hydroxy-aluminum montmorillonite," *Clays and Clay Minerals*, vol. 39, no. 2, pp. 158–166, 1991.

[28] R. P. T. Janssen, M. G. M. Bruggenwert, and W. H. Van Riemsdijk, "Interactions between citrate and montmorillonite-Al hydroxide polymer systems," *European Journal of Soil Science*, vol. 48, no. 3, pp. 463–472, 1997.

[29] J. L. Schroder, H. Zhang, D. Zhou et al., "The effect of long-term annual application of biosolids on soil properties, phosphorus, and metals," *Soil Science Society of America Journal*, vol. 72, no. 1, pp. 73–82, 2008.

[30] K. Sakurai and P. M. Huang, "Cadmium adsorption on the hydroxyaluminum-montmorillonite complex as influenced by oxalate," in *Environmental Impact of Soil Component Interactions: Vol. II, Metals, Other Inorganics, and Microbial Activities*, P. M. Huang, J. Berthelin, J.-M. Bollag, W. B. McGill, and A. L.

Page, Eds., pp. 39–46, Lewis Publisher, Boca Raton, Fla, USA, 1995.

[31] S. Taniguchi, N. Yamagata, and K. Sakurai, "Cadmium adsorption on hydroxyl-aluminosilicate-montmorillonite complex as influenced by oxalate and citrate," *Soil Science and Plant Nutrition*, vol. 46, no. 2, pp. 315–324, 2000.

[32] F. M. Eaton, "Automatically operated sand-culture equipment," *Journal of Agricultural Research*, vol. 53, pp. 433–444, 1936.

[33] H. Jenny and R. Overstreet, "Contact effects between plant roots and soil colloids," *Proceedings of the National Academy of Sciences of the United State of America*, vol. 24, no. 9, pp. 384–392, 1938.

[34] B.-J. Koo, A. C. Chang, D. E. Crowley, and A. L. Page, "Characterization of organic acids recovered from rhizosphere of corn grown on biosolids-treated medium," *Communications in Soil Science and Plant Analysis*, vol. 37, no. 5-6, pp. 871–887, 2006.

[35] L. M. Candalaria, *Interactions of citric acid and synthetic hydroxy-aluminum montmorillonite [Ph.D. dissertation]*, University of California, Riverside, Calif, USA, 1995.

[36] M. Tagliavini, A. Masia, and M. Quartieri, "Bulk soil pH and rhizosphere pH of peach trees in calcareous and alkaline soils as affected by the form of nitrogen fertilizers," *Plant and Soil*, vol. 176, no. 2, pp. 263–271, 1995.

[37] W. D. C. Schenkeveld and S. M. Kraemer, "Equilibrium and kinetic modelling of the dynamic rhizosphere," *Plant and Soil*, vol. 386, no. 1, pp. 395–397, 2015.

[38] P. H. Nye, "Changes of pH across the rhizosphere induced by roots," *Plant and Soil*, vol. 61, no. 1-2, pp. 7–26, 1981.

[39] H. Marschner, V. Romheld, and M. Kissel, "Different strategies in higher plants in mobilization and uptake of iron," *Journal of Plant Nutrition*, vol. 9, no. 3, pp. 695–713, 1986.

[40] T. S. Gahoonia, "Influence of root-induced pH on the solubility of soil aluminium in the rhizosphere," *Plant and Soil*, vol. 149, no. 2, pp. 289–291, 1993.

[41] H. Marschner, V. Römheld, and I. Cakmak, "Root-induced changes of nutrient availability in the rhizosphere," *Journal of Plant Nutrition*, vol. 10, no. 9, pp. 1175–1184, 1987.

[42] R. A. Youssef and M. Chino, "Root-induced changes in the rhizosphere of plants: I. pH changes in relation to the bulk soil," *Soil Science and Plant Nutrition*, vol. 35, no. 3, pp. 461–468, 1989.

[43] H. Marschner and V. Römheld, "*In vivo* measurement of root-induced pH changes at the soil-root interface: effect of plant species and nitrogen source," *Zeitschrift für Pflanzenphysiologie*, vol. 111, no. 3, pp. 241–251, 1983.

[44] P. R. Darrah, "The rhizosphere and plant nutrition: a quantitative approach," *Plant and Soil*, vol. 155-156, no. 1, pp. 1–20, 1993.

[45] H. T. Gollany and T. E. Schumacher, "Combined use of colorimetric and microelectrode methods for evaluating rhizosphere pH," *Plant and Soil*, vol. 154, no. 2, pp. 151–159, 1993.

[46] H.-N. Hyun, A. C. Chang, D. R. Parker, and A. L. Page, "Cadmium solubility and phytoavailability in sludge-treated soil: effects of soil organic carbon," *Journal of Environmental Quality*, vol. 27, no. 2, pp. 329–334, 1998.

[47] C. G. Millward and P. D. Kluckner, "Microwave digestion technique for the extraction of minerals from environmental marine sediments for analysis by inductively coupled plasma atomic emission spectrometry and atomic absorption spectrometry," *Journal of Analytical Atomic Spectrometry*, vol. 4, no. 8, pp. 709–713, 1989.

[48] W. R. Fischer, H. Flessa, and G. Schaller, "pH values and redox potentials in microsites of the rhizosphere," *Zeitschrift für*

Pflanzenernährung und Bodenkunde, vol. 152, no. 2, pp. 191–195, 1989.

[49] P. Hinsinger and R. J. Gilkes, "Mobilization of phosphate from phosphate rock and alumina-sorbed phosphate by the roots of ryegrass and clover as related to rhizosphere pH," *European Journal of Soil Science*, vol. 47, no. 4, pp. 533–544, 1996.

[50] T. C. Zhang and H. Pang, "Applications of microelectrode techniques to measure pH and oxidation—reduction potential in rhizosphere soil," *Environmental Science and Technology*, vol. 33, no. 8, pp. 1293–1299, 1999.

[51] V. Römheld and H. Marschner, "Plant–induced pH changes in the rhizosphere of 'Fe–efficient' and 'Fe–inefficient' soybean and corn cultivars," *Journal of Plant Nutrition*, vol. 7, no. 1–5, pp. 623–630, 2008.

[52] D. Riley and S. A. Barber, "Effect of ammonium and nitrate fertilization on phosphorus uptake as related to root-induced pH changes at the root-soil interface," *Soil Science Society of America Journal*, vol. 35, no. 2, pp. 301–306, 1971.

[53] D. A. Barber and K. B. Gunn, "The effect of mechanical forces on the exudation of organic substances by the roots of cereal plants grown under sterile conditions," *New Phytologist*, vol. 73, no. 1, pp. 39–45, 1974.

[54] D. L. Jones and P. R. Darrah, "Re-sorption of organic compounds by roots of *Zea mays* L. and its consequences in the rhizosphere. III. Characteristics of sugar influx and efflux," *Plant and Soil*, vol. 178, no. 1, pp. 153–160, 1996.

[55] B.-J. Koo, D. C. Adriano, N. S. Bolan, and C. D. Barton, "Root exudates and microorganisms," in *Encyclopedia of Soils in the Environment*, D. Hillel, Ed., vol. 3, pp. 421–428, Elsevier, Oxford, UK, 2005.

[56] D. C. Adriano, *Trace Elements in Terrestrial Environments: Biogeochemistry, Bioavailability, and Risks of Metals*, Springer, New York, NY, USA, 2nd edition, 2001.

[57] B.-J. Koo, W. Chen, A. C. Chang, A. L. Page, T. C. Granato, and R. H. Dowdy, "A root exudates based approach to assess the long-term phytoavailability of metals in biosolids-amended soils," *Environmental Pollution*, vol. 158, no. 8, pp. 2582–2588, 2010.

[58] C. Fernández-Ramos, O. Ballesteros, A. Zafra-Gómez et al., "Sorption and desorption of alcohol sulfate surfactants in an agricultural soil," *Environmental Toxicology and Chemistry*, vol. 33, no. 3, pp. 508–515, 2014.

[59] N. S. Bolan, D. C. Adriano, R. Natesan, and B.-J. Koo, "Effects of organic amendments on the reduction and phytoavailability of chromate in mineral soil," *Journal of Environmental Quality*, vol. 32, no. 1, pp. 120–128, 2003.

[60] S. P. Mishra, "Adsorption-desorption of heavy metal ions," *Current Science*, vol. 107, no. 4, pp. 601–612, 2014.

[61] S. A. Wasay, S. Barrington, S. Tokunagal, and S. Prasher, "Kinetics of heavy metal desorption from three soils using citric acid, tartaric acid, and EDTA," *Journal of Environmental Engineering and Science*, vol. 6, no. 6, pp. 611–622, 2007.

[62] J. Liu, J. Dai, R. Wang, F. Li, X. Du, and W. Wang, "Adsorption/desorption and fate of mercury (II) by typical black soil and red soil in China," *Soil and Sediment Contamination*, vol. 19, no. 5, pp. 587–601, 2010.

19

Variability of Soil Micronutrients Concentration along the Slopes of Mount Kilimanjaro, Tanzania

Mathayo Mpanda Mathew,[1] Amos E. Majule,[2] Robert Marchant,[3] and Fergus Sinclair[4]

[1] *World Agroforestry Centre (ICRAF), P.O. Box 6226, Dar es Salaam, Tanzania*
[2] *Institute of Resource Assessment, University of Dar es Salaam, P.O. Box 35097, Dar es Salaam, Tanzania*
[3] *Environment Department, York Institute for Tropical Ecosystems, University of York, Heslington, York, North Yorkshire YO10 5NG, UK*
[4] *World Agroforestry Centre (ICRAF), P.O. Box 30677, Nairobi 00100, Kenya*

Correspondence should be addressed to Mathayo Mpanda Mathew; m.mpanda@yahoo.com

Academic Editor: Claudio Cocozza

Soil micronutrients are important elements for plant growth despite being required in small quantities. Deficiency of micronutrients can result in severe crop failure while excess levels can lead to health hazards; therefore, investigating their status in agricultural land is crucial. Fifty plots were established along an altitudinal gradient from 680 to 1696 m a.s.l. on the slopes of Mount Kilimanjaro, Tanzania. Soils were sampled at the top- (0–20 cm) and subsoils (21–50 cm) in four locations within each plot. Fourier Transform Mid-Infrared (FT-MIR) spectroscopy and wet chemistry were used for soil analysis. Results indicated that the mean concentrations of the micronutrients in the topsoil were Fe (130.4 ± 6.9 mgkg^{-1}), Mn (193.4 ± 20.5 mgkg^{-1}), Zn (2.8 ± 0.2 mgkg^{-1}), B (0.68 ± 0.1 mgkg^{-1}), and Cu (8.4 ± 0.8 mgkg^{-1}). Variations of the micronutrients were not statistically different by elevation (df = 41, $p > 0.05$) and by soil depth (df = 49, $p > 0.05$). Correlations among micronutrients were significant for Fe *versus* Mn ($r = 0.46$, $p < 0.001$), B *versus* Zn ($r = 0.40$, $p = 0.003$), B *versus* Cu ($r = 0.34$, $p = 0.013$), and Cu *versus* Zn ($r = 0.88$, $p < 0.001$). The correlated micronutrients implied that they were affected by similar factors. Soil pH correlated positively with B, Fe, and Mn and negatively with Cu and Zn, hence probably influencing their availability. Therefore, the need for sustaining micronutrient at sufficient levels is crucial. Management interventions may include moderating soil pH by reducing acidity through liming in the higher elevations and incorporation of organic matter in the lowlands.

1. Introduction

Soil nutrients are important elements that support plant growth and crop productivity [1]. Maintenance of soil nutrients at sufficient levels for macro- and micronutrients remains prerequisite in ensuring sustained crop yields [2, 3]. Usually macronutrients, required in large quantities, are the focus of many interventions, unlike micronutrients that are required in small quantities [4–6]. In sub-Saharan Africa, soil infertility remains one of the key factors responsible for declining crop productions [7, 8]. Challenges of soil infertility caused by various factors such as reduction in crop diversity have led to application of various interventions including use of inorganic fertilizers and agroforestry practices that deploy leguminous species [9–11].

Micronutrients quantities required by plants are very small, and the thresholds for sufficient, deficient, and toxic levels are also very close. Several review studies have summarized and suggested the micronutrients range based on extraction methods [4, 12]. Major sources of soil micronutrients are inorganic forms from parent material and organic forms within humus, though deficiency or toxicity can mostly be attributed to the parent material [13, 14]. Furthermore, factors which play important roles in regulating micronutrients include soil pH, oxidation state, organic matter, mycorrhizae, organic compounds, and stability of chelates [15, 16].

Most soils vary in their micronutrient content, and deficiencies in supplying micronutrient are alarming [17]. Deficiency of micronutrients can result in severe crop failure; hence attempts to improve crop production and soil management [18–21] must be in line with micronutrients amendments [22, 23].

Normally, concentrations of soil nutrients are affected by soil types, climate, topography, and management practices [24–26]. For instance, declined vegetation cover and heavy precipitation may accelerate micronutrients leaching. Increased chances of leaching for micronutrients are due to their occurrence as free ions or soluble complexes in solution [27]. Therefore, translocation of micronutrients along the elevation due to surface runoff in sloping terrains and depositions in the valley bottoms calls for proper soil management practices to address both nutrient transfers and crop yield [28].

In Tanzania, few studies have attempted to assess concentration of soil micronutrients in relation to supporting crop productions. Such trend has led to partial understanding of the status and variability of micronutrients in various agricultural soils [18, 29, 30]. This study, therefore, aimed at determining concentration levels and variability of soil micronutrients along the slopes of Mount Kilimanjaro, Tanzania. The information generated can serve for planning soil management interventions to sustain soil micronutrients sufficient levels and addressing deficiencies in the study site.

2. Material and Methods

2.1. Study Site. The study was carried out on farmland along the southern slopes of Mount Kilimanjaro, in Moshi rural district, northern Tanzania (Figure 1). In general, soils in the study site originated from volcanic rocks which are rich in Ca and Mg [32, 33]. Mount Kilimanjaro is a stratovolcano found in the East Africa Rift Valley surrounded by the Precambrian rocks of the Mozambican Belt [32, 34]. Hydrological processes across the study area are very complex, comprising heavy precipitation and deep ground water infiltration [35]. The total population of the Kilimanjaro region is 1,640,087 with average household size of 4.3 [36]. The Chagga tribe forms the dominant inhabitants in the study site, with other ethnic groups including Pare and Taita.

2.2. Land Use Systems. Study transect was categorized into three land use zones based on altitude, climate, and soils. These land use zones exhibited variation in farming systems and were divided into upland (highland), midland (intermediate zone), and lowland.

The upland lies between 1438 and 1696 m a.s.l. Soils are dominated by Humic Nitisol [6, 37]. Mean annual temperature is 24°C and rainfall ranges between 1250 and 2000 mm per year [38–40]. The terrain is gentle slope. Chagga homegarden system is the dominant farming system, comprising multistrata agroforestry with banana plantations and coffee as main crops. Livestock keeping is done through zero grazing. For dairy cattle it includes Friesian, Jersey, Ayrshire, and crossovers between the improved and local breeds (Kilimanjaro Zebu). Other livestock include meat cattle, dairy and

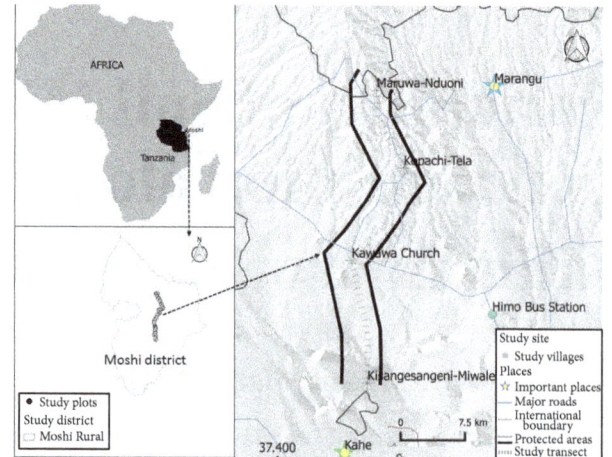

FIGURE 1: Location of the study site on the southern slopes of Mount Kilimanjaro, Tanzania. Insert map indicates the location of Tanzania within Africa continent [31].

meat goats, sheep, and pigs. Open spaces are also found for fodder and maize cultivation. Other crops spatially distributed on farms include yams, round potato, and vegetables [40, 41].

The midland forms the transition where highland and lowland converge. It lies between 900 and 1438 m a.s.l [40]. Major soils are Haplic Phaeozem [6, 37]. It has mean annual temperature of 26°C and rainfall range of 1000–1200 mm per year [38, 39]. The terrain is gentle slope. Mixture of Chagga homegarden and maize monocropping systems is the major farming system. As moving downslope, the midland, the maize is very predominant, such that the area is partly referred to as maize belt. Other crops found include coffee, banana, cardamom, and beans, which are intercropped together. Livestock keeping is a mixture of staff-fed and open field grazing, with dominant species being cattle, goats, and sheep.

The lowland zone extends below 900 m a.s.l. with an annual precipitation of 400–900 mm per year and a mean temperature of 33°C [40]. Major soils include Eutric Fluvisol [6, 37]. The terrain is plain and flat. Main annual crops include sunflower, cotton, maize, sorghum, cassava, paddy rice, and pigeon peas. Free livestock grazing mainly of indigenous breeds of cows (Kilimanjaro Zebu), goats, and sheep is commonly practiced on farms after the crops' harvest [40].

2.3. Soil Sampling and Analysis. Fifty plots were established for soil along 25 km long preselected transect running from 680 to 1696 m a.s.l: 12 plots in the upland, 14 in the midland, and 24 in the lowland. The African Soil Information System (AfSIS) protocol for soil sampling was adapted where inverted Y-shaped design was used in sampling 4 subplots within each plot [42]. Soils were sampled at the top- (0–20 cm) and subsoils (21–50 cm) using auger and sampling plate. Soils were mixed in buckets separately for the sub- and topsoils to prepare composite samples. Coning and quartering method was used to reduce each sample to 500 g, each from the top- and subsoils per plot [43]. Samples were packed

TABLE 1: Calibration results of soil properties on the southern slopes of Mount Kilimanjaro, Tanzania.

Soil property	Number of principal components	Calibrations	
		RMSEC	R-squared
Fe (mgkg^{-1})	5	0.54	0.31
Mn (mgkg^{-1})	5	0.74	0.34
Zn (mgkg^{-1})	5	0.57	0.46
B (mgkg^{-1})	5	0.76	0.78
Cu (mgkg^{-1})	5	0.90	0.32
Soil pH	5	0.06	0.93

Note. Fe: iron; Mn: manganese; Zn: zinc; B: boron; and Cu: copper.

in zip-lock bags and labelled. Samples were then air-dried, ground using a wooden rolling pin, and sieved through a 2 mm mesh.

2.3.1. Spectral Data Analysis. Air dried subsamples from all plots and soil depths each with approximately 20 g were loaded into four wells. The soils were then analysed using Fourier Transform Mid-Infrared Reflectance Spectroscopy at waveband range from 4001.6 to 601.7 cm, at World Agroforestry Centre (ICRAF) Soil-Plant Spectral Diagnostics Laboratory in Nairobi (Bruker Optik GmbH, Germany [44]). Soil samples were scanned 32 times and their four spectra averaged to account for variability within sample and differences in particle size and packaging in wells [43].

2.3.2. Reference Soil Analysis. About 30% of the soil subsamples were randomly selected for wet chemistry analysis at the Crop Nutrition Laboratory in Nairobi as calibration set [45]. Soil pH was analysed by standard potentiometric method using soil-to-water ratio of 1 : 2 on weight/volume basis [43]. Micronutrients (B, Cu, Fe, Mn, and Zn) were analysed using inductively coupled plasma atomic emission spectroscopy (ICP-AES) using Mehlich 3-Diluted ammonium fluoride-EDTA and ammonium nitrate [46].

2.3.3. Chemometric Analysis. Chemometric procedures were used in analysing data from soil spectra and measured values of the reference soil samples. Soil spectra were processed by cubic smoothing splines and, thereafter, the first derivatives were taken with a smoothing interval of 21 data points using "*trans*" function in the "*soil.spec*" in R-software. Measured soil properties were then calibrated using first derivative of the reflectance spectra by use of partial-least squares regression [47–49]. The regression model developed was used to predict the soil properties (Table 1) for the rest of the samples and their coefficient of correlation (R^2) and root mean standard errors of calibration (RMSEC):

$$RMSEC = \sqrt{\frac{\sum_{i=1}^{N}(y_i - X_i)^2}{N - A - 1}}, \qquad (1)$$

where y is the predicted reference value, X is the measured reference value, N is the number of samples, and A is the number of principal components used in the model.

2.4. Statistical Analysis. Descriptive statistics (maximum, minimum, and mean and standard error of the mean, standard deviation, skewness, kurtosis, and coefficient of variation) were computed for the soil properties. A nonparametric Kruskal-Wallis test (K-W test) was performed to determine the relationship of soil micronutrients with elevation and soil depths. Pearson's correlation was used to compare variables with different dimensional units [50], to determine relationships between soil pH and micronutrients, and to determine correlations among micronutrients. R-statistics software was used in all statistical analyses [48].

3. Results

Correlation coefficients (R^2) of calibration for the wet chemistry and MIR results (Table 1) were large for B and soil pH ($R^2 = 0.78$ and 0.93), indicating large correlation between wet chemistry and MIR analysis procedure. Results for Cu, Fe, Mn, and Zn showed medium correlations ($R^2 = 0.31$–0.46).

Concentration of B, Cu, Fe, Mn, and Zn varied with soil depth across the entire elevation range (Table 2). Variation of concentrations of micronutrients observed by this study (Table 2) ranges from deficient to sufficient, as required for plant growth as suggested by other studies [12].

Mean concentrations of Cu, Fe, Mn, and Zn were higher in the topsoil than in the subsoil except for boron (Table 2) across the elevation. However, there was no significant difference in variation of top- and subsoil concentrations (K-W test: df = 49, $p > 0.05$). Skewness was positive for topsoil (range of 1.2–2.0) and subsoils (0.99–2.05), with the exception of Fe which was close to symmetrical distribution. Kurtosis was positive in all soil micronutrients indicating a peaked distribution, with exception of Fe in the subsoil (Table 2) which was negative, indicating a flatter distribution.

Soil micronutrients indicated high variability especially for concentrations of B, Cu, Mn, and Zn (CV > 0.5). However, the concentration levels did not differ significantly with elevation (K-W test: df = 41, $p > 0.05$). This implied that the concentration levels varied within and between elevations, as further indicated by a scatter plot (Figure 2).

Soil pH was strongly acidic in the upland with estimated value of 5.2 and elevated to very strong alkaline with a value of 9 in the lowland (Table 2, Figure 2). Similarly, soil pH indicated positive correlation with B, Fe, and Mn and negative correlation with Cu and Zn (Table 3). This implied that soil pH influenced the availability of soil micronutrients in the study site.

Correlations among micronutrients were found to be statistically significant for Fe *versus* Mn, B *versus* Zn, B *versus* Cu, and Cu *versus* Zn, implying that these correlated micronutrients were affected by similar factors.

4. Discussion

Mean concentrations of B, Cu, Fe, Mn, and Zn ($n = 50$) were found to be in sufficient range (Table 2), while the minimum levels for B (0.000078 mgkg^{-1}), Cu (0.75 mgkg^{-1}), and Zn (0.92 mgkg^{-1}) indicated deficiencies. The deficiency,

TABLE 2: Descriptive statistics for the soil micronutrients on the southern slopes Mount Kilimanjaro, Tanzania.

Soil property	Max	Min	Mean (SE)	Std. dev.	Kurtosis	Skewness	CV
0–20 cm							
Fe (mgkg^{-1})	310.60	39.29	130.41 (6.9)	49.2	2.54	1.12	0.38
Mn (mgkg^{-1})	757.04	14.33	193.43 (20.56)	145.39	0.33	1.75	0.75
Zn (mgkg^{-1})	10.34	0.92	2.82 (0.27)	1.97	4.35	2.00	0.69
B (mgkg^{-1})	3.50	0.000078	0.68 (0.1)	0.72	4.24	1.92	1.06
Cu (mgkg^{-1})	24.67	0.75	8.49 (0.85)	6.03	0.70	1.23	0.71
Soil pH (1 : 2 soil : water)	9.03	5.21	6.58 (0.13)	0.93	−0.41	0.59	0.14
21–50 cm							
Fe (mgkg^{-1})	229.70	28.36	119.06 (6.12)	43.26	−0.03	0.31	0.36
Mn (mgkg^{-1})	827.58	8.9	185.45 (22.1)	156.3	5.26	2.05	0.84
Zn (mgkg^{-1})	7.24	0.51	2.18 (0.18)	1.24	3.8	1.64	0.57
B (mgkg^{-1})	3.74	0.00001	0.77 (0.12)	0.86	2.21	1.65	1.11
Cu (mgkg^{-1})	19.8	0.47	7.4 (0.72)	5.06	0.03	0.99	0.68
Soil pH (1 : 2 soil : water)	9.55	5.23	6.78 (0.15)	1.08	0.01	0.85	0.16

Note. max = maximum, min = minimum, SE = standard error of the mean, std. dev. = standard deviation, and CV = coefficient of variation.

TABLE 3: Pearson product-moment correlation coefficient among soil properties on the southern slopes of Mount Kilimanjaro, Tanzania.

Soil micronutrients	pH (1 : 2 soil : water)	Fe (mg/kg)	Mn (mg/kg)	Zn (mg/kg)	B (mg/kg)
0–20 cm					
Fe (mgkg^{-1})	0.05 (0.69)				
Mn (mgkg^{-1})	0.30 (**0.03**)	0.46 (**<0.001**)			
Zn (mgkg^{-1})	−0.12 (0.37)	0.20 (0.14)	0.08 (0.57)		
B (mgkg^{-1})	0.60 (**<0.001**)	−0.18 (0.2)	0.19 (0.169)	0.40 (**0.003**)	
Cu (mgkg^{-1})	−0.17 (0.21)	0.17 (0.23)	−0.06 (0.65)	0.88 (**<0.001**)	0.34 (**0.013**)
21–50 cm					
Fe (mgkg^{-1})	−0.04 (0.79)				
Mn (mgkg^{-1})	0.47 (**<0.001**)	0.26 (0.07)			
Zn (mgkg^{-1})	−0.0089 (0.95)	0.099 (0.49)	0.051 (0.72)		
B (mgkg^{-1})	0.64 (**<0.001**)	−0.4 (**0.004**)	0.22 (0.12)	0.29 (**0.04**)	
Cu (mgkg^{-1})	−0.14 (0.33)	0.096 (0.51)	−0.15 (0.3)	0.84 (**<0.001**)	0.23 (0.12)

Note. r (p value), significant level, $\alpha = 0.05$.

sufficiency, and toxicity range of micronutrients is very small [4, 12]; therefore, it is important to understand their concentration levels for proper land management. Overall, the concentration levels of micronutrients observed by this study falls under similar range with other studies in Tanzania [30, 51].

Soil pH was low in the high elevations (Table 2, Figure 2), contributed by higher concentration of Al due to the nature of the parent material as well as the higher mean annual precipitation which led to leaching of base cations. In the lower elevations, soil pH was alkaline; this was due to increased concentrations of exchangeable bases as a result of translocation and soil depositions and their exposure to the surface by evaporation. Therefore, soil pH increased with decreased elevation.

The pattern indicated by soil pH coincided with changes in the availability of the micronutrients (Figure 2), implying that it has direct influence. Other studies have indicated that soil pH influences micronutrients availability by favouring

conditions which accelerates oxidation, precipitation, and immobilization [5, 17]. Positive correlations were found for B, Mn, and Fe with soil pH (Table 3), therefore providing favourable conditions for their availability. Solubility of B, Mn, and Fe is known to increase with lowering soil pH [52].

Soil pH indicated negative correlation with Zn and Cu (Table 3). This implied that strong acidity in the higher altitude and alkaline conditions in the lower altitudes in the study area reduced the availability of the Zn and Cu (Figure 2). Saline soils tend to enhance formation of insoluble oxides and hydroxides of Cu and Zn, which limits their availability [4]. Furthermore, our observation indicated that Cu was much lower than 60 mgkg^{-1}. Any concentration of Cu above 60 mgkg^{-1} is considered to be toxic and that there was limited horizontal transmission of the Cu from the neighbouring farms unlike previous observation [30].

Correlation among micronutrients in the study site (Table 3) may explain their relationships in enhancing their availability. For instance, high concentration of Mn and Fe is

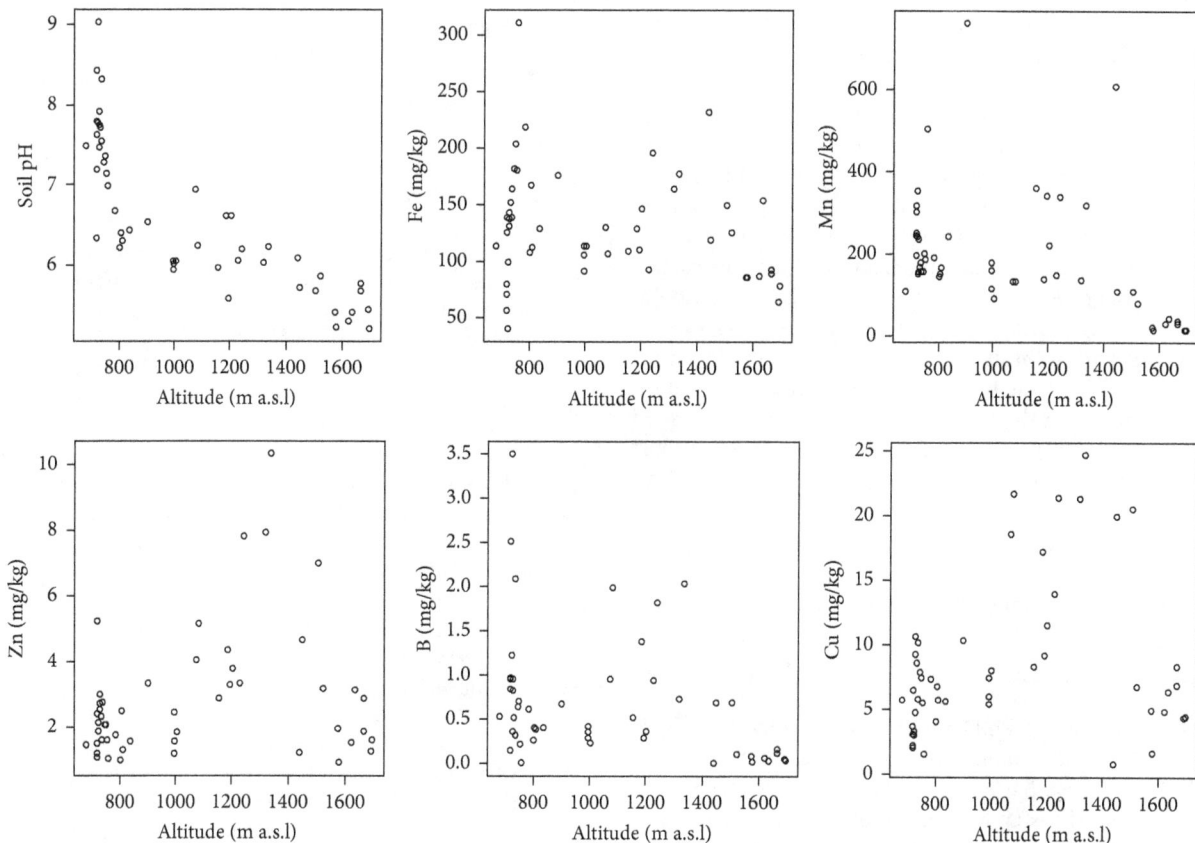

FIGURE 2: Patterns of soil properties along elevation gradient on Mount Kilimanjaro, Tanzania. *Note*. The upland (1438–1696 m a.s.l.); the midland (900–1438 m a.s.l.); and the lowland (extending below 900 m a.s.l.).

known to suppress extractable heavy metals like Zn and Cu [53]. However, this was not the case as there was poor and statistically insignificant correlation among these elements (Table 3). Therefore, other factors, including soil pH, remain responsible. Furthermore, a positive and significant correlation between Fe and Mn existed (Table 3), which underlines the fact that Mn influences availability of Fe. This implied that, under the same soil pH level, increase in concentration of Mn was likely to increase Fe availability as previously noted [54].

Topsoil indicated higher concentrations of soil micronutrients compared to subsoils (Table 2), though the difference was not statistically significant (K-W test: df = 49, $p > 0.05$). This implied that the depositions and decomposition of organic matter were higher in the topsoil, therefore contributing to the release of micronutrients [1, 55]. Furthermore, leaching did not remove the extractable micronutrients from the surface layer into the subsurface, across the three land uses. This can partly be explained by less soil drainage due to dry condition and high compaction below 900 m a.s.l. Similarly, above 900 m a.s.l., farms were composed of Chagga homegardens which retain higher vegetation cover estimated at above 10% [39], tending to reduce leaching through increase litter, mulch, and root production [56].

At the same elevation (Figure 2), some soil micronutrients have shown differences in their concentration levels. This can

probably be explained by differences in landforms especially in the mountainous areas which associated with localized management. It has been noted in another study that landforms and land use had impact on soil chemical properties [57]. Similarly, differences in soil micronutrients variability were observed in the Usambara Mountains in Tanzania due to differences in landforms [18].

On average, the sufficient concentration levels of soil micronutrients (Table 2) provide prospects for plant productions and human health. A study in Malawi noted that soil rich in micronutrients influenced their concentration in food crops [14]; therefore, observed concentrations of Fe (Table 2) could result in addressing Fe deficiency in diet in the study area. It has been established that Fe deficiency is a serious problem in Tanzania, affecting 30% of all women, and responsible for 50% of anaemia due to low consumption of Fe-rich foods [58].

5. Conclusion

Soil micronutrients in the study area varied with depth and elevation, though the variations were not statistically significant. The average concentrations of B, Cu, Fe, Mn, and Zn were in sufficient ranges for supporting plant growth. Soil pH increased as descending downslope from strong acidic in the high elevation to strong alkaline in the lowlands. Soil pH was shown to correlate positively with B, Fe, and Mn and negatively with Cu and Zn. Correlations among micronutrients were significant for Fe *versus* Mn, B *versus* Zn, B *versus* Cu, and Cu *versus* Zn. The observed soil micronutrients correlation implied that they were affected by similar factors. Soil pH has shown to influence the availability of soil micronutrients, including restricting B, Cu, and Zn. Improving crop production in the study area needs to take into account soil management to sustain micronutrient sufficient levels and addressing deficiencies in some parts for B, Cu, and Zn. Management interventions may include moderating soil pH by reducing acidity through liming in the higher elevations and addition of organic matter in the lowlands. Application of organic matter in the lowland may ensure slow release of the micronutrients at levels which are sufficient to support plant growth.

Competing Interests

The authors declare that they have no competing interests.

Acknowledgments

This work was part of Ph.D. program funded by the project titled "*The Climate Change Impacts on Ecosystem Services and Food Security in Eastern Africa (CHIESA)*." CHIESA was funded by the Ministry for Foreign Affairs of Finland and coordinated by the International Centre of Insect Physiology and Ecology (ICIPE) in Nairobi, Kenya. World Agroforestry Centre and CGIAR's CRP program on Humid Tropics supported Mathayo Mpanda Mathew in various capacities. Special thanks are due to Fortunatus Muya, Jimmy Sianga, Wilson Mchomvu, and Andrew Sila for assistance in field and laboratory work.

References

[1] E. W. Russell, *Soil Conditions and Plant Growth*, Longman Group Limited, London, UK, 1973.

[2] L. C. Campbell, "Managing soil fertility decline," *Journal of Crop Production*, vol. 1, no. 2, pp. 29–52, 1998.

[3] M. Kumar, A. K. Jha, S. Hazarika et al., "Micronutrients (B, Zn, Mo) for improving crop production on acidic soils of Northeast India," *National Academy Science Letters*, vol. 39, no. 2, pp. 85–89, 2016.

[4] N. C. Brady and R. R. Weil, *Elements of the Nature and Properties of Soils*, Pearson Education, New Jersey, NJ, USA, 3rd edition, 2010.

[5] D. C. Martens and D. T. Westermann, "Fertilizer applications for correcting micronutrient deficiencies," in *Micronutrients in Agriculture*, J. J. Mortvedt, F. R. Cox, L. M. Shuman, and R. M. Welch, Eds., pp. 549–592, Soil Science Society of America, Madison, Wis, USA, 1991.

[6] J. G. Mowo, J. Floor, F. B. S. Kaihurm, and J. P. Magoggo, *Review of Fertilizer Recommendations in Tanzania Part 2, S.F.R.N. 26*, National Soil Service, Tanga, Tanzania, 1993.

[7] R. Lal, "Enhancing crop yields in the developing countries through restoration of the soil organic carbon pool in agricultural lands," *Land Degradation and Development*, vol. 17, no. 2, pp. 197–209, 2006.

[8] S. Zingore, J. Mutegi, B. Agesa, L. Tamene, and J. Kihara, "Soil degradation in sub-saharan Africa and crop production options for soil rehabilitation," *Better Crops with Plant Food*, vol. 99, no. 1, pp. 24–26, 2015.

[9] F. K. Akinnifesi, W. Makumba, G. Sileshi, O. C. Ajayi, and D. Mweta, "Synergistic effect of inorganic N and P fertilizers and organic inputs from *Gliricidia sepium* on productivity of intercropped maize in Southern Malawi," *Plant and Soil*, vol. 294, no. 1-2, pp. 203–217, 2007.

[10] Z. Druilhe and J. Barreiro-Hurlé, "Fertilizer subsidies in sub-Saharan Africa," in *Working Paper*, Agricultural Development Economics Division, Ed., FAO, Rome, Italy, 2012.

[11] R. R. B. Leakey, "The role of trees in agroecology and sustainable agriculture in the tropics," *Annual Review of Phytopathology*, vol. 52, pp. 113–133, 2014.

[12] M. Sillanpaa, *Micronutrients and the Nutrient Status of Soil: A Global Study*, FAO Soils Bulletin, FAO, Rome, Italy, 1982.

[13] J. C. Ritchie, G. W. McCarty, E. R. Venteris, and T. C. Kaspar, "Soil and soil organic carbon redistribution on the landscape," *Geomorphology*, vol. 89, no. 1-2, pp. 163–171, 2007.

[14] E. J. M. Joy, M. R. Broadley, S. D. Young et al., "Soil type influences crop mineral composition in Malawi," *Science of the Total Environment*, vol. 505, pp. 587–595, 2015.

[15] L. S. Murphy, R. Ellis Jr., and D. C. Adriano, "Phosphorus-micronutrient interaction effects on crop production," *Journal of Plant Nutrition*, vol. 3, no. 1–4, pp. 593–613, 2008.

[16] F. Ali, "Effect of applied phosphorus on the availability of micronutrients in alkaline-calcareous soil," *Journal of Environment and Earth Science*, vol. 4, no. 15, pp. 143–147, 2014.

[17] J. G. White and R. J. Zasoski, "Mapping soil micronutrients," *Field Crops Research*, vol. 60, no. 1-2, pp. 11–26, 1999.

[18] J. L. Meliyo, B. Massawe, L. Brabers et al., "Status and variability of soil micronutrients with landforms in the plague focus of western usambara mountains, Tanzania," *International Journal of Plant & Soil Science*, vol. 4, no. 4, pp. 389–403, 2015.

[19] H. Foroughifar, A. A. Jafarzadeh, H. Torabi, A. Pakpour, and M. Miransari, "Using geostatistics and geographic information system techniques to characterize spatial variability of soil properties, including micronutrients," *Communications in Soil Science and Plant Analysis*, vol. 44, no. 8, pp. 1273–1281, 2013.

[20] A. K. Shukla, P. C. Srivastava, P. K. Tiwari et al., "Mapping current micronutrients deficiencies in soils of Uttarakhand for precise micronutrient management," *Indian Journal of Fertilisers*, vol. 11, no. 7, pp. 52–63, 2015.

[21] M. A. Wani, J. A. Wani, M. A. Bhat, N. A. Kirmani, Z. M. Wani, and S. N. Bhat, "Mapping of soil micronutrients in kashmir agricultural landscape using ordinary kriging and indicator approach," *Journal of the Indian Society of Remote Sensing*, vol. 41, no. 2, pp. 319–329, 2013.

[22] A. García-Ocampo, "Fertility and soil productivity of Colombian soils under different soil management practices and several crops," *Archives of Agronomy and Soil Science*, vol. 58, supplement 1, pp. S55–S65, 2012.

[23] A. Kabata-Pendias and H. Pendias, *Trace Elements in Soils and Plants*, CRC Press LLC, New York, NY, USA, 2001.

[24] M. Zhang, X.-K. Zhang, W.-J. Liang et al., "Distribution of soil organic carbon fractions along the altitudinal gradient in Changbai Mountain, China," *Pedosphere*, vol. 21, no. 5, pp. 615–620, 2011.

[25] A. C. Hamilton, "Vegetation, climate and soil: altitudinal relationships on the East Usambara Mountains, Tanzania," *Journal of East African Natural History*, vol. 87, no. 1, pp. 85–89, 1998.

[26] J. M. Maitima, S. M. Mugatha, R. S. Reid et al., "The linkages between landuse change, land degradation and biodiversity across East Africa," *African Journal of Environmental Science and Technology*, vol. 3, no. 10, pp. 310–325, 2009.

[27] P. M. Huang, Y. Li, and M. E. Sumner, *Handbook of Soil Sciences: Resource Management and Environmental Impacts*, CRC Press, London, UK, 2011.

[28] Z. Mbaga-Semgalawe and H. Folmer, "Household adoption behaviour of improved soil conservation: the case of the North Pare and West Usambara Mountains of Tanzania," *Land Use Policy*, vol. 17, no. 4, pp. 321–336, 2000.

[29] S. B. Mwango, B. Msanya, P. Mtakwa et al., "The influence of selected soil conservation practices on soil properties and crop yields in the usambara mountains, Tanzania," *Advances in Research*, vol. 3, no. 6, pp. 558–570, 2015.

[30] Y. H. Senkondo, E. Semu, and F. M. G. Tack, "Vertical distribution of copper in copper-contaminated coffee fields in Kilimanjaro, Tanzania," *Communications in Soil Science and Plant Analysis*, vol. 46, no. 10, pp. 1187–1199, 2015.

[31] M. M. Mathew, A. E. Majule, F. Sinclair, and R. Marchant, "Effect of soil properties on tree distribution across an agricultural landscape on a tropical mountain, Tanzania," *Open Journal of Ecology*, vol. 6, no. 5, pp. 264–276, 2016.

[32] B. Le Gall, P. Nonnotte, J. Rolet et al., "Rift propagation at craton margin. Distribution of faulting and volcanism in the North Tanzanian Divergence (East Africa) during Neogene times," *Tectonophysics*, vol. 448, no. 1–4, pp. 1–19, 2008.

[33] M. Schrumpf, W. Zech, J. C. Axmacher, and H. V. M. Lyaruu, "Biogeochemistry of an afrotropical montane rain forest on Mt. Kilimanjaro, Tanzania," *Journal of Tropical Ecology*, vol. 22, no. 1, pp. 77–89, 2006.

[34] P. Nonnotte, H. Guillou, B. Le Gall, M. Benoit, J. Cotten, and S. Scaillet, "New K–Ar age determinations of Kilimanjaro volcano in the North Tanzanian diverging rift, East Africa," *Journal of Volcanology and Geothermal Research*, vol. 173, no. 1-2, pp. 99–112, 2008.

[35] P. C. Røhr and Å. Killingtveit, "Rainfall distribution on the slopes of Mt Kilimanjaro," *Hydrological Sciences Journal*, vol. 48, no. 1, pp. 65–77, 2003.

[36] URT, *2012 Population and Housing Census: Population Distribution by Administrative Areas*, National Bureau of Statistics and Office of Chief Government Statistician, Dar es Salaam, Tanzania, 2013.

[37] E. De-Pauw, *Soils, Physiography and Agroecological Zones of Tanzania*, 1984.

[38] E. Soini, "Land use change patterns and livelihood dynamics on the slopes of Mt. Kilimanjaro, Tanzania," *Agricultural Systems*, vol. 85, no. 3, pp. 306–323, 2005.

[39] A. Hemp, "The banana forests of Kilimanjaro: biodiversity and conservation of the Chagga homegardens," *Biodiversity and Conservation*, vol. 15, no. 4, pp. 1193–1217, 2006.

[40] URT, *Kilimanjaro Region Socio-Economic Profile*, The Planning Commission and Regional Commissioner's Office Kilimanjaro, Kilimanjaro, Tanzania, 1998.

[41] M. J. Mambo, "Soil nutrient flow dynamics in the slopes of Mount Kilimanjaro, a case study of North Ward in Hai District, Tanzania," in *Geography and Environmental Management*, University of Dar es Salaam, Dar es Salaam, Tanzania, 2005.

[42] UNEP, *Land Health Surveillance: An Evidence-Based Approach to Land Ecosystem Management. Illustrated with a Case Study in the West Africa Sahel*, United Nations Environment Programme, Nairobi, Kenya, 2012.

[43] E. K. Towett, K. D. Shepherd, A. Sila, E. Aynekulu, and G. Cadisch, "Mid-infrared and total X-ray fluorescence spectroscopy complementarity for assessment of soil properties," *Soil Science Society of America Journal*, vol. 79, no. 5, pp. 1375–1385, 2015.

[44] L. Raphael, "Application of FTIR spectroscopy to agricultural soils analysis," in *Fourier Transforms—New Analytical Approaches and FTIR Strategies*, G. Nicolic, Ed., pp. 385–404, InTech, Shanghai, China, 2011.

[45] B. S. Waswa, P. L. G. Vlek, L. D. Tamene, P. Okoth, D. Mbakaya, and S. Zingore, "Evaluating indicators of land degradation in smallholder farming systems of western Kenya," *Geoderma*, vol. 195-196, pp. 192–200, 2013.

[46] A. Mehlich, "Mehlich 3 soil test extractant: a modification of Mehlich 2 extractant," *Communications in Soil Science and Plant Analysis*, vol. 15, no. 12, pp. 1409–1416, 1984.

[47] A. Sila and T. Hengl, *Soil.Spec Package: Soil Spectroscopy Tools and Reference Models*, 2014.

[48] R-Core-Team, *R: A Language and Environment for Statistical Computing*, R Foundation for Statistical Computing, Vienna, Austria, 2013.

[49] A. Savitzky and M. J. E. Golay, "Smoothing and differentiation of data by simplified least squares procedures," *Analytical Chemistry*, vol. 36, no. 8, pp. 1627–1639, 1964.

[50] G. V. Belle, L. D. Fisher, P. J. Heagerty, and T. Lumley, *Biostatistics: A Methodology for the Health Sciences*, John Wiley & Sons, Hoboken, NJ, USA, 2nd edition, 2004.

[51] S. T. Klopfenstein, D. R. Hirmas, and W. C. Johnson, "Relationships between soil organic carbon and precipitation along a climosequence in loess-derived soils of the Central Great Plains, USA," *Catena*, vol. 133, pp. 25–34, 2015.

[52] H. D. Foth, *Fundamentals of Soil Science*, John Wiley & Sons, New York, NY, USA, 8th edition, 1990.

[53] K. M. B. Kitundu and J. P. Mrema, "The status of Zn, Cu, Mil, and Fe in the soils and tea-leaves of-Kibena-tea. Estates-Njombe, Tanzania," *Tanzania Journal of Agricultural Science*, vol. 7, no. 1, pp. 34–41, 2006.

[54] M. R. Reddy, M. R. Tucker, and S. J. Dunn, "Effect of manganese on concentrations of Zn, Fe, Cu and B in different soybean genotypes," *Plant and Soil*, vol. 97, no. 1, pp. 57–62, 1987.

[55] L. M. Shuman and W. L. Hargrove, "Effect of tillage on the distribution of manganese, copper, iron, and zinc in soil fractions," *Soil Science Society of America Journal*, vol. 49, no. 5, pp. 1117–1121, 1985.

[56] J. Lehmann and G. Schroth, "Nutrient leaching," in *Trees, Crops, and Soil Fertility: Concepts and Research Methods*, G. Schroth and E. L. Sinclair, Eds., pp. 151–166, CAB International, Wallingford, UK, 2003.

[57] Y. Hao, Q. Chang, L. Li, and X. Wei, "Impacts of landform, land use and soil type on soil chemical properties and enzymatic activities in a Loessial Gully watershed," *Soil Research*, vol. 52, no. 5, pp. 453–462, 2014.

[58] A. Temu, B. Waized, and S. Henson, "Mapping value chains for nutrient-dense foods in Tanzania," in *Reducing Hunger and Undernutrition, E.r. 76*, Institute of Development Studies, Brighton, UK, 2014.

PERMISSIONS

LIST OF CONTRIBUTORS

Mufeed Batarseh
Chemistry Department, Mu'tah University, Mútah 61710, Jordan Abu Dhabi Polytechnic, Abu Dhabi, UAE

Cécile Noel
BRGM, 3 avenue Claude Guillemin, 45060 Orléans, France
LPC2E, CNRS, 3 avenue de la Recherche Scientifique, 45071 Orléans, France

Jean-Christophe Gourry, Jacques Deparis, Michaela Blessing and Ioannis Ignatiadis
BRGM, 3 avenue Claude Guillemin, 45060 Orléans, France

Christophe Guimbaud
LPC2E, CNRS, 3 avenue de la Recherche Scientifique, 45071 Orléans, France

Marco Alves and Jacyra Soares
Department of Atmospheric Science, IAG, University of São Paulo, 05508-090 São Paulo, SP, Brazil

John Pichtel
Natural Resources and Environmental Management, Ball State University, Muncie, IN 47306, USA

Yulia Markunas, Vadim Bostan, Andrew Laursen and Lynda McCarthy
Department of Chemistry and Biology, Ryerson University, 350 Victoria Street, Toronto, ON, Canada M5B 2K3

Michael Payne
Black Lake Environmental, 246 Black Lake Route, Perth, ON, Canada K7H 3C5

J. Ryschawy
Université de Toulouse, INRA, INP-ENSAT, UMR 1248 AGIR, 31324 Castanet-Tolosan, France

M. A. Liebig, S. L. Kronberg, D. W. Archer and J. R. Hendrickson
USDA-ARS, Northern Great Plains Research Laboratory, Mandan, ND 58554 0459, USA

Pasicha Chaikaew
Department of Environmental Science, Faculty of Science, Chulalongkorn University, Bangkok, Thailand

Suchana Chavanich
Department of Marine Science, Faculty of Science, Chulalongkorn University, Bangkok,Thailand

José Maria Filippini Alba and Carlos Alberto Flores
Empresa Brasileira de Pesquisa Agropecuária, Centro de Pesquisa Agropecuária de Clima Temperado, Pelotas, RS, Brazil

Alberto Miele
Empresa Brasileira de Pesquisa Agropecuária, Centro Nacional de Pesquisa de Uva e Vinho, Bento Gonçalves, RS, Brazil

Jamal M. A. Alsharef, Ali Akbar Firoozi and Panbarasi Govindasamy
Department of Civil and Structural Engineering, Universiti Kebangsaan Malaysia (UKM), 43600 Bangi, Selangor, Malaysia

Mohd Raihan Taha
Department of Civil and Structural Engineering, Universiti Kebangsaan Malaysia (UKM), 43600 Bangi, Selangor, Malaysia
Institute for Environment and Development (LESTARI), Universiti Kebangsaan Malaysia (UKM), 43600 Bangi, Selangor, Malaysia

He Zhang, Baiyu Zhang and Bo Liu
Northern Region Persistent Organic Pollution Control (NRPOP) Laboratory, Faculty of Engineering and Applied Science, Memorial University, St. John's, NL, Canada A1B 3X5

N. E. Kosheleva and E. M. Nikiforova
Faculty of Geography, Moscow State University, Leninskie Gory, Moscow 119991, Russia

Tri Joko
Doctoral Program of Environmental Science, School of Postgraduate Studies, Diponegoro University, Semarang City, Indonesia
Public Health Faculty, Diponegoro University, Semarang City, Indonesia

Sutrisno Anggoro
Doctoral Program of Environmental Science, School of Postgraduate Studies, Diponegoro University, Semarang City, Indonesia
Faculty of Fisheries and Marine Science, Diponegoro University, Semarang City, Indonesia

Henna Rya Sunoko
Doctoral Program of Environmental Science, School of Postgraduate Studies, Diponegoro University, Semarang City, Indonesia
Faculty of Medicine, Diponegoro University, Semarang City, Indonesia

Savitri Rachmawati
Public Health Faculty, Diponegoro University, Semarang City, Indonesia

Mohamed Mekkaoui
Groupe Physicochemistry of Materials, Nanomaterials and Environment, Faculty of Sciences, Mohammed V University, Ibn Battuta Avenue, Rabat, Morocco

Malika Laghrour
Groupe Physicochemistry of Materials, Nanomaterials and Environment, Faculty of Sciences, Mohammed V University, Ibn Battuta Avenue, Rabat, Morocco
National Institute of Agricultural Research (INRA), BP 6356, Rabat, Morocco

Rachid Moussadek, Rachid Mrabet, Rachid Dahan and Abdelmajid Zouahri
National Institute of Agricultural Research (INRA), BP 6356, Rabat, Morocco

Mohammed El-Mourid
International Center for Agricultural Research in the Dry Areas (ICARDA), Rabat Instituts, North-Africa Platform, 10112 Rabat, Morocco

Masahiko Katoh
Department of Agricultural Chemistry, School of Agriculture, Meiji University, 1-1-1 Higashi- Mita, Tama-ku, Kanagawa 214-8571, Japan
Department of Civil Engineering, Faculty of Engineering, Gifu University, 1-1 Yanagido, Gifu 501-1193, Japan

Takeshi Sato
Department of Civil Engineering, Faculty of Engineering, Gifu University, 1-1 Yanagido, Gifu 501-1193, Japan

Wataru Kitahara
Department of Civil Engineering, Graduate School of Engineering, Gifu University, 1-1 Yanagido, Gifu 501-1193, Japan
In Situ Solutions Co., Ltd., 2-5-2 Kandasudamachi, Chiyoda-ku, Tokyo 101-0041, Japan

Abraham Jera, France Ncube and Artwell Kanda
Department of Environmental Science, Bindura University of Science Education, P. Bag 1020, Bindura, Zimbabwe

Panbarasi Govindasamy, Jamal Alsharef and Kowstubaa Ramalingam
Department of Civil and Structural Engineering, Universiti Kebangsaan Malaysia (UKM), 43600 Bangi, Selangor, Malaysia

Alden D. Smartt, Kristofor R. Brye, Richard J. Norman and Trenton L. Roberts
Department of Crop, Soil, and Environmental Sciences, University of Arkansas, Fayetteville, AR 72701, USA

Christopher W. Rogers
Department of Plant, Soil, and Entomological Sciences, Aberdeen Research and Extension Center, University of Idaho, Aberdeen, ID 83210, USA

Edward E. Gbur
Agricultural Statistics Laboratory, University of Arkansas, Fayetteville, AR 72701, USA

Jarrod T. Hardke
Department of Crop, Soil, and Environmental Sciences, Rice Research and Extension Center, University of Arkansas, Stuttgart, AR 72160, USA

Mathayo Mpanda Mathew
World Agroforestry Centre (ICRAF), Dar es Salaam, Tanzania

Amos E. Majule
Institute of Resource Assessment, University of Dar es Salaam, Dar es Salaam, Tanzania

Robert Marchant
Environment Department, York Institute for Tropical Ecosystems, University of York, Heslington, York, North Yorkshire YO10 5NG, UK

Fergus Sinclair
World Agroforestry Centre (ICRAF), P.O. Box 30677,Nairobi 00100, Kenya

Won-Pyo Park, Thomas E. Ferko, Jonathan R. Parker, Tracy H.Ward, Stephanie V. Lara and ChauM. Nguyen
Department of Natural and Mathematical Sciences, California Baptist University, Riverside, CA 92504-3297, USA

Bon-Jun Koo
Department of Natural and Mathematical Sciences, California Baptist University, Riverside, CA 92504-3297, USA
Department of Environmental Sciences, University of California, Riverside, CA 92521-0001, USA

Andrew C. Chang
Department of Environmental Sciences, University of California, Riverside, CA 92521-0001, USA

Index